Christopher H. T. Lee
Editor

Emerging Technologies for Electric and Hybrid Vehicles

Editor
Christopher H. T. Lee
School of Electrical and Electronic Engineering
Nanyang Technological University
Singapore, Singapore

ISSN 1865-3529 ISSN 1865-3537 (electronic)
Green Energy and Technology
ISBN 978-981-99-3059-3 ISBN 978-981-99-3060-9 (eBook)
https://doi.org/10.1007/978-981-99-3060-9

This Springer imprint is published by the registered company Springer Nature Singapore Pte Ltd.
The registered company address is: 152 Beach Road, #21-01/04 Gateway East, Singapore 189721, Singapore

Paper in this product is recyclable.

Contents

Electric Vehicle Systems: State of the Art and Emerging Technologies

Shuangxia Niu and Yuan Mao

1 Preface

To deal with the energy and environmental crisis, the world's major automotive powers are sparing no effort in the research of electric vehicles (EVs). Electrification, intellectualization, network share have become the new development direction of the automobile industry chain. Promoting technological innovation around this trend is the fundamental means for EVs to break through to the high end. Electric vehicles use electricity as the whole or part of the power source. In comparison to vehicles driven by the traditional combustion engine, EVs have advantages that include fast start, high efficiency, low energy consumption, zero emissions, and no energy consumption in traffic congestion. Furthermore, electricity consumed by EVs can be obtained by renewable power generation, which can also reduce the damage to the natural world by over-exploitation.

According to the energy source, there are three types of EVs, which are classified into battery EV (BEV), hybrid EV (HEV) as well as fuel-cell EV (FEV) namely [2]. BEV is powered entirely by electricity and driven by an electric motor. HEV adopts both electricity and gasoline as energy sources. The operating system is driven by an electric motor and engine. FEV is powered by hydrogen and propelled by an electric motor. In this chapter, the outline of EV systems including power systems, propulsion systems, and the dynamical system is discussed. Especially, the hybridization and control of HEV are introduced in detail.

S. Niu (✉) · Y. Mao
Department of Electrical and Electronic Engineering, The Hong Kong Polytechnic University, Hong Kong, China
e-mail: eesniu@polyu.edu.hk

C. H. T. Lee (ed.), *Emerging Technologies for Electric and Hybrid Vehicles*,
Green Energy and Technology, https://doi.org/10.1007/978-981-99-3060-9_1

2 Introduction to EV Systems

2.1 EV Systems: Outline

Making a comparison with traditional internal combustion engine vehicles (ICEVs), the transmission structure of EVs has been changed, with new systems such as a power supply system and a drive motor. In order to exploit greater flexibility of electric propulsion, present EV technology has the following characteristics:

(a) The energy flow is generally transmitted employing flexible electrical cables instead of bolted flanges or rigid shafts.
(b) There are diverse types of EV propulsion arrangements with different configurations.
(c) Due to the different heaviness, dimensions, and appearance of EV energy sources, there are various types of hardware and mechanism in EV refueling systems.

The general outline of the EV system mainly consists of three modules: energy source subsystem, electric propulsion subsystem, and auxiliary subsystem (Fig. 1). The electric propulsion subsystem is the nucleus of the EV system, which converts the electric power into mechanical power to provide energy for the EV. The role of the energy source subsystem is to convert the chemical energy of batteries and other fuels into electricity. The function of the auxiliary subsystem involves temperature control, auxiliary power supply, and power steering control. Although EVs on the market come in a variety of constructions, they can be broadly divided into several categories.

According to variations in the manner of electric propulsion, EVs can be classified into the following types (Fig. 2). Among them, (a), (b), and (c) can be categorized as centralized propulsion constructions. (d), (e) and (f) may be sorted into in-wheel

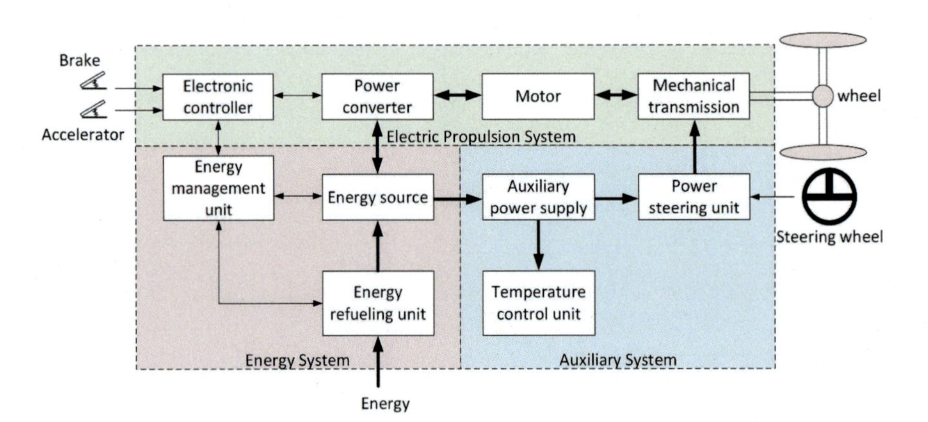

Fig. 1 The general outline of the EV system

propulsion constructions. The centralized propulsion system is developed based on the traditional automobile, which possesses the advantages of simple structure, uncomplicated control strategy, and easy maintainability.

Figure 2a depicts the traditional centralized propulsion system, which is a modification of the traditional ICEVs. In a traditional centralized propulsion system, the internal combustion engine is substituted by the drive motor, however, the clutch, transmission, and differential devices are organized much the same as in traditional

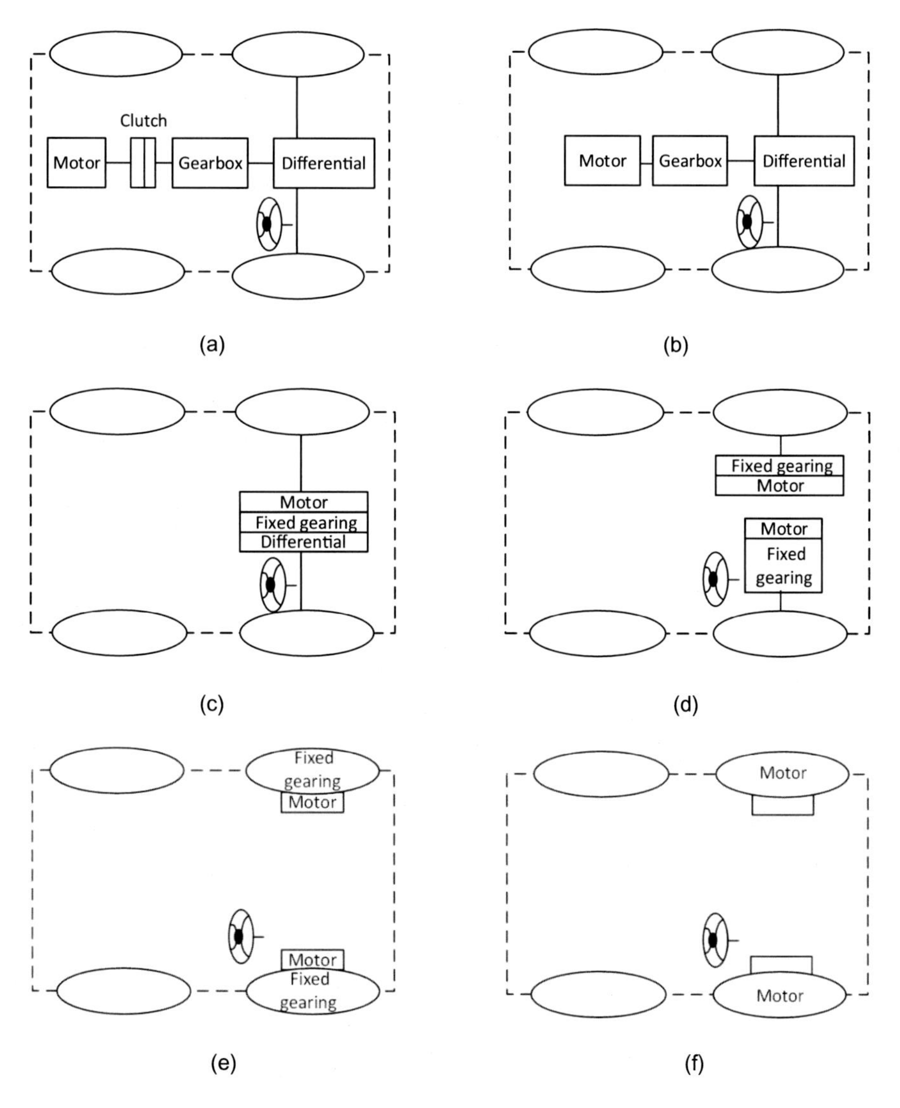

Fig. 2 **a–c** Centralized propulsion system. **d–f** in-wheel propulsion system

ICEVs'. Figure 2b shows a centralized propulsion system with a continuously variable transmission (CVT) system. Compared with the traditional mechanical transmission structure, the CVT propulsion system can save weight and volume and reduce the control difficulty caused by shifting gears. The integrated centralized drive system is shown in Fig. 2c and has similar modules to the CVT propulsion system. The drive shaft accommodates the electric motor and fixed gearing integrated with the differential.

Compared with the centralized propulsion system, the in-wheel propulsion system has a more compact structure, smaller mass, and higher transmission efficiency. In Fig. 2d, the wheel is bridged with the electric motor by a fixed gearing whereas in Fig. 2e, the planetary gear embedded in the wheel replaces the fixed gearing. It is worth mentioning that the transmission modules can be removed as depicted in Fig. 2f. In this design, gear is not required as the outer rotor of the electric motor equals the wheel rim, making the rotor velocity equivalent to the wheel velocity.

Based on variation in energy source, EVs can be also divided into six categories (Fig. 3).

2.2 *EV Parameters*

2.2.1 Vehicle Weight and Size Parameters

Weight and size parameters are significant indexes, which relate to the stability, trafficability characteristic, and comfortability of the EV. Table 1 lists the key weight and size parameters of EVs.

Because of the battery weight, even without the internal combustion engine and its relevant components, the curb weight of an EV is higher than that of a similar size ICEV. The increased weight of the EV will inevitably affect the load on the suspension and tires. It will also cause an increase in the overall power consumption. On this account, it is an essential issue to trim the EVs' curb weight.

2.2.2 Performance Parameters

(1) Maximum speed

It simply refers to the maximum speed that the vehicle can reach on a horizontal and well-conditioned road (usually a concrete or asphalt road).

(2) Acceleration rate

The acceleration capacity of a vehicle refers to the maximum acceleration that the vehicle can achieve on a level road. Since the acceleration during acceleration is constantly changing and difficult to express, it generally indicates the shortest time needed for the EV to accelerate from 0 km/h to a certain velocity.

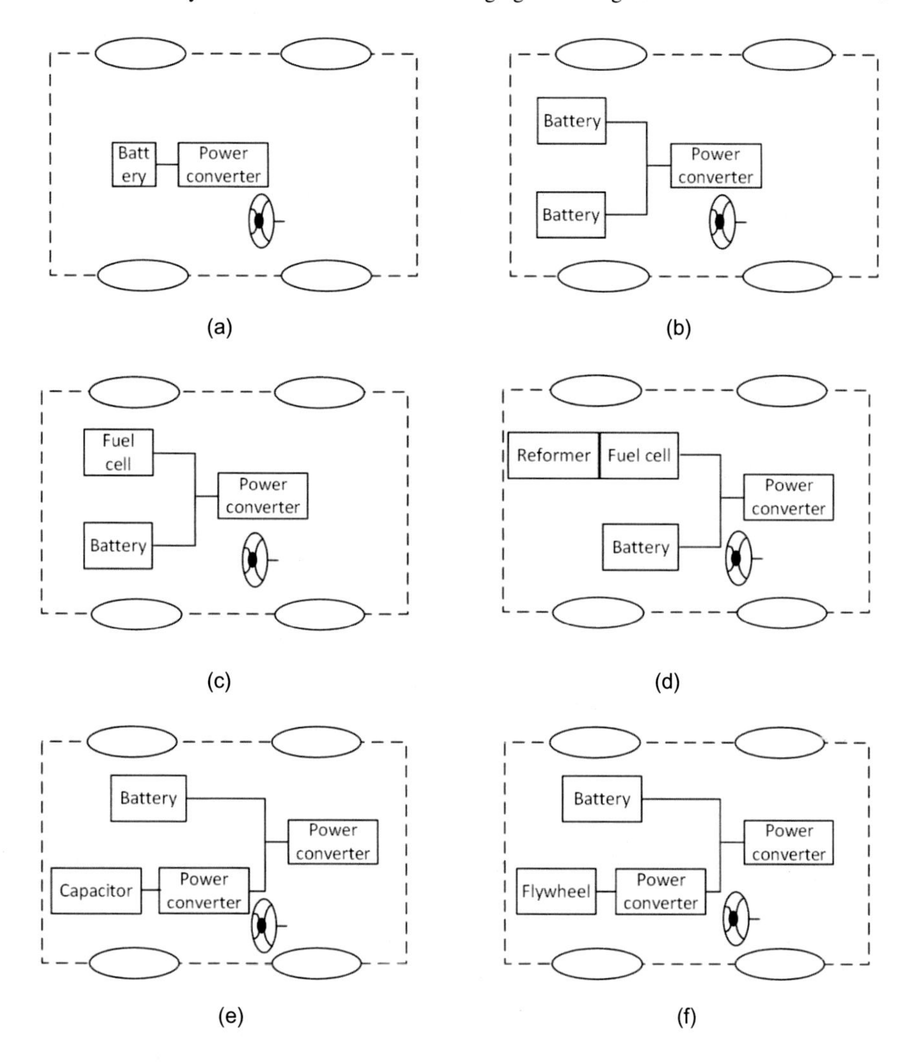

B: Battery
C: Capacitor
FC: Fuel cell
FW: Flywheel
P: Power converter
R: Reformer

Fig. 3 Six categories of EV energy source systems

Table 1 Vehicle weight and size parameters

Gross weight	Weight including payload
Curb weight	Weight excluding payload
Payload	Weight of passengers and cargo
Maximum weight	Maximum gross weight for safety operation
Battery weight	Weight of the whole battery pack in the EV
Vehicle dimensions	Length, width, height, and ground
Frontal area	Equivalent frontal area affecting the vehicle aerodynamic drag
Seating capacity	The number of passengers, sometimes adult/child is also specified
Cargo capacity	Volume of cargo
Wheelbase	Distance from the center of the front axle to the center of the rear axle
Wheel track	The distance between the front wheels or rear wheels
Front overhang	The horizontal distance between the front-wheel center and front end
Rear overhang	The horizontal distance from the end to the center of the rear axle

(3) Range per charge

As the name implies, range per charge describes the mileage that the EV can travel after being completely charged up. This standard can also be extended to fuel vehicles. It is greatly noticeable that the range of each charge may be greatly different because it is claimed in various ways. It could depend on a designated driving at a constant speed on a flat path, or for a certain kind of driving period.

2.2.3 Energy Parameters

(1) Fuel economy evaluation index

The fuel economy evaluation index is a measurement of the number of miles that a unit of energy can support a vehicle. The energy consumption evaluation units for ICEVs and EVs are different due to the energy resources they expended. It is usually donated by l/km for ICEVs and kWh/km for battery-powered EVs. Nevertheless, the unit of l/km can also be used to express fuel efficiency in some EVs. This situation generally applies to the EV supported by fuel cells, whose corresponding fuel includes but is not limited to compressed gaseous hydrogen, liquid hydrogen, and liquid methanol.

(2) Energy efficiency and power efficiency

The energy efficiency of EVs can be formulated as

$$\eta_e = \frac{E_{\text{out}}}{E_{\text{in}}} \tag{1}$$

Also, the power efficiency can be formulated as

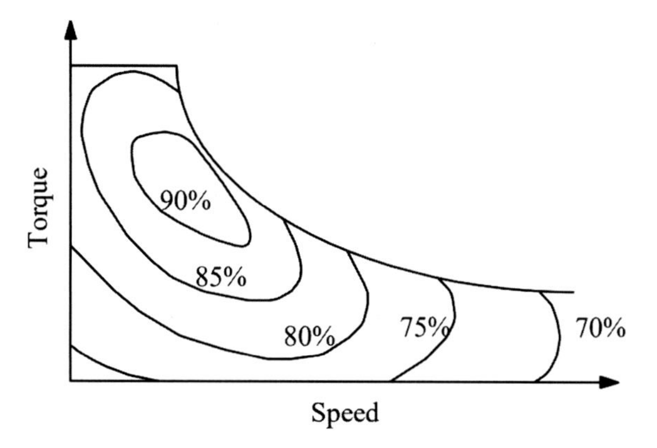

Fig. 4 Typical power efficiency map of an EV induction motor

$$\eta_p = \frac{P_{\text{out}}}{P_{\text{in}}} \tag{2}$$

For industrial operation, there may not be a necessarily distinguishable point between energy efficiency and power efficiency. However, the continuous variation of power efficiency is a significant difference in the operation of vehicles. Instead of using a particular linear chart, it is widely adopted to use the efficiency map to measure the energy efficiency performance of EVs. Figure 4 is a typical efficiency map.

2.2.4 Load Parameters

Since the EV may operate in different application scenarios, the power rating is bound to be affected by different loads. There are generally three types of load duties: continuous load, temporary load, and intermittent load.

The continuous load indicates that maintenance can be conducted on the specified power continuously or lasting for a practical time much longer than the thermal time constant [3]. The temporary load signifies that within a specified duration, the specified power will not make the motor temperature reach thermal equilibrium, and then continue for a while until the motor temperature does not exceed 2 K from the initial temperature [6]. Intermittent load means that within a specified time, the specified power will not make the motor temperature reach thermal equilibrium, and then continue for some time until the temperature of the motor goes back to the primary cycle temperature.

The power rating of EV motor under temporary load is the highest in the group of these loads. In addition, the temporary power of motors gets higher along with operating time becomes shorter. Among all three kinds of loads, the power rating with intermittent load is in the middle. The reason is that the time of the motor is limited to gradual cooling to the initial temperature. In general, EV motors have the lowest power rating at continuous loads.

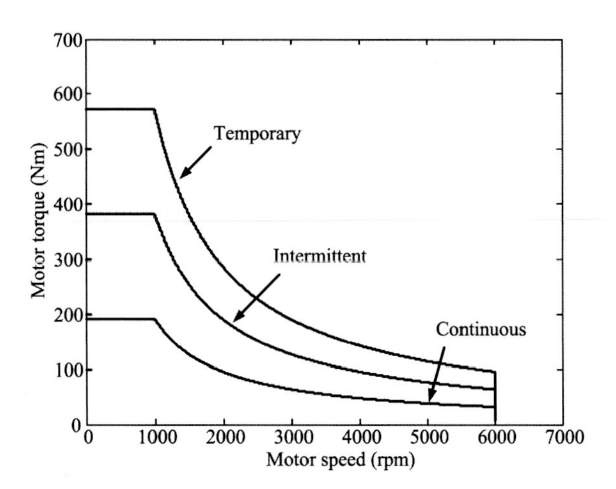

Fig. 5 Torque-speed curve of a typical induction motor [3]

Figure 5 depicts the torque-speed characteristics of a proper induction motor with different loads. The conclusion can be made that it is most economical to operate under continuous load. Besides, that is especially crucial for the dynamic system to operate under long-term hill-climbing. In addition, the temporary capability of the dynamic system is the most aggressive, which is especially worthwhile to describe the overtaking capability of EVs. Furthermore, the intermittent capability most affects the dynamic system to meet the immutable daily way of urban drive.

2.3 Force Act of EVs

Road load is the force that vehicles have to overcome to travel. As Fig. 6 shows, this road load is composed of three types of components: F_d for the aerodynamic drag force, F_r for the resistance force, and F_c for the climbing force. In addition, they should fulfill the following equation:

$$F_l = F_d + F_r + F_c \tag{3}$$

(1) Aerodynamic drag force

Aerodynamic drag force is caused by the dragging force put on a vehicle's body when traveling through air. Aerodynamic drag force mainly comes from the following three parts: i. the skin friction drag of air against the surface of a vehicle; ii. drag caused by the downshoot of the vortex at the rear of the vehicle; iii. the normal pressure on the vehicle itself,

Typically, the value of aerodynamic drag force can be expressed in a form that is proportional to the dynamic pressure of the relative velocity of the airflow, i.e.

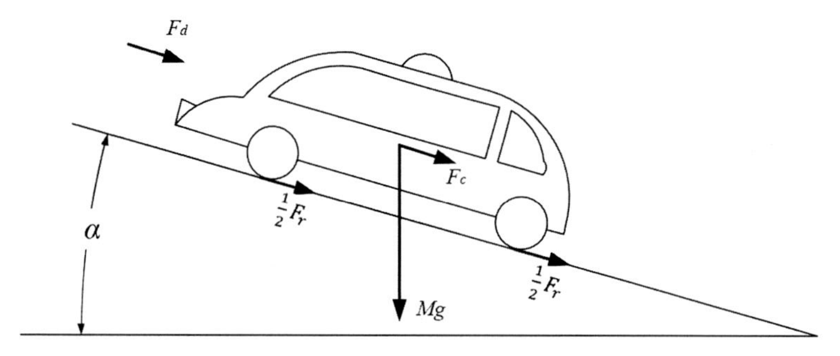

Fig. 6 Road load analysis diagram

$$F_d = 0.5C_D A\rho(v + v_0)^2 \tag{4}$$

where C_D represents the aerodynamic drag coefficient (dimensionless), which changes its value significantly between 0.2 and 1.5 based on different vehicle models. For instance, C_D value ranges [0.2,0.3] for purposely streamlined vehicles, [0.3, 0.5] for coaches, [0.5, 0.6] for vans, [0.6, 0.7] for buses and [0.8, 1.6] for trucks. ρ represents air density with measurement unit kg/m^3, generally $\rho = 1.2258$ N s^2/m^4, A represents the frontal area in m^2, v and v_0 are the vehicle speed and the headwind speed respectively with measurement unit m/s. Obviously, if v_0 use measurement unit km/h, the above formula can be transformed to

$$F_d = 0.0386C_D A\rho(v + v_0)^2 \tag{5}$$

(2) Rolling resistance force

Rolling resistance force emerges as a consequence of distortion on the surface of the wheel and road, yet the former prevails excessively. Factors that may affect the rolling resistance force include tire parameters, vehicle speed, quantity of charging piles, and torque conversion extent. Among all the factors, the tire parameters are relatively dominant.

The expression of rolling resistance force is

$$F_r = M\mathrm{g}C_r\cos\alpha \tag{6}$$

where M represents vehicle mass in kg, g represents gravitational acceleration (9.81 m/s^2) and C_r Indicates the rolling resistance coefficient (dimensionless). The tire types will affect the value of C_r. For example, C_r for radial-ply tires is around 0.013 while C_r for cross-ply tires is around 0.018. In addition, C_r changes in reverse to the air pressure of the tire.

(3) Climbing force

The component force of the vehicle gravity along the ramp is the climbing force, which can be expressed as

$$F_c = M\mathrm{g}\sin\alpha \tag{7}$$

where α represents the inclined angle in degree or radian. In usual, the incline can be expressed by the percentage grade ability *p*:

$$p = \left(\frac{h}{l}\right)100\% \tag{8}$$

where *h* indicates the vertical height above *l*, which is the horizontal distance. Thus, the relationship of *p* and sinα can be written as:

$$\alpha = \tan^{-1}\left(\frac{p}{100}\right) \tag{9}$$

It is worth noting that the maximum grade ability indicates the utmost incline when the climbing speed of the vehicle is essentially zero.

(4) Acceleration

The acceleration of a vehicle can be formulated in

$$\mathrm{a} = \frac{(F - F_l)}{k_r M} \tag{10}$$

where *F* denotes the power applied by the wheels, which is necessary to overcome the road load F_l described above. k_r denotes a correction factor of the fluctuation in weight of the vehicle as a result of the inertia produced by a rotational mass, (like the wheel, the gear, and the shaft), in scope between around 1.01–1.40.

The equation indicates that if the motive force is lower than road load, it will become deceleration and become a negative value.

3 Introduction to HEV Systems

3.1 HEV Systems: Outline

When breakthroughs were made in EV energy sources, HEVs employing both internal combustion engines and electric motors have been recommended as a changeable solution until pure EVs are fully developed and promoted. There are several positive features of HEVs. The most apparent advantage would be that the original EV driving range can be greatly extended by two or four times. As well as these further favorable aspects would be that HEVs provide fast liquid petrol or diesel refueling and it only needs a few transformations on the energy supply base station. Compared with ICEVs, HEVs generate remarkably less pollution and consume less fuel while having the same range. The most impractical characteristics of HEVs

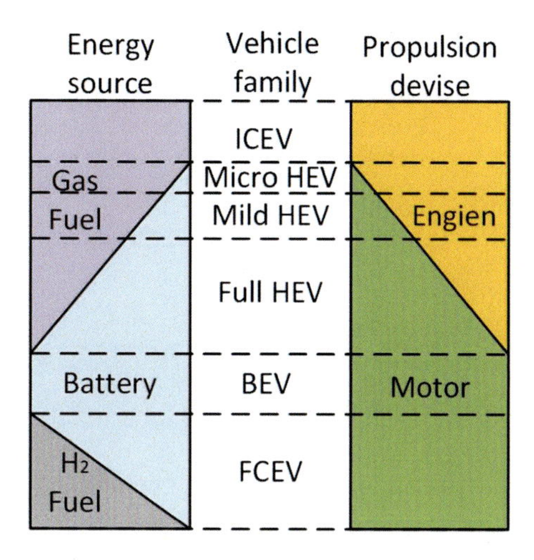

Fig. 7 Spectrum of various EVs

might be the increased complexity in construction and loss of the zero-emission concept.

Considering the classifications of power producers (engines and motors) and energy sources (gasoline fuel, batteries, and hydrogen fuel) and their utilization levels, the entire spectrum of vehicles is presented in Fig. 7.

From the perspective of energy sources, all of these HEVs are blended with gasoline fuel and batteries. Similarly, the main energy sources of FCEVs are the hybridization of gasoline fuel and batteries. For the BEV, the only battery is adopted as an energy source. According to the share of battery in the overall energy source of HEVs, they can be further classified as micro HEV, mild HEV, and full HEV.

Micro HEVs, as the name suggests, have a very low proportion of electric power sources. A starter motor is usually absent in the micro HEV, and the generator section is also replaced by a belt-driven integrated starter generator (ISG). Power levels of these kinds of ISG are usually between 3 and 5 kW. The ISG is not used for propulsion but to provide two significant blended functions for HEVs. The first one is shutting down the engine when the vehicle is not propelling, the idling stop function, thereby achieving the purpose of energy conservation. Vehicles can work in an idling stop state that does not consume fuel when parked. The second function is regenerative braking, which aims to use the electric motor for energy recovery when the vehicle is decelerating or going downhill. The battery voltage for micro HEVs is 12 V. The micro HEV model we are familiar with is Citroën C3.

For the mild HEVs, the power levels of ISG are usually between 7 and 12 kw. The acceleration cannot be provided by electricity solely because the ISG and the engine are connected by the same shaft. Therefore, mild HEVs are usually not electrically launched. However, it can offer all other hybrid features, involving regenerative braking and idling stop. In addition, ISG is characterized by reduced engine size because it helps propel the vehicle. The levels of battery voltage for the mild HEVs

Table 2 Features of Different Levels of HEVs

	Hybrid features				Battery rating (V)	Motor rating (kW)	Project cases
	Electric launch	Idling stop	Regenerative braking	Downsized engine			
Micro HEV		√	√		12	3–5	Citroën C3
Mild HEV		√	√	√	36–144	7–12	GM Silverado, GM Sierra, Honda Civic, and Honda Accord
Full HEV	√	√	√	√	200–500	30–50	Toyota Prius, Ford Escape

are typically between 36 and 144 V. Common mild HEVs on the market include the GM Silverado, GM Sierra, Honda Civic, and Honda Accord.

For the full HEVs, all the hybrid functions involving regenerative braking, downsized engine, electric launch, and idling stop are provided. The respective motor and battery ratings are generally between 200–500 V and 30–50 kW. Common full HEV in daily life include the Toyota Prius and Ford Escape. In full HEV, the engine and the motor can be operated simultaneously, which may provide additional torque. Compared with conventional vehicles which have the same-sized engine, full HEV can offer better acceleration performance to cope with different working conditions.

To sum up, the functions and characteristics of each level of HEVs are compared and listed in Table 2.

3.2 Classification of HEV Systems

Compared with conventional ICE vehicles and BEVs, HEV drive trains have increased vehicle energy management systems and control systems. Its main function is to optimize the efficiency of the generator, distribute the power flow between the generator and the motor, and perform the energy management of the battery and the fuel tank at the same time. According to the difference in the ways of construction, HEVs can be classified into series hybrid, parallel hybrid, series–parallel hybrid, and complex hybrid.

(1) Series hybrid system

Figure 8 exhibits the fundamental architecture of a typical series hybrid system. The fuel tank-ICE-generator and battery form the onboard energy source. The power generated by the generator driven by the ICE and the electric power streamed from the battery are sent to the motor together. Electric driving is the only driving mode in a series hybrid system. As a result of omitting clutches among mechanical connection parts, it is flexible to locate the ICE-generator set. Although the transmission system is not complex, it demands three propulsion devices—an ICE, a generator, and an electric motor. When parallel EVs are used to climb long slopes, some problems need to be solved urgently that the above-mentioned devices are necessary to be dimensioned to achieve an optimal level of sustained power.

(2) Parallel hybrid system

As exhibited in Fig. 9, parallel HEVs have two complementary propulsion systems, the ICE propulsion system and the electric motor propulsion system. The power of the two propulsion systems can be superimposed on each other to meet the needs of different working conditions of the vehicles. Compared with series HEVs, parallel HEVs need fewer motive devices. Besides, a smaller volume of ICE and electric motors are employed, which may also achieve similar performance. For endurance runs, the maximum stained power of ICE is required to be rated, while the motor is only about half that. From this perspective, parallel HEV saves energy more than series HEV.

(3) Series–parallel hybrid system

As Fig. 10 shows, the series–parallel hybrid system integrates the traits of both the series and parallel HEVs, which includes a separate mechanical link between the ICE and the transmission system in comparison with the series hybrid system. In addition, the series–parallel hybrid system owes one more generator than the parallel hybrid system. It is clear that although a series–parallel hybrid system absorbs merits from

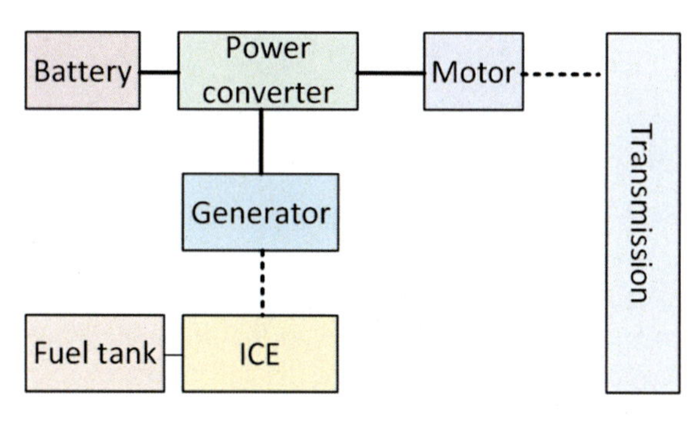

Fig. 8 Schematic diagram of the series hybrid system

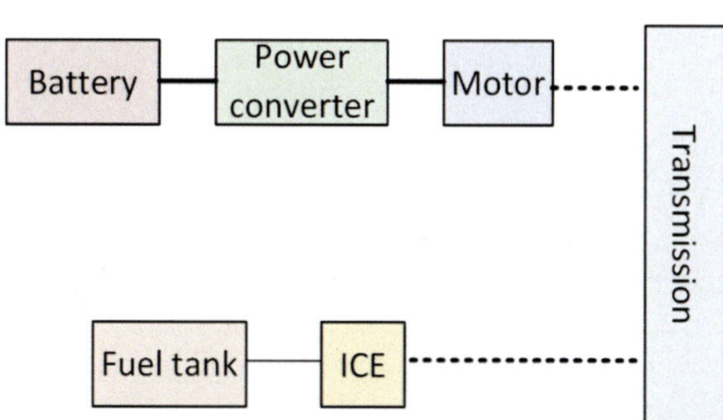

Fig. 9 Schematic diagram of the parallel hybrid system

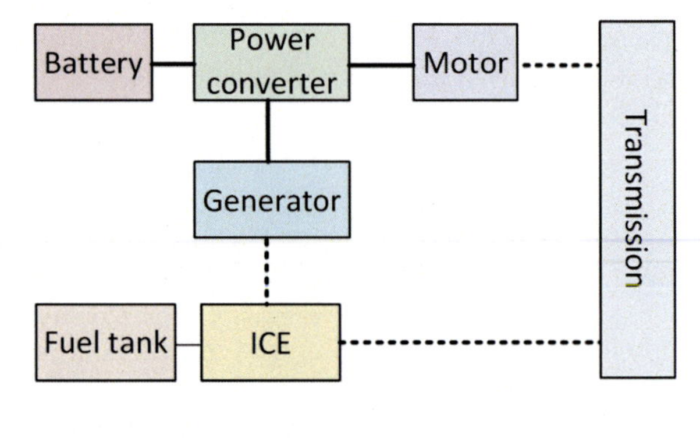

Fig. 10 Schematic diagram of the series–parallel hybrid system

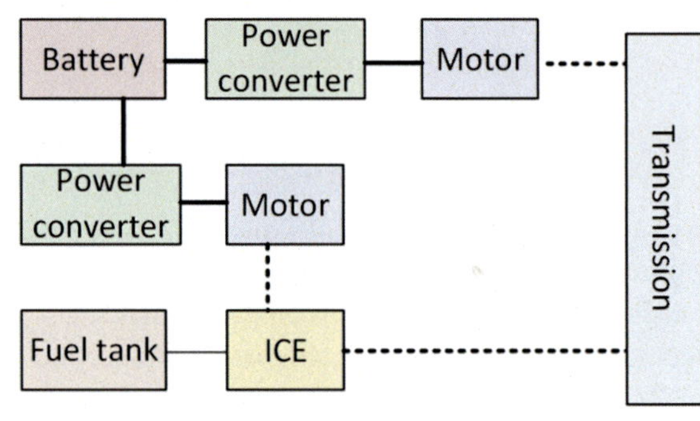

Fig. 11 Schematic diagram of the complex hybrid system

pure series and parallel hybrid systems yet the design and manufacturing costs may increase.

(4) Complex hybrid system

As depicted in Fig. 11, a complex hybrid system provides versatile power flow to the electric motor. It can allow multiple operating modes that series–parallel hybrids cannot provide, particularly the three-power propulsion mode (actuated by ICE and two electric motors). Closed to the series–parallel HEV, the complex HEV was damaged from greater complexity and expenses.

3.3 Power Flow Control of HEV Systems

(1) Series hybrid control

Series HEVs have realized the diversification of onboard energy sources, which can take advantage of various sorts of energy sources throughout, and achieve their best combination through appropriate control to meet various special needs of vehicle driving. For example, as shown in Fig. 12, when the vehicle is in a startup, normal driving or acceleration, the ICE and the battery supply power to the electric vehicle at the same time. While at light load working conditions, deceleration or braking, the extra electricity can be used to charge the battery.

(2) Parallel hybrid control

Different from the series HEV, the engine in the parallel HEV directly supplies mechanical power to the drive wheels. As shown in Fig. 13, the driving power of the parallel HEV is provided by the engine, generator/motor separately or jointly through a mechanical coupling device. Through the electromechanical coupling device, the engine can transmit all or part of the power to the electric motor. The electric motor can work in generating mode to charge the battery. It can also implement regenerative braking during vehicle braking to recover braking energy.

(3) Series–parallel hybrid control

Control of this kind of hybrid system contains advantages from both series and/or parallel hybrids. Thus, various operating modes become possible to perform its power flow control. It usually manifests itself in two modes: the ICE-heavy one (Fig. 14) and the electric-heavy one (Fig. 15). As the name suggests, the ICE-heavy one represents a heavier proportion of ICE in the entire system, while the electric-heavy one represents the motor playing a major propulsion role.

(4) Complex hybrid control

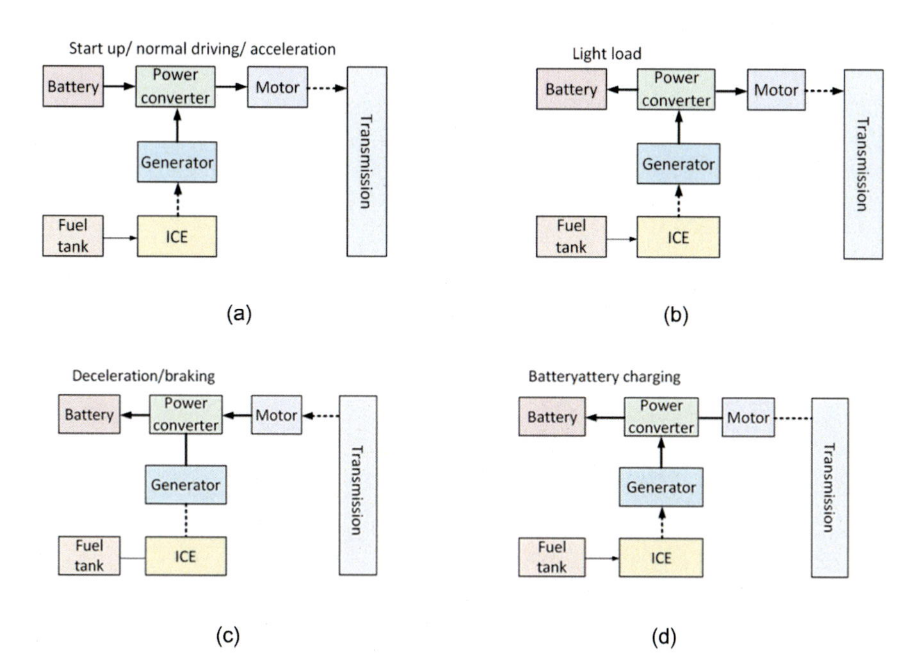

Fig. 12 Schematic diagram of the series hybrid system

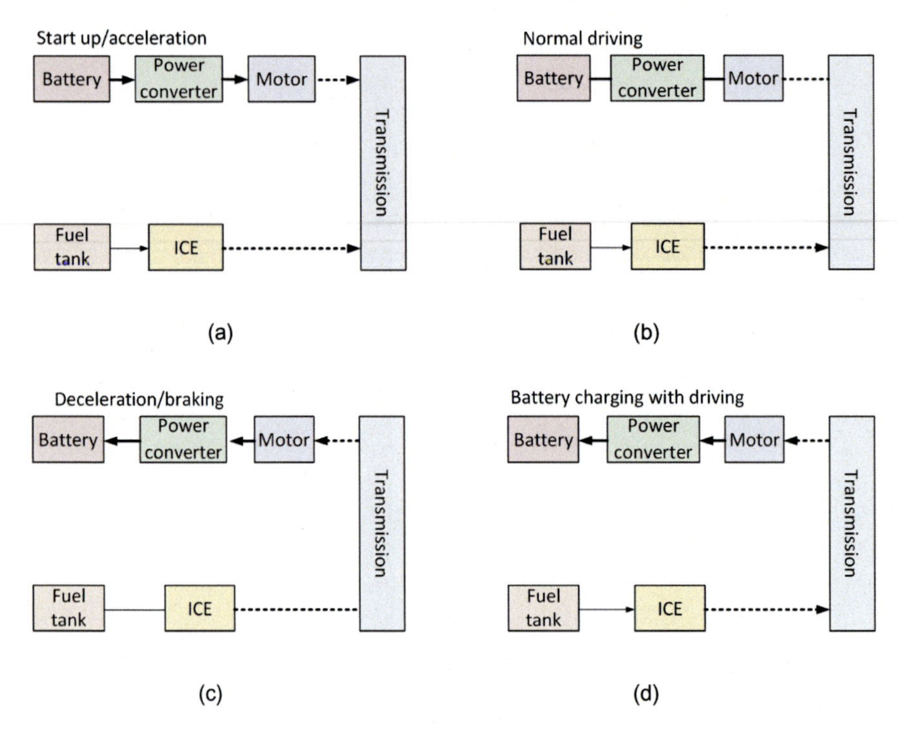

Fig. 13 Schematic diagram of the parallel hybrid system

For this, recent trends have led to a proliferation of studies about dual-axle propulsion systems. The rear-wheel axle and the front-wheel system are driven independently inside this system. Figure 16 illustrates the schematic diagram of the front-hybrid rear-electric complex hybrid system while that of the front-electric rear-hybrid one is shown in Fig. 17. In this system, the shaft or transfer that connects the front and rear wheels is eliminated, which may contribute to a lighter propulsion system and more elastic packaging of cars. Furthermore, the vehicle fuel efficiency can be improved drastically by regenerative braking on the four-wheel scale, thereby improving fuel economy.

4 Emerging Technologies of EV Systems

4.1 Electronic Differential

The electronic differential is an emerging technology for the transmission system. When a vehicle turns, the inner and outer wheels have different wheelbases. For the vehicle to turn accurately and smoothly, a differential mechanism is needed to rotate

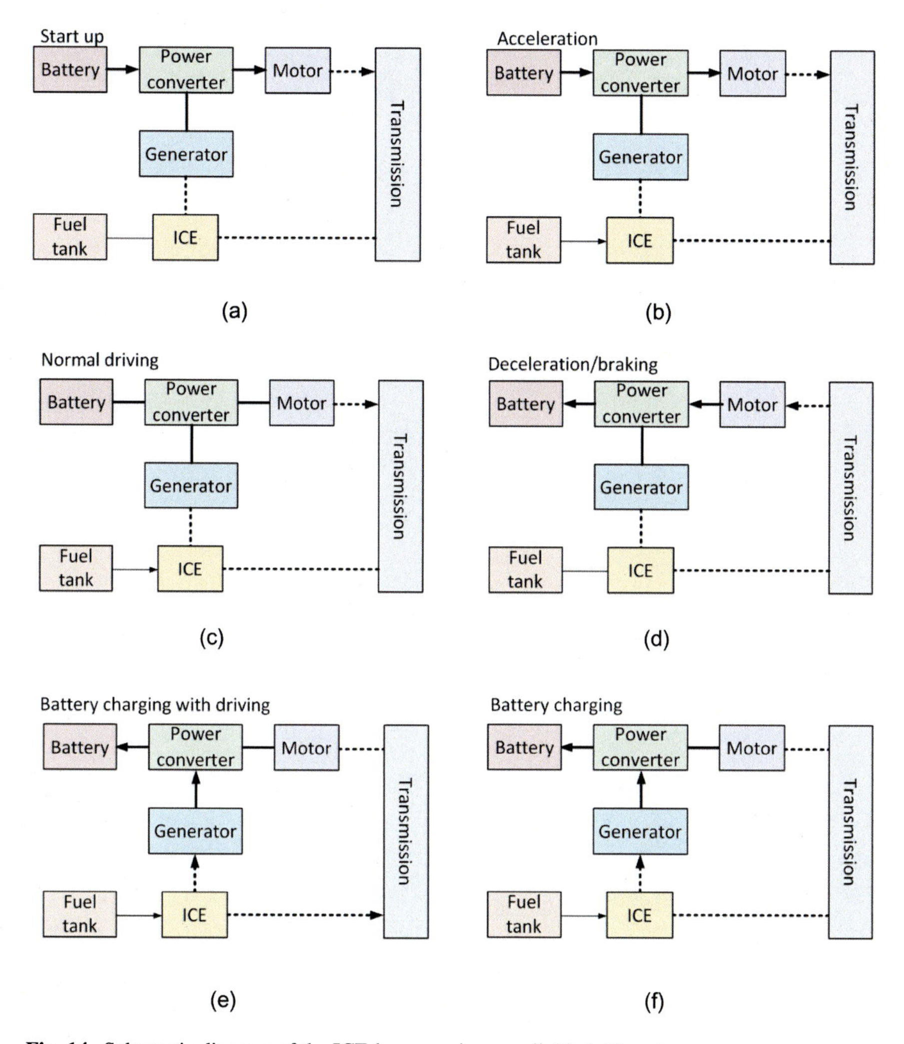

Fig. 14 Schematic diagram of the ICE-heavy series–parallel hybrid system

the inner and outer wheels at different rates. The core of the mechanical differential is the left and right half gears, the two planetary gears, and the gear frame (Fig. 18a). The mechanical differentials are widely used in ICEVs. Recently, the application of electronic differential in EVs has become a hot research issue. As mentioned before, EVs can be powered by the in-wheel propulsion system. As shown in Fig. 18b, each wheel is connected to an individual motor with fixed gearing. By this means, the speed of every wheel is able to be controlled independently to achieve equilibrium in corning.

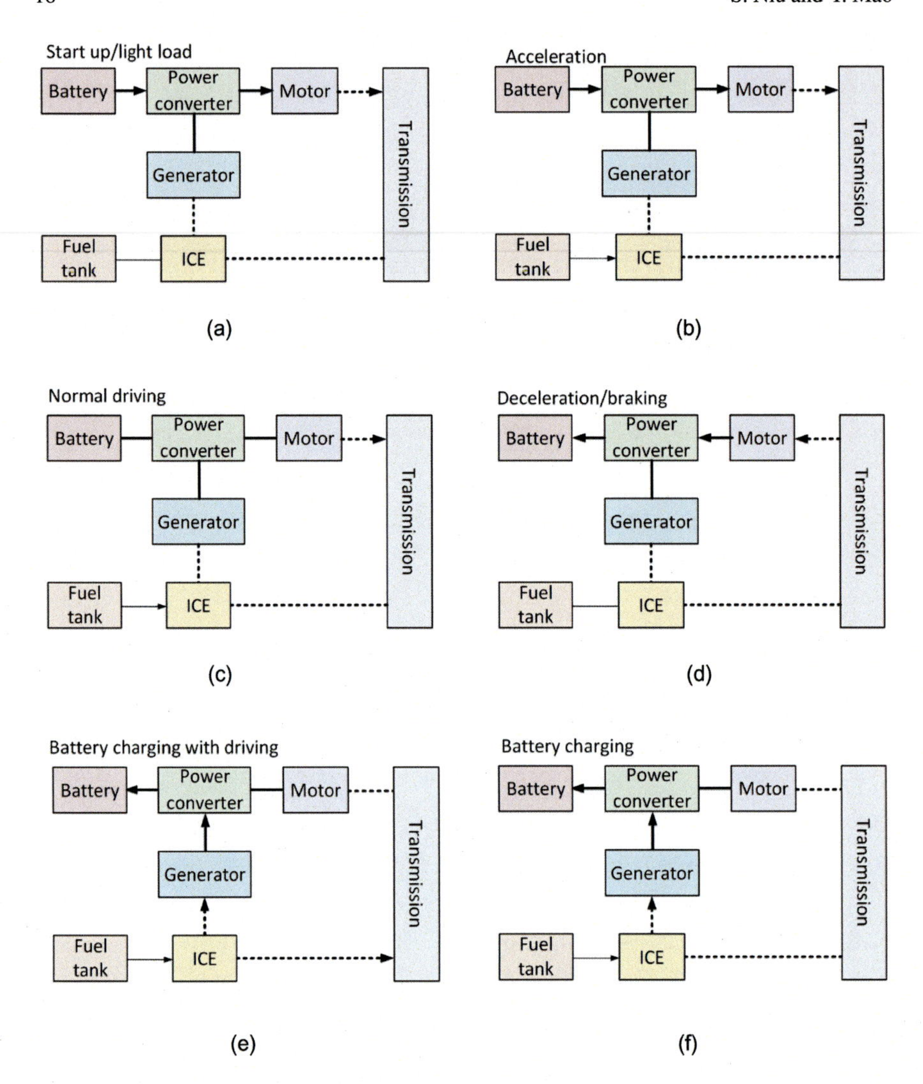

Fig. 15 Schematic diagram of the electric-heavy series–parallel hybrid system

Compared with mechanical differential, electronics differential has higher control precision and smaller overall mass. However, additional motors as well as power converters are needed, which may inevitably cause a rise in cost.

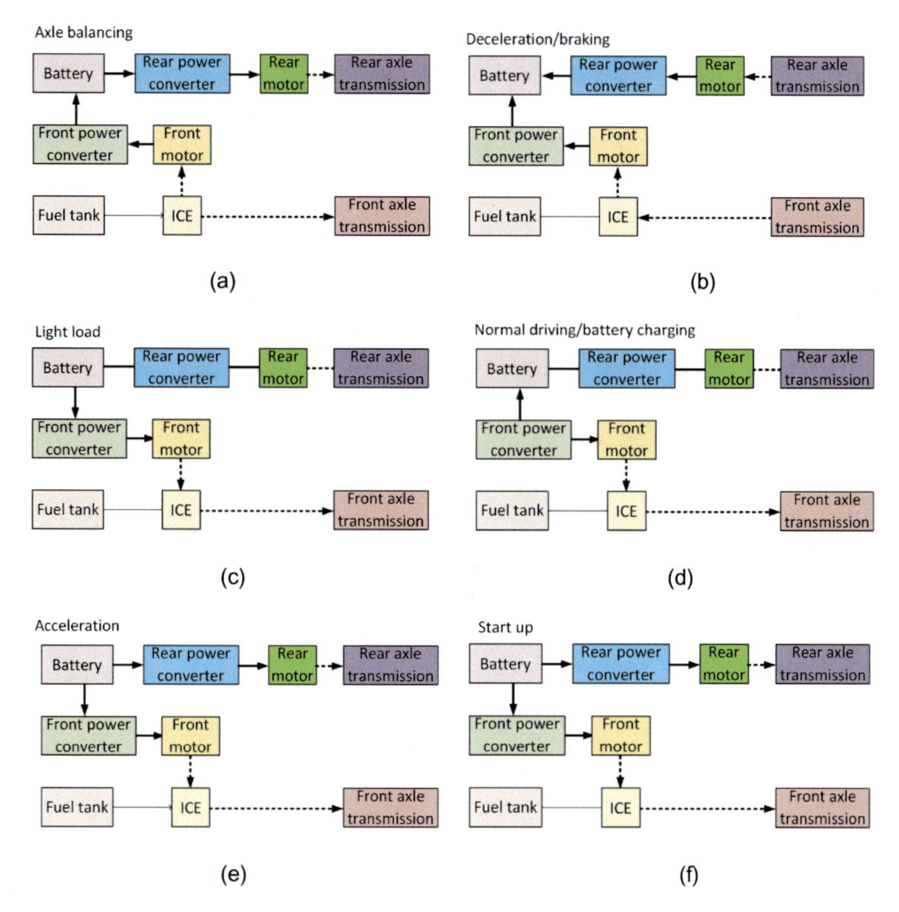

Fig. 16 Schematic diagram of the front-hybrid rear-electric complex hybrid system

4.2 In-Wheel Motor

The current findings add to a growing body of literature on the in-wheel motor propulsion system of EVs. In-wheel motor propulsion system not only can omit a large number of transmission parts but also can realize a variety of complex driving modes. The common in-wheel motor genre includes the high-speed inner-rotor in-wheel motor and in-wheel motor conversely with low-speed and outer-rotor.

The shaft of the high-speed inner-rotor in-wheel motor is connected to the wheel drive shaft through a high-speed-reduction planetary gear set so that the motor bearing does not directly sustain the wheel and road load. Typically, the design of this motor allows it to operate up to 1000 rpm in order to realize higher power density.

In the low-speed outer-rotor in-wheel motor propulsion system, inside the wheel-mounted corresponding outer rotor directly, there is no need to use a reduction mechanism [5].

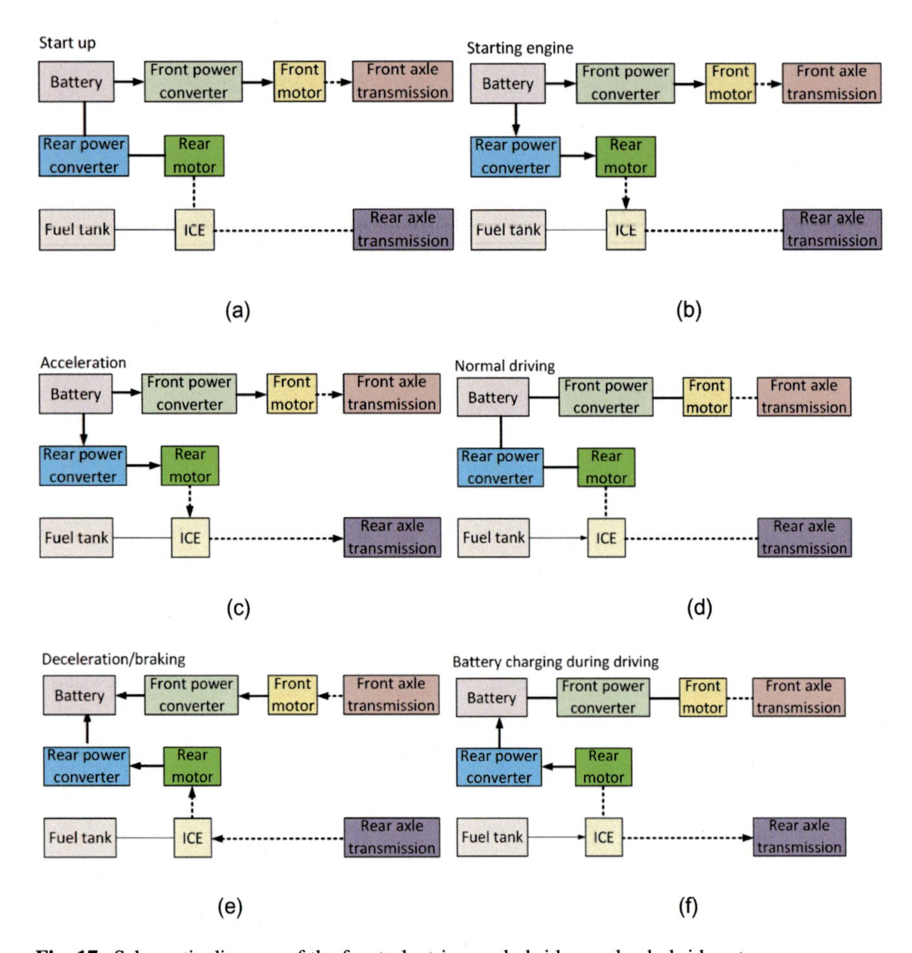

Fig. 17 Schematic diagram of the front-electric rear-hybrid complex hybrid system

4.3 E-CVT Propulsion Systems

E-CVT system that is short for the electric continuously variable transmission system was initially proposed for use in the Toyota Prius in 1997 (Fig. 19). E-CVT system can be regarded as a turning point in the development of HEV. After Toyota Prius introduced E-CVT, various automakers followed suit.

The core of the E-CVT propulsion system is planetary gear set. The innermost is the sun gear, the small gears in the middle are the planetary gears, and the outermost is the ring gear. Hybrid based on planetary gear sets can split power from one output/input shaft to two input/output shafts freely.

Figure 20 is a planetary gear-linked double electric machine E-CVT system. The ring gear is coupled with the planetaries gears through teeth. The engine and the motor are both connected to the ring gear. Besides, the engine is coupled with the

Fig. 18 **a** Mechanical differential **b** Electronic differential

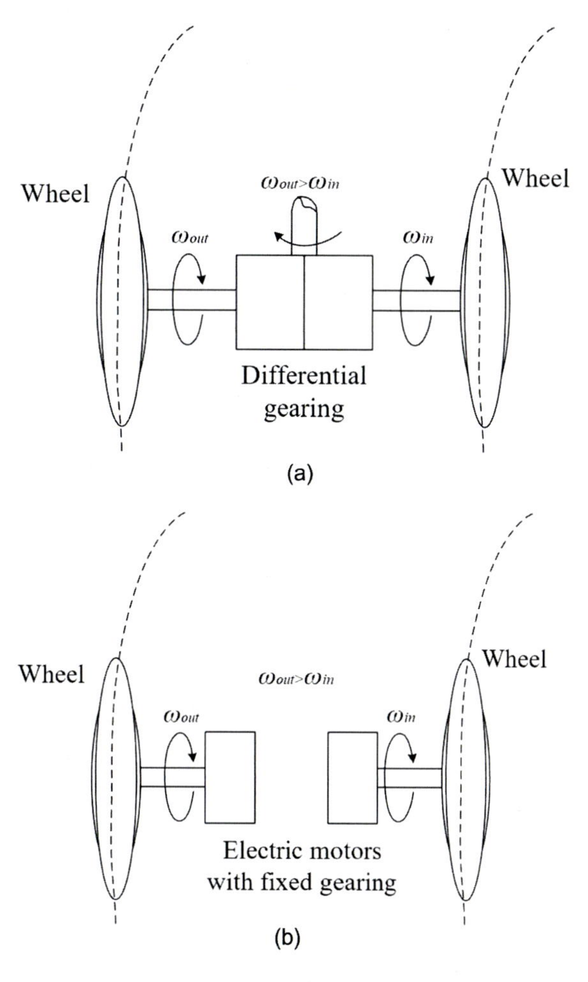

Fig. 19 E-CVT propulsion system by Toyota Prius. *Source* Carfolio [1]

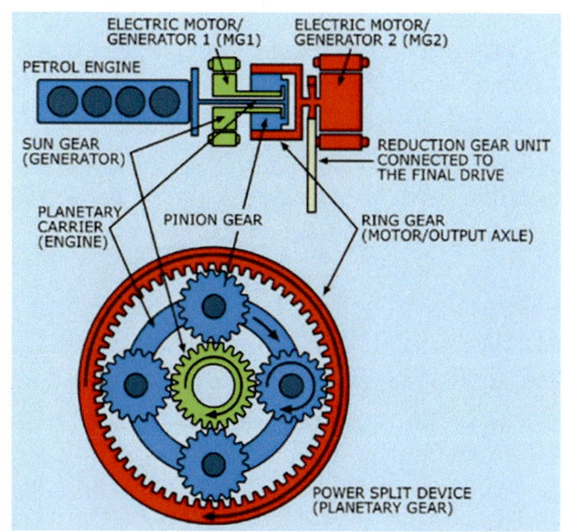

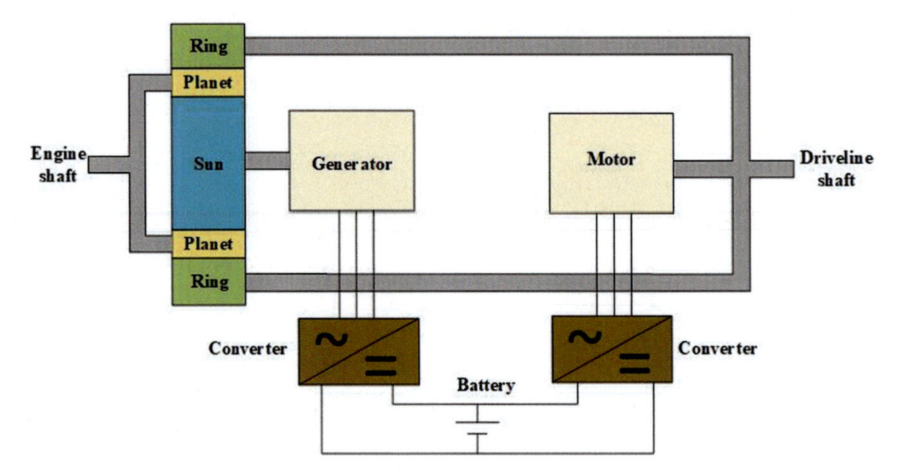

Fig. 20 Planetary gear-linked double electric machine E-CVT system

planet carrier and the motor is linked to the driveline shaft. The sun gear is assembled to the generator.

With the help of the E-CVT system, the power acquired by the generator can be then fed back to the motor. When the speed of the transmission system changes, the engine speed can be kept constant. Therefore, there can be a realization of a continuously variable ratio between the speed of the engine and the wheel [4].

The advantages of E-CVT advancement include the following points: i. The overall size of the transmission system can be greatly cut down due to the absence of clutches, shifting gears, and torque converters. ii. The transmission efficiency, energy efficiency as well as power density can be improved. iii. The mechanically simple structure may result in better reliability. iv. With the continuously variable ratio between the motion of the wheel and engine, the engine can operate in a position of maximum energy efficient, thereby greatly reducing fuel consumption. v. The E-CVT system helps to enable the idle stop function and electric start function. In this way, the engine can be conveniently turned off when the vehicle is stopped, and it can also provide a greater driving force when the vehicle is started. vi. The E-CVT device can help to activate regenerative braking to a large extent when the vehicle is decelerating or coasting down a slope. The engine and electric motor then run simultaneously to provide the power needed for full-throttle acceleration.

A major challenge associated with existing E-CVT propulsion systems is to eliminate the sun gear, which may help to avoid the negative effects caused by mechanical transmissions such as gear noise, transmission loss, and regular lubrication. To this end, scholars are committed to researching E-CVT propulsion systems without mechanical planetary gears in recent years.

A breakthrough idea is to concentrically combine two machines both electrically and mechanically to achieve a power split. As shown in Fig. 21, although the two motors are separated in space, they indirectly interact with each other structurally and electrically. The machine I is a double-rotor induction machine while Machine II

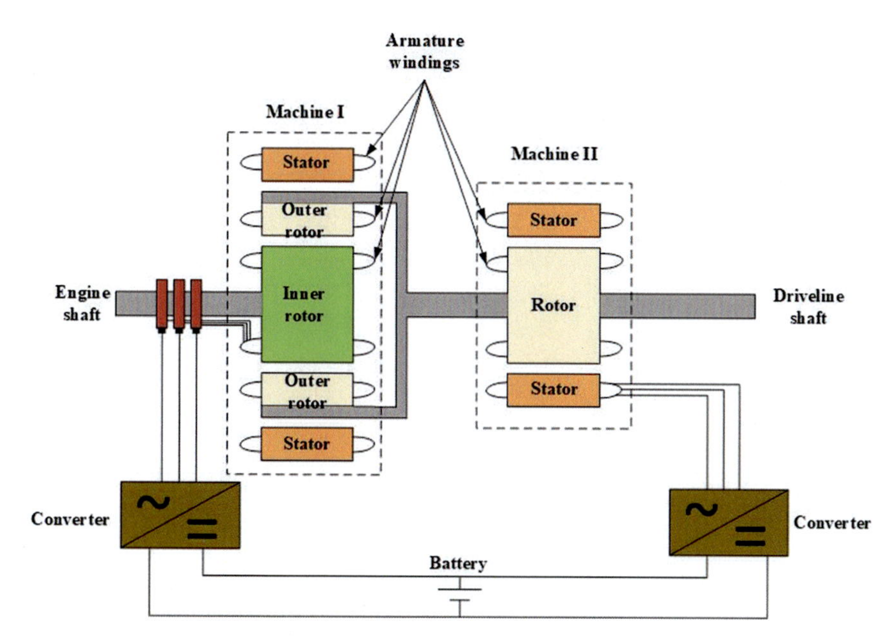

Fig. 21 Concentrically settled dual rotor machine E-CVT system

is a traditional induction machine. The outer rotor of Machine I is linked to Machine II while its inner rotor is mechanically coupled with the engine through the shaft. Meanwhile, the rotor of Machine II is linked to the driveline shaft. Subject to the structural characteristics of induction motors, the inner rotor of Machine I is bridged with the power converter through the slip ring and carbon brush. Machine II is powered by the battery through another three-phase converter. The above-mentioned power converters are electrically bridged to the battery. The power flow in this system can be regarded as an integration of mechanical power flow and electrical flow. The former directly passes through the outer rotor of Machine I to the rotor of Machine II. The latter penetrates two power converters and one slip ring. The continuously variable speed ratio is realized by regulating the electrical power flow, thus the engine can work at a constant speed when the driveline speed is changing.

More in-depth research found that the integrated double-rotor induction motor E-CVT system can be more compact in structure (Fig. 22). The most important point is to integrate the outer rotor of Machine I with the rotor of Machine II. The operation principle of the integrated type is similar to the concentrically settled type while taking the specific benefits of being highly compact and lightweight.

Further research on E-CVT has been fast and encouraging, combined with the permanent magnet synchronous machine. As depicted in Fig. 23, the outer rotor is replaced by a PM ring. By adopting the PM brushless structure, the mechanical stability and power density of the E-CVT system can be further improved.

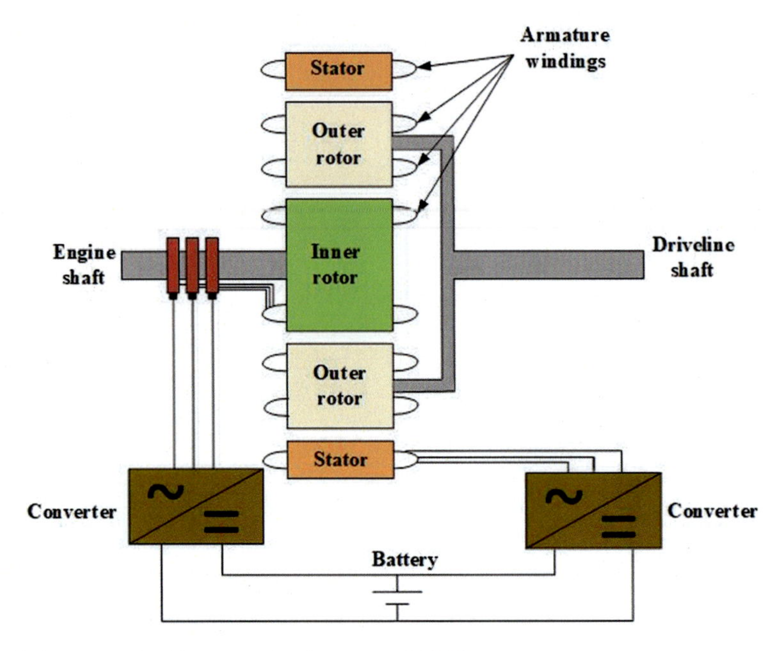

Fig. 22 Integrated double-rotor induction motor E-CVT system

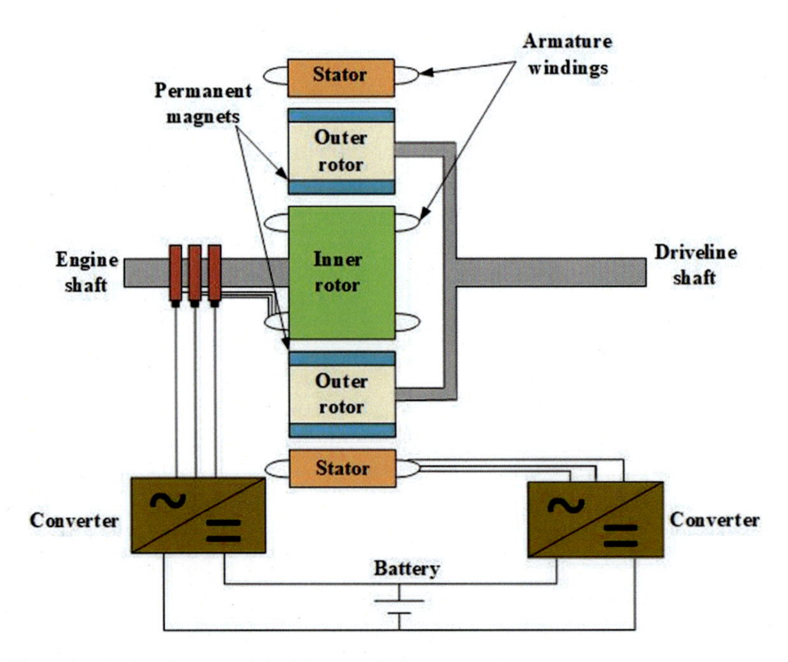

Fig. 23 E-CVT using integrated double-rotor PM motor

References

1. Carfolio (2014) 1997 Toyota Prius technical specifications, Carfolio.com.
2. Chan CC, Chau KT (2001) Modern electric vehicle technology. Oxford University Press, Oxford
3. Chau KT (2014) EV powertrain parameters. In: Crolla D, Foster DE, Kobayashi T, Vaughan N (eds) Encyclopedia of automotive engineering
4. Chau KT, Chan CC (2007) Emerging energy-efficient technologies for hybrid electric vehicles. Proc IEEE 95(4):821–835
5. Ehsani M, Wang FY, Brosch GL (2013) Transportation technologies for sustainability. Springer, New York Inc.
6. Rockwell Automation (1996) Application basics of operation of three-phase induction motors, Sprecher + Schuh AG Rockwell Automation, Aarau

Classical Electric Machines for Electric and Hybrid Vehicles

Wenlong Li and Yuan Wan

Abstract Classical electric machines, including DC machines, induction or asynchronous machines, and synchronous machines, play a key role in modern electric vehicle (EV) propulsion. Because of the maturity and simplicity, DC machines have been the favorable candidate for early EV traction motors. Due to the disadvantages of carbon brushes and commutators, their reliability and efficiency limit the wide application. Nowadays, AC machines are mainly applied for modern EV propulsion, primarily squirrel-cage induction machines (SCIMs), wound-rotor synchronous machines (WRSMs), and permanent magnet synchronous machines (PMSMs). Especially, SCIMs and PMSMs are becoming the predominant traction motors in the market share for modern EVs recently. In this chapter, the classical electric machines, namely, DC machines, SCIMs, WRSMs, and PMSMs are presented. Their configurations, operating principle, mathematical modeling, and control methods are briefly discussed.

1 Background

Transportation electrification such as the implementation of electric vehicles (EVs) is considered as an effective way to achieve carbon neutrality goals and help to tackle the climate change issue. To combat internal combustion engine vehicles, EVs should keep improving their technology and attract more and more consumers. In an EV propulsion system, one of the key components is the electric machine. Since the operating conditions for EV traction applications are different from those for industrial applications, the requirements for EV traction machines are much stricter: (1) high torque and power density, (2) wide speed range for both low- and high-speed regions, (3) high efficiency for the full speed range, (4) high overload capability, (5)

W. Li (✉) · Y. Wan
Nanjing University of Science and Technology, Nanjing 210094, China
e-mail: wlli@njust.edu.cn

Y. Wan
e-mail: 12019097@njust.edu.cn

C. H. T. Lee (ed.), *Emerging Technologies for Electric and Hybrid Vehicles*,
Green Energy and Technology, https://doi.org/10.1007/978-981-99-3060-9_2

high robustness and reliability under various operating scenarios, and (6) low acoustic noise. According to these requirements, different types of classical electric machines have been successfully applied in recent commercial EVs, as shown in Table 1.

Based on the commutating method, the classical electric machines can be divided into two main categories as shown in Fig. 1. According to the machines, whether have commutators, the classical electric machines can be categorized into DC machines and AC machines. DC machines are a kind of classical electric machines that were invented almost two centuries ago. Owning to the advantages of mature technology and simplicity of control, DC machines have been adopted as EV traction motors in the early stage. However, due to well-known problems such as lower efficiency and requirements for the routine maintenance of the mechanical commutator and carbon brushes, the DC machines are no more popular for the most recent EVs but are still applied in some low-end and off-road EVs. AC machines mainly include induction machines and synchronous machines. Induction machines, especially squirrel-cage induction machines (SCIMs), are one of the most popular candidates for recent EV traction applications, because of their low cost and ruggedness. Wound-rotor synchronous machines (WRSMs) have been adopted by BWM in the recent EV type of iX3, due to the rear-earth-free feature and higher performance compared to SCIMs. Permanent magnet synchronous machines (PMSMs) are another attractive candidate for EV traction applications. Because of the utilization of high-energy rear-earth permanent magnet (PM) materials, PMSMs exhibit distinct advantages such as higher power density and efficiency than other counterparts. Therefore, PMSMs for EV traction motors are steadily increasing the market share in recent years. As shown in Table 1, PMSMs are adopted as the traction motors for the recent EV models by world-renowned automakers such as Audi, Tesla, Porsche, Ford, GM, Jaguar, etc. In the following sections, the machine configurations, operating principles, mathematical modeling, and control strategies will be discussed for these classical electric machines.

Table 1 Application of classical electric machines for EV propulsion

Machine type	Typical application
DC	Fiat Panda Elettra [1], Citroën Berlingo Electrique [1]
IM	2022 Audi e-tron S Sportback quattro [2], 2021T Model 3 (front motor) [3], 2021T Model S (rear motor) [4]
WRSM	BMW iX3 [5]
PMSM	2022 Audi RS e-tron GT [6], 2021T Model 3 (rear motor) [3], 2021T Model S (front motor) [4], 2020 Porsche Taycan 4S [7], 2021 Ford Mustang Mach-E GT [8], 2021 Chevrolet Bolt [9], 2020 Jaguar I-Pace S EV400 [10]

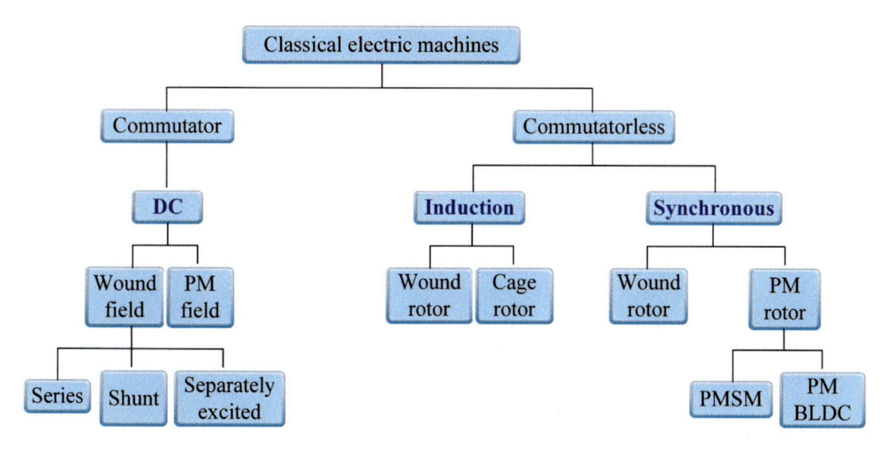

Fig. 1 Classification of the classical electric machines

2 DC Machines

The DC machine operates with DC voltage and current but its armature works on AC with the mechanical rectifier to change it into DC. Different from AC machines, the magnetic field produced by the stator is not rotating but has a steady direction. Figure 2 depicts the configuration of a DC machine with one pole pair. Its stator consists of two PM poles generating a constant magnetic field. Its rotor consists of a simple single-turn coil which is called the armature. The armature coil is connected to a mechanical commutator with two segments contacted to two carbon brushes. Finally, two terminals of a DC source are connected to the carbon brushes to energize the armature coil. The commutator is the mechanical rectifier that reverses the current direction of the armature coil. Using the mechanical commutator, during the rotor spinning, a fixed spatial distribution of the current flowing in the same direction in the armature coil is produced. Due to the use of the commutator, the current keeps flowing in only one direction in the external circuit which is connected to the armature coil through the carbon brushes, and the carbon brushes make the current flow from the DC source to the armature coil via the commutator. Based on the formula of Ampere's force, the torque generated on the side coil is calculated by:

$$T_e = F_e D \cos\theta = BILD \cos\theta \tag{1}$$

where F_e is the Ampere's force created on each coil side, D is the width of the armature coil, θ is the angle between the magnetic field and the armature coil surface and varies along with the rotation, B is the magnetic flux density generated by the PM poles or field coil in the stator, I is the amplitude of the current flowing in the armature coil, and L is the length of the rotor armature coil.

To establish the mathematical model for an actual DC machine, an equivalent circuit of the separately excited DC machines is taken for an example which is

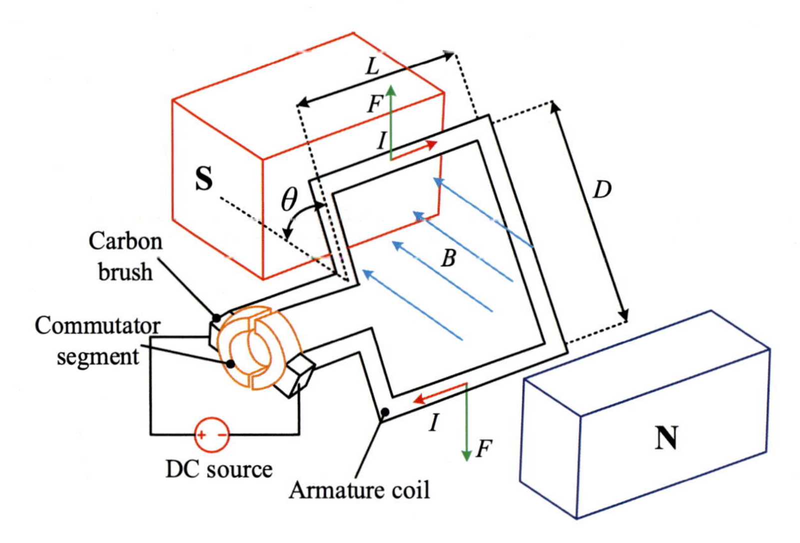

Fig. 2 Configuration of a DC machine

shown in Fig. 3, the machine operating principle is governed by:

$$\begin{cases} E_A = K_e \phi \omega_m \\ \phi = K K_F I_F \\ V_A = E_A + R_A I_A \\ VF = RFIF \\ T_{\text{dev}} = K \phi I_A \end{cases} \tag{2}$$

where E_A is the induced back electromotive force (EMF) in the armature winding, K_e is the constant of the back EMF, ϕ is the flux linkage of the armature winding, ω_m is the rotor mechanical speed, K is the magnetic field constant, K_F is a constant under unsaturated condition representing the slope of the magnetization characteristic, I_F is the field current, V_A is the terminal voltage of the armature winding, R_A is the internal resistance of armature winding, I_A is the armature current, V_F is the terminal voltage of the field winding, R_F is the internal resistance of the field winding, and T_{dev} is the developed torque. In addition, K_F depends on the number of field winding turns, parameters of the equivalent magnetic circuit, and B-H characteristics of the iron core.

Based on the different connections of the field winding and armature winding, the DC machines can be divided into several types, namely, series, shunt, separately excited, cumulative compound, differential compound, and PM DC machines. Due to different relationships between the field and armature currents, various torque-speed curves of these DC machines can be obtained. For shunt DC machines, the armature and field winding are connected in parallel to the same DC source, and thus both the field and armature currents can be regulated simultaneously. For series DC machines, the field and armature windings are connected in series, hence the field and

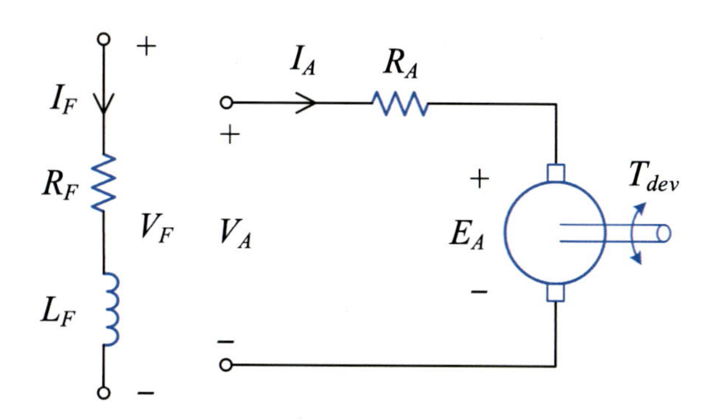

Fig. 3 The equivalent circuit for a separately excited DC machine

armature currents are identical and can be regulated at the same time. For separately excited DC machines, the field and armature windings are fed by different voltage sources individually, therefore the field and armature currents are decoupled totally and can be regulated independently. The cumulative compound DC machines have two types of field windings. One field winding is connected in series to the armature winding, and the other one is connected in parallel to the armature winding which can be regarded as the combination of a series and shunt DC machines. The flux components produced by the series field winding and the shunt field winding have the same direction. Compared to the cumulative compound DC machine, the flux components produced by the series field winding and the shunt field winding of the differential compound DC machine have the reversed direction hence opposing each other. By using PMs to replace the field winding in the stator, the PMDC machine is achieved. Compared to other electrically excited DC machines, the power density and efficiency of the PMDC machines are higher because of the magnetic field produced by high-energy-density PM materials without additional ohmic losses in the field winding. Nevertheless, since the magnetic field produced by the PMs in the PMDC machine is not adjustable, without the flux control it cannot achieve the operating characteristics, especially for high-speed operation.

Because of the utilization of the commutator and carbon brushes, the electrically excited and PM DC machines have the same inherent disadvantages, such as tear and wear. In addition, the commutator limits the rotor speed and also causes torque ripples. The carbon brushes induce the problems such as friction and radio frequency interference. Furthermore, routine maintenance for the carbon brushes and the mechanical commutator is required due to the problem of tear and wear. Due to these limitations and disadvantages, DC machines are not unsuitable for high-performance EV drives, which hinders the wide application of DC machines in EV propulsion systems.

In this chapter, the operating principle of several classical DC machines, namely the shunt, series, and separately-excited DC machines, is also discussed in detail.

Figure 5 shows the equivalent circuit and torque-speed characteristics of a shunt DC machine. The governed equations are given by:

$$\begin{cases} V_T = R_A I_A + E_A \\ I_A = \dfrac{T_{\text{dev}}}{K\phi} \\ T_{\text{dev}} = \dfrac{K\phi}{R_A}(V_T - K\phi\omega_m) \end{cases} \tag{3}$$

where V_T is the armature winding terminal voltage.

According to the torque equation in (3), when keeping the field constant K, field flux linkage ϕ, the armature internal resistance $R_{A,}$ and the armature terminal voltage V_T unchanged, the torque and speed have a linear relationship, as shown in Fig. 5b. The stall torque is $K\phi\ V_T/R_A$ and the slop is $(K\phi)^2/R_A$.

Figure 6 shows the equivalent circuit and torque-speed characteristics of a series DC machine. The governed equations are given by:

$$\begin{cases} \phi = K_F I_F = K_F I_A \\ E_A = K\phi\omega_m = K K_F I_A \omega_m \\ V_T = R_F I_F + R_A I_A + E_A \\ T_{\text{dev}} = \frac{K K_F V_T^2}{(R_F + R_A + K K_F \omega_m)^2} \end{cases} \tag{4}$$

According to the torque equation in (4), when keeping the field constant K, field flux linkage ϕ, the armature internal resistance $R_{A,}$ and the armature terminal voltage V_T unchanged, the torque and speed have a nonlinear relationship and there is no no-load speed, as shown in Fig. 6b. The stall torque is $\frac{K K_F V_T^2}{(R_A + R_F)^2}$.

Figure 7 shows the equivalent circuit and torque-speed characteristics of a series DC machine. The torque-speed curves of a series and shunt DC machines are similar because the current control for both the field and armature windings is independent. As illustrated in Fig. 7b, the torque and speed have a linear relationship, the stall torque is $K\phi\ V_T/R_A$, and the slop is $(K\phi)^2/R_A$.

Speed control is important for DC machines, especially for EV traction applications. Take the shunt DC motor for example, according to (3), there are three ways to vary the machine speed: (1) to insert resistance connected to the armature winding in series which can change the resistance of the armature branch, and hence the slope of the torque-speed curve; (2) to vary the field current while keeping the armature supply voltage constant, hence varying the field flux ϕ; (3) to vary the voltage supplied to the armature winding while keeping the field constant, namely changing the stall torque. The torque-speed curves under these three methods are illustrated in Fig. 8.

Figure 9 presents a comparison of torque-speed characteristics for different DC machines, namely shunt, series, and cumulative compound DC machines. It should be noted that among three DC machines, the shunt DC motor has very less variation in the speed under different load conditions. Thus, shunt DC machines are known as constant-speed machines. Furthermore, the series DC machine has the largest variation in speed as the load torque varies. The cumulative compound DC machine illustrated in Fig. 4 can be regarded as a combination of a shunt and series DC machine and has the average speed variation under different load conditions. When the load is light, the shunt and compound DC machines can work safely at a definite speed. Under heavy load, the speeds for these two DC machines drop less than that of a series DC machine.

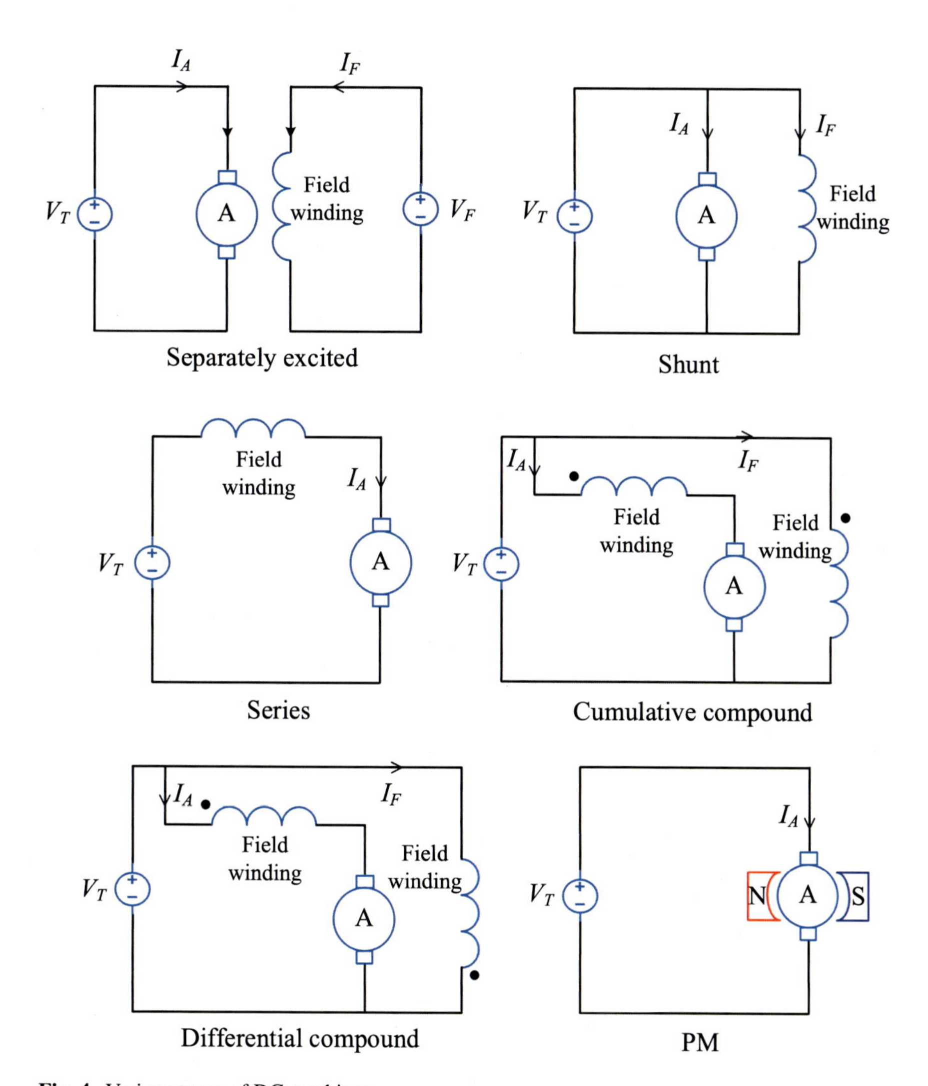

Fig. 4 Various types of DC machines

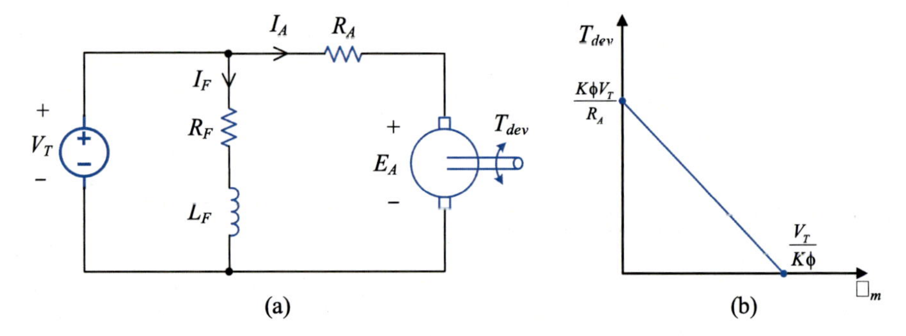

Fig. 5 Shunt DC machine. **a** Equivalent circuit. **b** Torque-speed characteristic

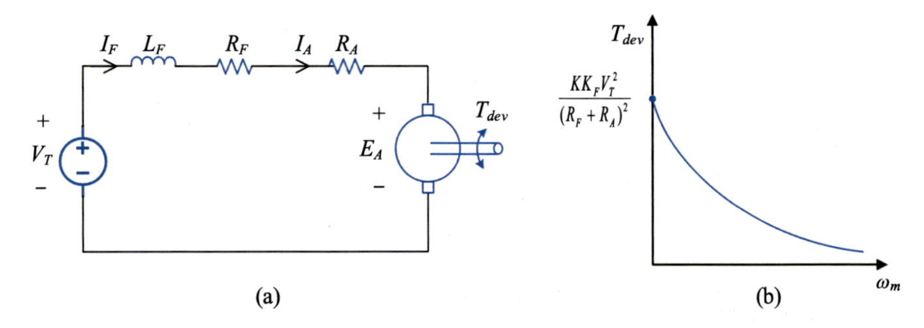

Fig. 6 Series DC machine. **a** Equivalent circuit. **b** Torque-speed characteristic

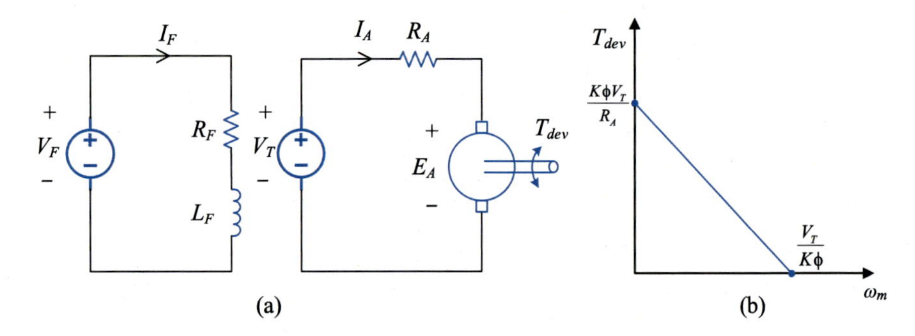

Fig. 7 Separately excited DC machine. **a** Equivalent circuit. **b** Torque-speed characteristic

3 Induction Machines

Compared to DC machines presented before, induction machines are commutator-less and brushless which makes them more reliable, hence they are widely adopted by the automobile company. According to the configuration of the rotor, induction

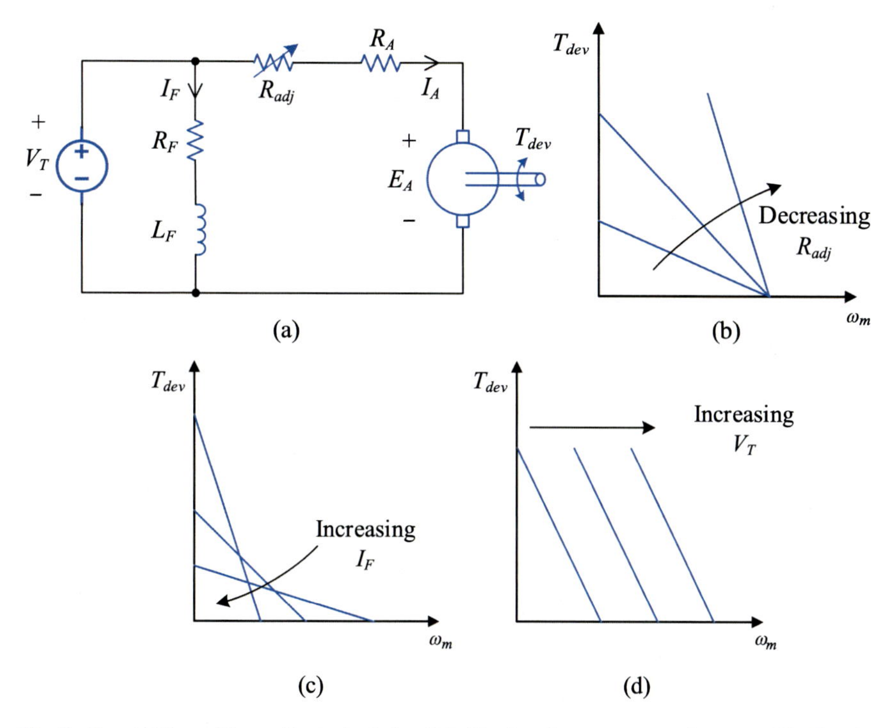

Fig. 8 Shunt DC machine. **a** Equivalent circuit. **b** Varying the armature resistance. **c** Varying the field current. **d** Varying the armature voltage

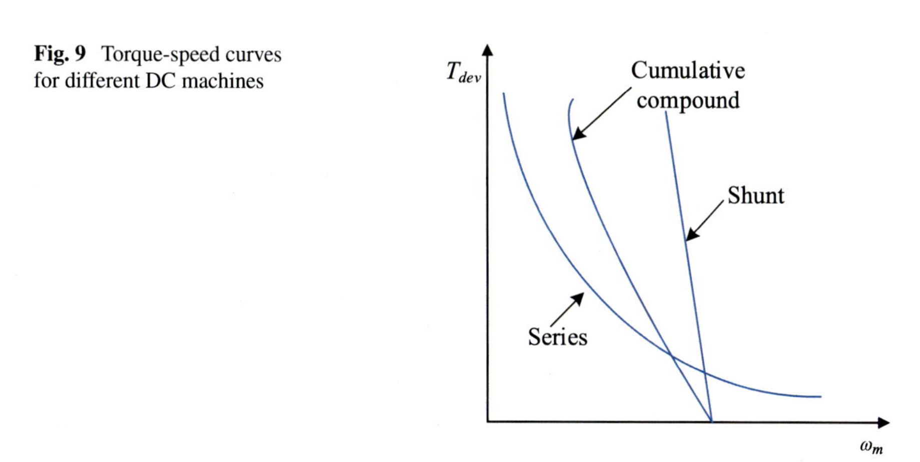

Fig. 9 Torque-speed curves for different DC machines

machines can be classified into two main categories, namely, wound-rotor induction machines (WRIMs) and squirrel-cage induction machines (SCIMs). Due to the simple structure and lower cost, and maintenance-free feature, the SCIM is more popular than the squirrel-cage WRIM for EV traction applications. Figure 10 shows

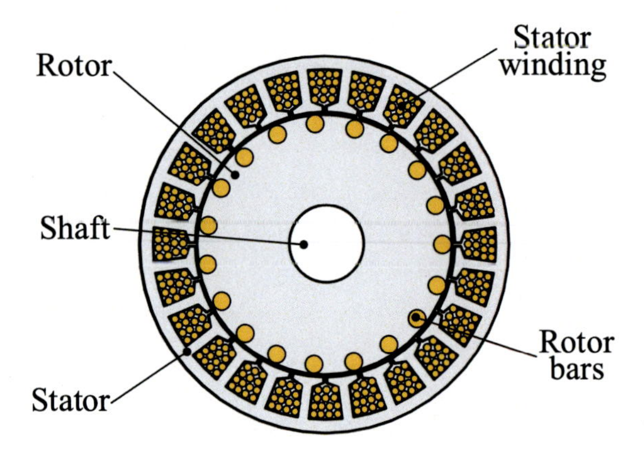

Fig. 10 Configuration of a SCIM

a typical configuration of a SCIM, which consists of a stator with polyphase armature windings and a rotor with cage bars. This configuration makes the rotor circuit not accessible from the outside but more robust and more complex for control.

The three-phase windings are assumed to be sinusoidally distributed along the stator circumference. Each phase is energized by an AC varying in sinusoidal magnitude with time. Without loss of generality, the balanced three-phase currents are given as follows:

$$\begin{cases} i_a = I_0 \cos \omega_e t \\ i_b = I_0 \cos(\omega_e t - 2\pi/3) \\ i_c = I_0 \cos(\omega_e t + 2\pi/3) \end{cases} \tag{5}$$

where I_0 is the current amplitude and ω_e is the current angular frequency.

The fundamental harmonics of magnetomotive force (MMF) are generated by each phase and it can be regarded as the superposition of two MMF waveforms traveling in the positive and negative direction, respectively.

$$\begin{cases} F_a = F_{a+} + F_{a-} = F_0[\cos(\theta_0 - \omega_e t) + \cos(\theta_0 + \omega_e t)]/2 \\ F_b = F_{b+} + F_{b-} = F_0[\cos(\theta_0 - \omega_e t) + \cos(\theta_0 + \omega_e t + 2\pi/3)]/2 \\ F_c = F_{c+} + F_{c-} = F_0[\cos(\theta_0 - \omega_e t) + \cos(\theta_0 + \omega_e t - 2\pi/3)]/2 \end{cases} \tag{6}$$

where F_0 is the amplitude and θ_0 is the initial angle, "+" and "−" signs represent the MMF travels in the positive and negative direction, respectively.

Therefore, the resultant MMF superposed by three-phase windings from phases *abc* is calculated by:

$$F_s = F_a + F_b + F_c = 3F_0 \cos(\theta_0 - \omega_e t)/2 \tag{7}$$

From (7), it shows that the fundamental air-gap MMF wave is a rotating wave whose angular velocity is ω_e. For a machine having p pole-pair number, the synchronous velocity ω_s is ω_e/p. When the rotor rotates at a mechanical velocity of ω_r, the rotor slip s is defined as:

$$s = \frac{\omega_s - \omega_r}{\omega_s} \tag{8}$$

In general, since the three-phase winding is not perfectly distributed in a sinusoidal manner along the stator circumference, there are other harmonic components of MMFs produced by the three-phase sinusoidal currents, notably the 3rd, 5th, and 7th harmonic components. The MMFs produced by these harmonic components are discussed in detail to unveil the effect on the machine performance due to these harmonics.

The MMF produced by the 3rd harmonic component can be expressed as:

$$\begin{cases} F_{a3} = F_3 \cos 3\theta_0 \cos \omega_e t \\ F_{b3} = F_3 \cos 3(\theta_0 - 2\pi/3) \cos(\omega_e t - 2\pi/3) = F_3 \cos 3\theta_0 \cos(\omega_e t - 2\pi/3) \\ F_{c3} = F_3 \cos 3(\theta_0 + 2\pi/3) \cos(\omega_e t + 2\pi/3) = F_3 \cos 3\theta_0 \cos(\omega_e t + 2\pi/3) \end{cases} \tag{9}$$

where F_3 is the amplitude and the subscript "3" denotes the 3rd harmonic component.

The resultant MMF F_{s3} of these three components is given by:

$$F_{s3} = F_{a3} + F_{b3} + F_{c3} = 0 \tag{10}$$

From (10), it can be found that the resultant MMF due to the 3rd harmonic component is canceled off in the symmetrical three-phase stator winding. The MMFs produced by the harmonic order of $3k$ ($k = 1, 2, 3\ldots$), such as the 6th, 9th, and 15th, etc. have the same result.

Similarly, the MMF produced by the 5th harmonic component can be calculated by:

$$\begin{cases} F_{a5} = F_{a5+} + F_{a5-} = F_5[\cos(5\theta_0 - \omega_e t) + \cos(5\theta_0 + \omega_e t)]/2 \\ F_{b5} = F_{b5+} + F_{b5-} = F_5[\cos(5\theta_0 - \omega_e t - 2\pi/3) + \cos(5\theta_0 + \omega_e t)]/2 \\ F_{c5} = F_{c5+} + F_{c5-} = F_5[\cos(5\theta_0 - \omega_e t + 2\pi/3) + \cos(5\theta_0 + \omega_e t)]/2 \end{cases} \tag{11}$$

where F_5 is the amplitude and the subscript "5" denotes the 5th harmonic component.

The resultant MMF F_{s5} of these three components is given by:

$$F_{s5} = F_{a5} + F_{b5} + F_{c5} = 3F_5 \cos(5\theta_0 + \omega_e t)/2 \tag{12}$$

From (12), it shows that the air-gap MMF produced by the 5th harmonic component rotates in the reversed direction as that of the fundamental MMF. It can be obtained that the MMFs produced by the harmonic order of $6k - 1$ ($k = 1, 2, 3\ldots$) have the same rotating direction as that of the 5th harmonic component.

The MMF produced by the 7th harmonic component can also be calculated by:

$$\begin{cases} F_{a7} = F_{a7+} + F_{a7-} = F_7[\cos(7\theta_0 - \omega_e t) + \cos(7\theta_0 + \omega_e t)]/2 \\ F_{b7} = F_{b7+} + F_{b7-} = F_7[\cos(7\theta_0 - \omega_e t) + \cos(7\theta_0 + \omega_e t + 2\pi/3)]/2 \\ F_{c7} = F_{c7+} + F_{c7-} = F_7[\cos(7\theta_0 - \omega_e t) + \cos(7\theta_0 + \omega_e t - 2\pi/3)]/2 \end{cases} \quad (13)$$

where F_7 is the amplitude and the subscript "7" denotes the 7th harmonic component.

The resultant MMF F_{s7} of these three components is given by:

$$F_{s7} = F_{a7} + F_{b7} + F_{c7} = 3F_7 \cos(7\theta_0 - \omega_e t)/2 \quad (14)$$

From (14), it shows that the air-gap MMF produced by the 7th harmonic component rotates in the same direction as that of the fundamental MMF. It is also easily deduced that the MMFs produced by the harmonic order of $6k + 1$ ($k = 1, 2, 3\ldots$) have the same rotating direction as that of the 7th harmonic component.

The magnetic field produced by the rotor currents rotates at $s\omega_s$ with respect to the rotor, and hence the rotor magnetic field rotates at ω_s which is superimposed on the rotation of the rotor. Therefore, a steady torque is produced by the stator and rotor magnetic fields.

The per-phase stator and rotor equivalent circuit for SCIMs are shown in Fig. 11. In the circuit, V_s is the terminal voltage of the stator per phase; i_s and i_r are the currents flowing in the stator and rotor windings; $i_r{'}$ is the rotor current referred to the stator side; R_s and X_s are the resistance and leakage reactance per phase in the stator, respectively; R_r and X_r are the resistance and leakage reactance per phase in the rotor, respectively; E_s and E_r are the back EMF in the stator and rotor, respectively, R_c and X_m are the equivalent core loss resistance and magnetizing reactance; n_s and n_r are the numbers of turns of the stator winding and rotor winding, respectively. Compared to a transformer where the frequencies for the primary and secondary are equal, the stator current frequency f and the rotor current frequency f_r are not the same motion between the stator and the rotor. The rotor frequency f_r equals sf. The rotor back EMF and reactance are sE_r and sX_r, respectively.

Based on the electromagnetic law, the flux linkage and voltage equations for SCIMs can be derived based on the inductances:

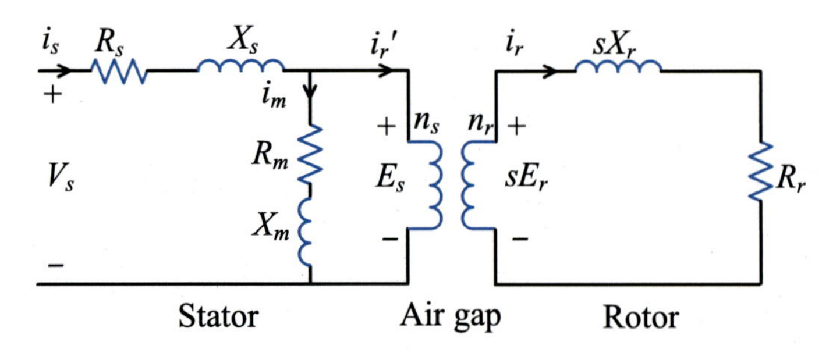

Fig. 11 Stator and rotor equivalent circuit for SCIMs

$$\begin{cases} \phi^s_{abs} = L_{os} i^s_{abc} + M_{osr} i^r_{abc} \\ \phi^r_{abs} = L_r i^r_{abc} + M_{sr} i^s_{abc} \end{cases} \tag{15}$$

$$\begin{cases} v^s_{abs} = R_s i^s_{abc} + d\phi^s_{abc}/dt \\ v^r_{abs} = R_r i^r_{abc} + d\phi^r_{abc}/dt \end{cases} \tag{16}$$

where the quantities ϕ, i, and v denote the flux linkage, current, and voltage, respectively, the superscripts s and r denote the stator and rotor, respectively, R_s and R_r are resistances of the stator and the rotor, L_{os} and L_r are self-inductances of the stator and the rotor, and M_{sr} is the mutual inductance between the stator and the rotor, respectively.

The above mathematical modeling is based on the stator reference frame, and the mutual inductance is rotor position-dependent. To transform it into the rotor reference frame, the following transformation matrix is applied:

$$T = k \begin{bmatrix} \cos\theta & \cos(\theta - 2\pi/3) & \cos(\theta + 2\pi/3) \\ \sin\theta & \sin(\theta - 2\pi/3) & \sin(\theta + 2\pi/3) \\ 1/2 & 1/2 & 1/2 \end{bmatrix} \tag{17}$$

where k is the transformation coefficient. For amplitude-invariant transformation, k equals 2/3 and for power-invariant transformation k is $\sqrt{2/3}$.

By applying the transformation matrix, the modeling of the flux linkage and voltage equations based on *abc*-domain (stator reference frame) can be transformed into *qd0*-domain (rotor reference frame):

$$\begin{bmatrix} \phi^s_d \\ \phi^s_q \\ \phi^s_0 \end{bmatrix} = L_s \begin{bmatrix} i^s_d \\ i^s_q \\ i^s_0 \end{bmatrix} + M_{sr} \begin{bmatrix} i^r_d \\ i^r_q \\ i^r_0 \end{bmatrix} \tag{18}$$

$$\begin{bmatrix} \phi^r_d \\ \phi^r_q \end{bmatrix} = L_r \begin{bmatrix} i^r_d \\ i^r_q \end{bmatrix} + M_{sr} \begin{bmatrix} i^s_d \\ i^s_q \end{bmatrix} \tag{19}$$

$$\begin{cases} v^s_d = r_s i^s_d + d\phi^s_d/dt - \omega_s \phi^s_q \\ v^s_q = r_s i^s_q + d\phi^s_q/dt + \omega_s \phi^s_d \end{cases} \tag{20}$$

$$\begin{cases} v^r_d = r_r i^r_d + d\phi^r_d/dt - (\omega_s - p\omega_m)\phi^r_q \\ v^r_q = r_r i^r_q + d\phi^r_q/dt + (\omega_s - p\omega_m)\phi^r_d \end{cases} \tag{21}$$

where the superscripts s and r denote the stator and rotor quantities, p is the pole-pair number of the magnetic field, and ω_s and ω_m are the synchronous and rotor angular velocities.

The synchronous and rotor angular velocities are determined by:

$$\omega_s = d\theta_s/dt, \omega_m = d\theta_m/dt \tag{22}$$

The developed torque can be expressed based on the rotor reference frame:

$$T_{\text{dev}} = \frac{3}{2}P\frac{M_{sr}^2}{L_r}i_d^s i_q^s \tag{23}$$

where P is the magnetic field pole-pair number. The developed torque formula (23) is similar to the torque equation of a separately excited DC machine presented in (2). The stator d-axis current i_d^s resembles ϕ while the stator q-axis current i_q^s resembles I_A.

For induction machine control, there are mainly two types of methods, namely, scalar control and vector control. The scalar control is much simpler than the vector control method and can maintain relatively high efficiency throughout a wide range. It is based on the constant volts/hertz principle, which is also called the variable voltage and variable frequency (VVVF) method. At sufficiently high speeds, the stator reactance is much greater than the resistance and the voltage drop across the stator resistance is negligible, and the back EMF is approximately equal to the terminal voltage. To keep the flux density constant, the terminal voltage of the stator should also be adjusted to maintain the same ratio as the voltage frequency. By keeping the ratio V_s/f, the peak torque of the machine seemingly stays constant regardless of the frequency. This implies that the machine will be able to generate its peak torque at any frequency, which allows the machine to accelerate as fast as possible from a standstill to the desired speed.

However, it is limited to frequencies above the rated frequency. Beyond base speed, the stator terminal voltage is generally maintained constant at its rated value. For the low-speed region, the stator terminal voltage should be improved to compensate for the large voltage drop across the stator resistance. Therefore, in the low-speed operating region, the voltage drop across the stator impedance can be compensated by the boosting voltage based on the constant volts/hertz strategy. The open-loop scalar control is much more attractive in the industrials when a small offset in speed and air-gap flux due to fluctuation are of no significance.

Although the VVVF control is relatively easy to be implemented, the dynamic performance of the motor drive is not acceptable. To meliorate the dynamic performance of the motor drive, the field-oriented control (FOC) is proposed for the induction machine control.

Figure 12 shows the control diagram of the FOC for SCIMs [11]. When applying FOC, at first the motor model should be transformed from the stator frame to the synchronously spinning rotor frame with speed ω_e. Based on the FOC, from the operating principle respect, the SCIMs can be converted to separately excited DC machines using a similar control strategy. By using the reference frame transformation presented in (17), the motor model of SCIMs can be transformed from the stator reference frame presented in (15) and (16) to the general synchronously rotating frame presented in (18)–(21). Based on FOC, the rotor flux is regulated by the stator

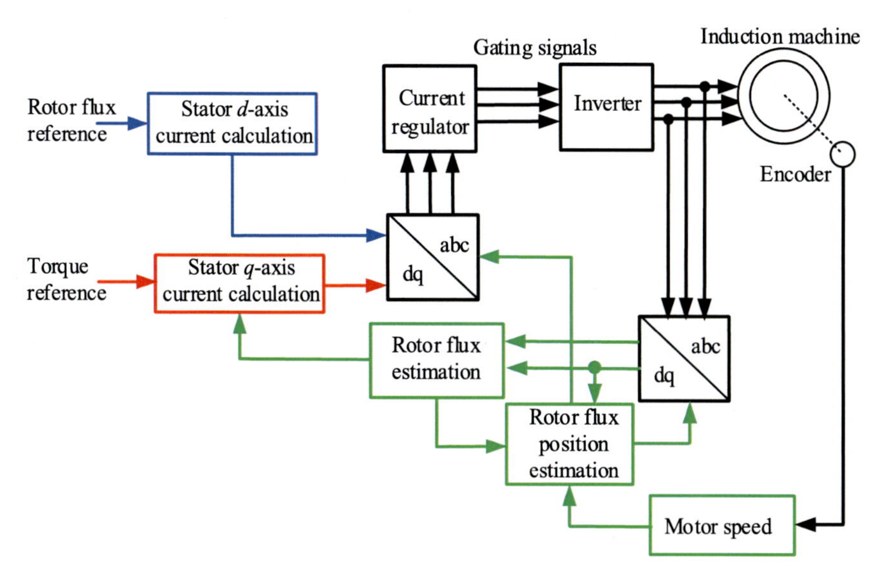

Fig. 12 FOC control block diagram for SCIMs

d-axis current and the developed torque is controlled by the stator q-axis current effectively to provide a satisfactory transient response with a short transition time. With the development of high-performance power electronic hardware, FOC-based induction machine drives have been more and more popular for EV propulsion systems.

The direct torque control (DTC) is another popular control strategy for high-performance induction machine drives. Compared to the FOC, the objectives of the DTC are to control the stator flux and machine torque, and the machine model is established based on the stationary reference frame. Figure 13 shows the control diagram of the DTC for induction machine drives [11]. The instantaneous stator flux and machine torque are estimated by the stator voltage and current vectors. The stator flux and machine torque are controlled by two respective regulators. The outputs of two regulators along with the stator flux angle determine the voltage vector applied to the inverter and control the stator flux and machine torque to the desired values. Since the transformation from the stationary reference frame to the rotating reference frame needs additional trigonometric calculation, the complexity of the FOC implementation is higher than that for the DTC. It also reported that the DTC possesses a higher dynamic performance since the torque is indirectly controlled by the stator currents in the FOC [11].

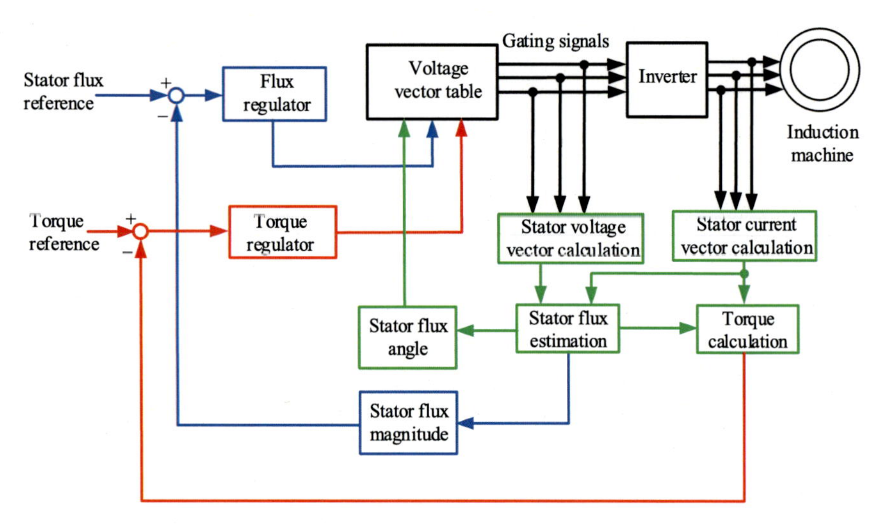

Fig. 13 DTC control block diagram for SCIMs

4 Synchronous Machines

The operating principle of synchronous machines is much similar to that of induction machines, and the developed torque is produced based on the interaction of the magnetic fields produced by the stator and rotor. The difference is that for a synchronous machine, the rotor rotating velocity equals the synchronous speed which means that the slip in the induction machine is zero. The magnetic field from the rotor is generated from the field winding or permanent magnets for which the machines are named wound-rotor synchronous machines (WRSMs) and permanent magnet synchronous machines (PMSMs). As shown in Fig. 14, the rotor magnetic field of a WRSM is generated by the field winding wound in the rotor core. The field winding is energized through an external circuit connected to brushes on the collector rings. Compared to PMSMs, WRSMs have the advantages of flexible adjustment of the rotor magnetic field in the air gap and no requirement for the rear-earth permanent magnets. However, due to the utilization of the PMs, the torque density and power density of PMSMs are much higher than those of WRSMs.

Compared to the WRSMs, the PMSMs are currently selected by the EV companies for the EV traction application. Since the high-energy PM material is applied in these machines, the power density, torque density, and efficiency are higher than the other classical electric machines. Their major advantages make them much more suitable for EV propulsion [12, 13]. Firstly, the overall weight and size are greatly decreased under the same power rating, due to the magnetic field excited by high-energy PMs; secondly, since there is no copper loss in the rotor, the overall efficiency of PMSMs is much higher; thirdly, because the heat primarily is produced by the stator, it is easily dissipated from the machine; finally, as the manufacturing defects, temperature rise or

Fig. 14 Configuration of a WRSM

mechanical integrity of PMs are improved due to the mature technology, the PMSMs exhibit higher robustness and reliability.

As that of a WRSM shown in Fig. 14, the stator of the PMSMs is also identical to that of the SCIMs, which consists of a stator core lamination equipped with polyphase winding. Compared with SCIMs, the rotor is incorporated with PM poles, which is relatively simpler since there are no rotor bars and end rings. Moreover, since the generated heat from the rotor is not substantial, fan blades are not always required to be mounted on the shaft to cool the rotor.

Similar to the WRSMs, the stator of a PMSM also generally adopts a three-phase distributed winding. By injecting balanced three-phase sinusoidal currents into the stator winding, a rotating sinusoidal magnetic field can be produced. By replacing the field winding of the WRSM with PM poles having the same pole-pair number as the stator field produced by the stator winding, the WRSM can be converted to a PMSM. Therefore, the rotor of a PMSM with the same pole-pair number as the rotating stator field revolves synchronously with the stator field. Based on the PM locations in the rotor, the PMSM can be roughly categorized into four types, namely, the surface-mounted, surface-inset, interior-radial, and spoke-type topologies.

Figure 15a shows the topology of a surface-mounted PMSM where PMs are fixed on the rotor surface directly. Therefore, the machine structure is easy and simple to be fabricated. Since the PM permeability is similar to the permeability of air, the effective air-gap length is the summation of the PM thickness and the physical air-gap length. The stator inductance is relatively low due to the large effective air-gap length; hence the armature reaction field is weak. In addition, the *d*- and *q*-axis inductances of stator winding are approximately equal, and there is no reluctance torque component. However, for high-speed operation, PMs may be separated from the rotor due to the centrifugal force, and the carbon sleeve is required to secure the PMs in position.

Figure 15b depicts the configuration of a surface-inset PMSM. Unlike the surface-mounted one, PMs for this topology are placed into the rotor surface. The effective air-gap length is not constant and varies along with the rotor position. The *d*- and

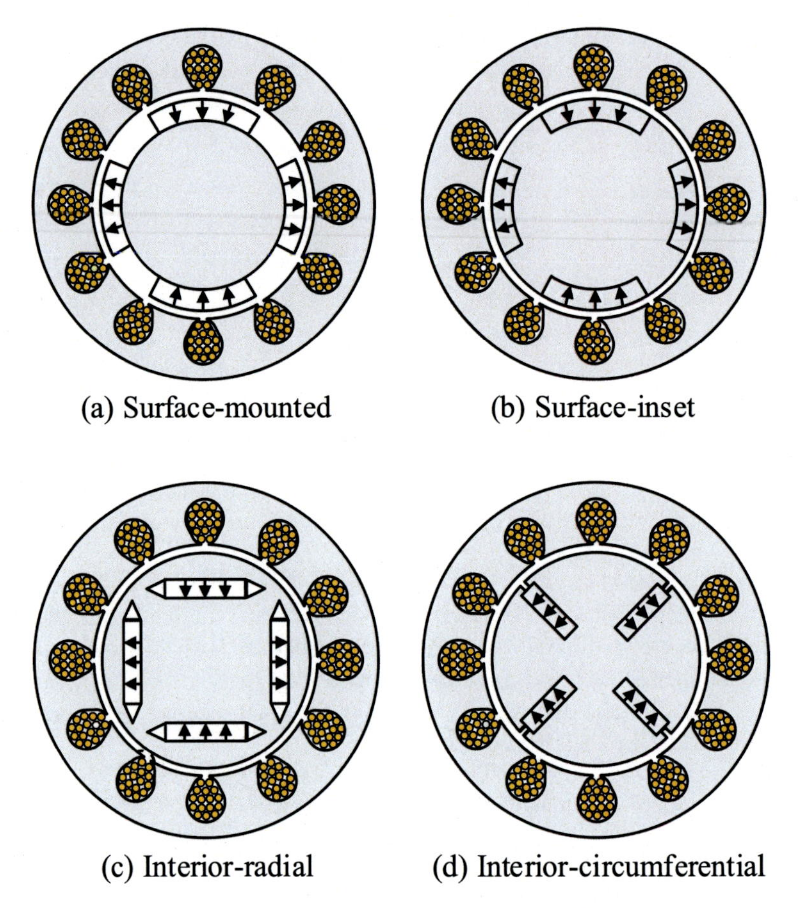

Fig. 15 Different topologies for PMSMs. **a** Surface-mounted. **b** Surface-inset. **c** Interior-radial. **d** Spoke-type

q-axis inductances are not equal and the q-axis inductance is higher. Thus, the reluctance torque is generated due to the salient structure. Compared to the surface-mounted topology, the surface-inset structure exhibits better mechanical ruggedness for holding the PMs under high-speed operation.

Figure 15c shows the topology of an interior-radial topology PMSM. The PMs are magnetized radially and embedded inside the rotor. Since PMs are totally inside the rotor, the mechanical integrity is better than that of the surface-inset topology. In addition, like the surface-inset PMSM, due to the salient structure, the reluctance torque is not zero. However, compared to the surface-inset PMSM, the interior-radial structure uses flat PMs, which are much easier in terms of manufacture and assembly.

Figure 15d illustrates the structure of a spoke-type PMSM. The PMs are circumferentially magnetized and also embedded inside the rotor. Compared to the interior-radial topology, it has a flux-concentration effect and the air-gap flux density is higher. Furthermore, like the surface-inset and interior radial PMSMs, spoke-type PMSMs also show good mechanical ruggedness and a better torque density due to the reluctance torque component. On the other hand, since there is a large amount of flux leakage at one end, the shaft should be nonmagnetic to avoid magnetic flux leakage. This may deteriorate the shaft torsional stiffness.

The operating principle of WRSMs and PMSMs is similar, which depends on two rotating magnetic fields produced by the stator armature winding and PMs on the rotor or the rotor field winding. These two magnetic fields rotate at the same angular velocity in the same direction, namely, the synchronous speed. In the stator reference frame, some parameters such as inductances and flux linkage are rotor position-dependent. For high-performance drives, the reference frame transformation is proven to be useful, which changes the PMSMs into DC machines and eases the synchronous machines and control. When the flux linkage, voltage, and currents are assumed to be sinusoidal and the magnetic saturation effect is ignored, the flux linkage and voltage of WRSMs in the rotor reference frame are expressed as:

$$\begin{cases} \phi_d^s = L_d^s i_d^s + M_{sr} i_f^r \\ \phi_q^s = L_q^s i_q^s \\ \phi_f^r = L_f^r i_f^r + M_{sr} i_d^s \end{cases} \tag{24}$$

$$\begin{cases} v_d^s = R_s i_d^s + d\phi_d^s/dt - \omega_e \phi_q^s \\ v_q^s = R_s i_q^s + d\phi_q^s/dt + \omega_e \phi_d^s \\ v_f^r = R_r i_r^s + d\phi_f^r/dt \end{cases} \tag{25}$$

where the quantities ϕ, I, and v denote the flux linkage, current, and voltage, respectively, the superscripts s and r denote the stator and rotor, and R_s and R_r are the stator and rotor resistances, L_d, L_q and L_f are the stator d-axis and q-axis inductances and rotor inductances, and M_{sr} is the mutual inductance, respectively.

The developed torque can be computed from the *d*- and *q*-axis currents and flux linkages:

$$T_{\text{dev}} = \frac{3P}{2}\left[M_{sr} i_f^r i_q^s + i_d^s i_q^s \left(L_d^s - L_q^s\right)\right] \tag{26}$$

where P is the magnetic field pole-pair number.

Similar to the WRSMs, for PMSMs, the rotor reference frame-based machine model is established by replacing the flux linkage of the rotor field winding with the PM flux linkage:

$$\begin{cases} \phi_d^s = L_d^s i_d^s + \phi_{PM} \\ \phi_q^s = L_q^s i_q^s \end{cases} \tag{27}$$

$$\begin{cases} v_d^s = R_s i_d^s + d\phi_d^s/dt - \omega_e \phi_q^s \\ v_q^s = R_s i_q^s + d\phi_q^s/dt + \omega_e \phi_d^s \end{cases} \tag{28}$$

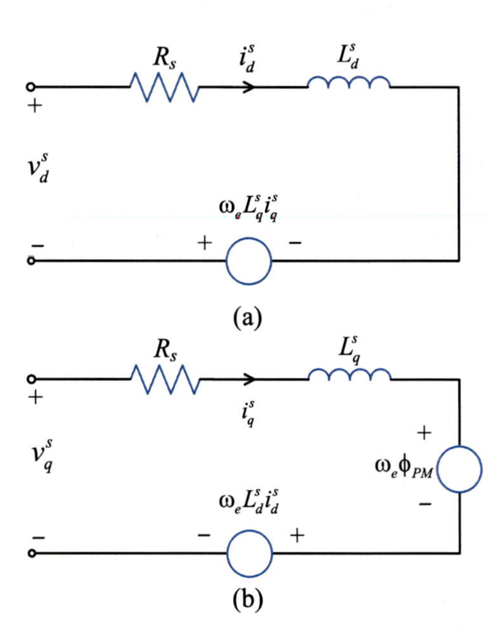

Fig. 16 Dynamic equivalent circuit of a PMSM. **a** *d*-axis based circuit. **b** *q*-axis based circuit

where ϕ_d and ϕ_q are the *d*- and *q*-axis components of flux linkages; v_d and v_q are the *d*- and *q*-axis stator voltages; i_d and i_q are the *d*- and *q*-axis stator currents. Based on the rotor reference frame model, the dynamic equivalent circuit of this PMSM can be represented by a *d*-axis-based circuit and a *q*-axis-based circuit as shown in Fig. 16.

Similar to (26), the developed torque of a PMSM can be calculated by:

$$T_{\text{dev}} = \frac{3P}{2}\left[\phi_{PM} i_q^s + i_d^s i_q^s \left(L_d^s - L_q^s\right)\right] \tag{29}$$

where P is the magnetic field pole-pair number and ϕ_{PM} is the PM flux linkage.

As shown in Fig. 17, the implementation of the FOC required the sampled stator currents converted from the stator reference frame (*abc*-domain) into the rotor reference frame (*dq0*-domain) [13]. For this transformation, the rotor flux position is required and is obtained from a position sensor. The discrepancy between the speed reference and measured speed is applied to calculate the reference torque based on a speed regulator, such as a proportional-integral (PI) controller. Then, the torque reference and the machine speed are fed into the current reference generator, such as a look-up table to generate the *d*- and *q*-axis current references. After that, two independent current regulators are implemented to control the *d*- and *q*-axis currents to generate the *d*- and *q*-axis voltage references. From the inverse transformation, the reference voltages for three phases are obtained. Finally, the PWM gating signals are generated from the PWM generator to feed into the three-phase inverter. The implementation of the DTC for PMSMs is much similar to that for the SCIMs as

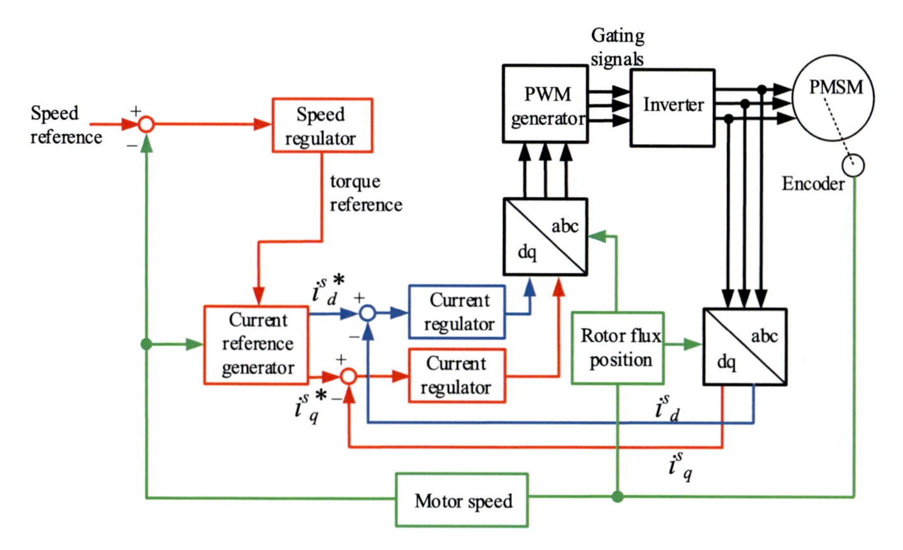

Fig. 17 FOC control block for PMSMs

shown in Fig. 13. The machine torque and stator flux are controlled in the DTC. The actual torque and stator flux are estimated from the stator voltage and current vectors. Finally, the voltage vectors generated from these regulators are applied to control the torque and stator flux in their predetermined directions.

Another type of PMSM is the permanent magnet brushless DC (PMBLDC) machine. This kind of machine has a similar configuration as the PMSMs, which consists of the stator three-phase windings and a PM rotor. Compared to PMSMs, the stator winding of a PMBLDC machine usually adopts the concentrated configuration, and the back EMF waveform is trapezoidal other than sinusoidal. Due to the utilization of concentrated-winding configuration in PMBLDC machines, the overhang winding and the copper material are greatly decreased. Furthermore, PMs in PMBLDC machines are usually surface-mounted, and thus both rotor structure and control are simple. In addition, other types for PM location such as the surface-inset, interior-radial, and spoke-type configurations can also be applied. For the PMBLDC machine, the torque generation will be degraded, when the air-gap flux waveform and back EMF are not nearly trapezoidal.

Two types of switching schemes to drive this PMBLDC machine are proposed, namely, a two-phase 120° conduction strategy and a three-phase 180° conduction strategy, as shown in Fig. 18 [13]. For the first strategy, there are only two phases carrying current simultaneously within the conduction period of 120°. During the conduction, the stator current waveform is approximately rectangular as depicted in Fig. 18a. For the second one, currents flow in all three phases during the conduction interval of 180°. Under this situation, the stator current waveform is nearly quasi-square as shown in Fig. 18b.

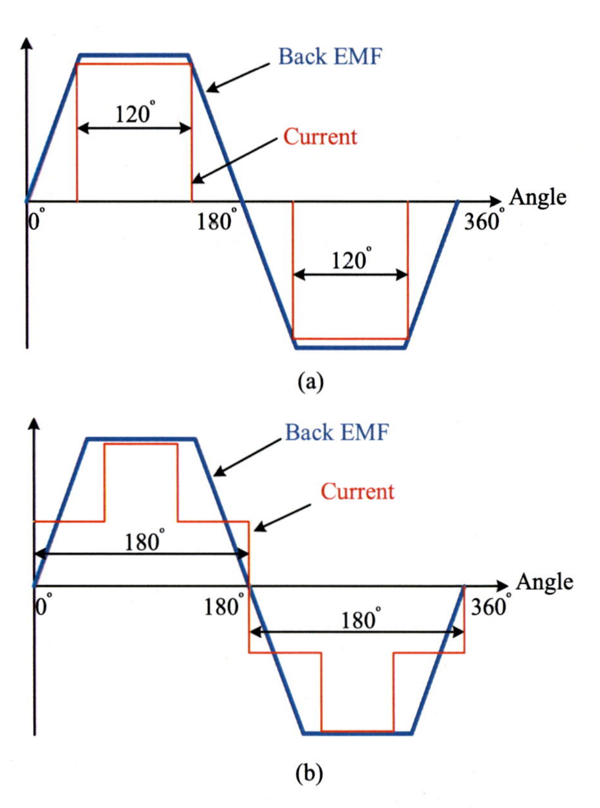

Fig. 18 BLDC operation. **a** 120° conduction strategy. **b** 180° conduction strategy

5 Summary

Although classical electric machines are a mature technology, they still play a key role in modern EV propulsion. Owning to the disadvantages such as low efficiency, reliability, and also the wear-and-tear problem for the carbon brushes and commutator, the DC machines are no more popular for the most recent EVs but still applied in some low-end and off-road EVs. The commutatorless machines, such as SCIMs, PMSMs, and PMBLDC machines, are the main players in the EV market. Owing to the definite merits of rugged structure and low cost, the SCIMs are more attractive than their wound-rotor counterpart. Although the control scheme for SCIMs is complicated, these advantages are superior to the disadvantages for EV traction machines and hold a significant market share in the future under the policy of green vehicles. Moreover, PMs have the demerits of high cost, shortage of supply, and risk of demagnetization, SCIMs are still promising candidates for EV propulsion applications.

On the other hand, PM machines exhibit higher power density, higher torque density, and higher efficiency than induction machines, which are highly promising for EV traction applications. They are becoming predominant in the EV market share. The WRSMs also have the definite merits of low cost, free of rear-earth

PMs, and high controllability for the rotor flux, and they are also selected as the potential traction motors for EV propulsion. Due to the high-performance requirements, the multi-phase PMSMs exhibit a higher power density, efficiency, lower torque ripple, and better fault-tolerant capability than the three-phase machines [14, 15]. The multi-phase machines are becoming a potential candidate for future EV propulsion applications [16].

References

1. Chau KT, Li W (2014) Overview of electric machines for electric and hybrid vehicles. Int J Veh Des 64(1):46
2. 2022 Audi e-tron S Sportback quattro—specifications and price. EVSpecifications. https://www.evspecifications.com/en/model/37981a9. Accessed 14 Mar 2022
3. 2021 tesla model 3 performance AWD—specifications and price. EVSpecifications. https://www.evspecifications.com/en/model/1af7111. Accessed 14 Mar 2022
4. "2021 tesla model S long range plus—specifications and price. EVSpecifications. https://www.evspecifications.com/en/model/37c6108. Accessed 14 Mar 2022
5. BMW iX3: everything we know about the all-new electric SUV. https://insideevs.com/news/434058/everything-we-know-bmw-ix3/. Accessed 14 Mar 2022
6. Scherr E, Sutton M (18 Nov 2021) Tested: 2022 Audi RS e-tron GT whirs to 60 MPH in 2.9 seconds. Car and Driver. https://www.caranddriver.com/reviews/a35834678/2022-audi-rs-e-tron-gt-us-drive/. Accessed 14 Apr 2022
7. 2020 Porsche Taycan 4S—specifications and price. EVSpecifications. https://www.evspecifications.com/en/model/0c44ce. Accessed 14 Mar 2022
8. 2021 ford mustang Mach-E GT—specifications and price. EVSpecifications. https://www.evspecifications.com/en/model/454de7. Accessed 14 Mar 2022
9. 2021 Chevrolet bolt EV Premier—specifications and price. EVSpecifications. https://www.evspecifications.com/en/model/97d7138. Accessed 14 Mar 2022
10. 2020 Jaguar I-Pace S EV400 AWD automatic—specifications and price. EVSpecifications. https://www.evspecifications.com/en/model/913dfd. Accessed 14 Mar 2022
11. Le-Huy H (Oct 1999) Comparison of field-oriented control and direct torque control for induction motor drives. In: Conference record of the 1999 IEEE industry applications conference. Thirty-Forth IAS annual meeting, vol 2, pp 1245–1252
12. Chau KT, Chan CC (2007) Emerging energy-efficient technologies for hybrid electric vehicles. Proc IEEE 95(4):821–835
13. Zhu ZQ, Howe D (2007) Electrical machines and drives for electric, hybrid, and fuel cell vehicles. Proc IEEE 95(4):746–765
14. Li W, Feng G, Li Z, Tjong J, Kar NC (2021) Multireference frame based open-phase fault modeling and control for asymmetrical six-phase interior permanent magnet motors. IEEE Trans Power Electron 36(10):11712–11725
15. Li W, Song P, Li Q, Li Z, Kar NC (2022) Open-phase fault modeling for dual three-phase PMSM using vector space decomposition and negative sequence components. IEEE Trans Magn 58(8). Art no 8204106
16. Hussain M, Ulasyar A, Zad HS, Khattak A, Nisar S, Imran K (2021) Design and analysis of a dual rotor multiphase brushless DC motor for its application in electric vehicles. Eng Technol Appl Sci Res 11(6). Art no 6

Advanced Electric Machines for Electric and Hybrid Vehicles

Hao Chen, Zhou Shi, and Ayman M. EL-Refaie

Abstract In this chapter, more advanced electric machines for electric and hybrid vehicles are introduced. Even though not all the performance metrics of these advanced electric machines are as good as conventional electric machines, they have some distinct advantages surpassing the conventional counterparts. Hence, these advanced electric machines have been gaining attention. As the rotor permeant magnet (PM) machines suffer from the risk of irreversible demagnetization due to poor thermal dissipation and/or faulty conditions as well as demanding magnet retention, several stator PM machines are introduced including doubly-salient PM machines, flux-switching PM machines, and flux-reversal PM machines. In order to improve the torque density, magnetic gear-based machines are introduced including vernier PM machines and magnetically-geared PM machines. Both the stator PM machines and the magnetic gear-based machines belong to the family of flux-modulated machines, since their original PM flux is modulated by the flux modulator in principle. Furthermore, in order to achieve both high torque density and wide speed range, the concept of flux-controllable/variable-flux machines is introduced including hybrid-excited machines and memory PM machines. The principle, development, and novel topologies of all these aforementioned advanced electric machines will be demonstrated in detail in this chapter. The trends and tradeoffs of these machines for electric and hybrid vehicle applications are highlighted.

H. Chen (✉)
College of Electrical Engineering, Zhejiang University, Hangzhou 310027, China
e-mail: hao.chen-ee@ieee.org

Z. Shi
Xiaomi EV Technology Co., Ltd., Beijing 100000, China
e-mail: shizhoujiangda@163.com

A. M. EL-Refaie
Department of Electrical and Computer Engineering, Marquette University, Milwaukee, WI 53233, USA
e-mail: ayman.el-refaie@marquette.edu

C. H. T. Lee (ed.), *Emerging Technologies for Electric and Hybrid Vehicles*,
Green Energy and Technology, https://doi.org/10.1007/978-981-99-3060-9_3

1 Background

As one of the most efficient energy conversion devices that interconvert mechanical and electrical powers, electric machines are the key enabling technology for EVs and HEVs, since the operation of electric machines directly affects the overall vehicle performance. As aforementioned in previous chapters, two types of electric machines have been dominantly adopted for EVs and HEVs, i.e., induction machines and PM synchronous machines (PMSMs).

Induction machines are widely used due to their high reliability and low production cost. However, because of their low efficiency, low power factor, and low inverter-usage factor, induction motors are facing fierce competition from PMSMs [1]. PMSMs are becoming more and more attractive for EVs and HEVs due to their high torque/power density, high efficiency, and high power factor. However, they suffer from the drawbacks of relatively high PM material cost and relatively low efficiency at high-speed partial-load operation region (flux-weakening region) [2]. Nowadays, in automotive market, the applications of DC machines and switched reluctance machines are phasing out, while the induction machines and PMSMs dominate the EVs/HEVs market. However, as aforementioned, induction machines and PMSMs are not without problems. Hence, in addition to the achieved performance of conventional electric machines, many advanced electric machines are continuously being proposed and developed to achieve higher torque/power density, higher efficiency, wider speed range, etc.

2 Stator-PM Machines

In conventional PM machines, e.g., surface-mounted PMSMs and interior PMSMs, the PMs are mounted on or inserted into the rotor. Such machines usually suffer from the risk of irreversible demagnetization due to poor thermal dissipation and/or faulty conditions. In addition, for high-speed EV/HEV applications, an additional retaining sleeve is required for surface-mounted PMSMs to retain the PMs against the centrifugal force or rely on thick bridges and/or center posts in case of interior PMSMs. By contrast, in stator-PM machines, the PMs are located in the stator instead of the rotor, which provides potential solutions for the issues mentioned above. Since there are neither PMs nor field windings on the rotor side, the rotor of such machines is mechanically simple and robust, which is more attractive for high-speed EV or HEV applications. There are three kinds of stator-PM machines, namely, doubly-salient PM machines, flux-switching PM machines, and flux-reversal PM machines, which will be introduced in detail in the following sections.

It should be noted that the number of rotor teeth, N_r, of stator PM machines is equivalent to the pole pair number of rotor magnets of conventional PMSMs. In other words, the fundamental frequency of stator PM machines is based on the number of poles/rotor teeth rather than the number of pole pairs. This is due to the fact that a complete electrical period of stator PM machines is associated with one rotor-tooth-pitch, and during one rotor-tooth-pitch, the direction of flux-linkage switches twice (one complete electrical period).

2.1 *Doubly-Salient PM Machines*

The doubly-salient PM machine was first proposed by Yuefeng Liao in 1992, which aims to improve the torque density of switched reluctance machines [3]. A 12-slot/8-pole doubly-salient PM machine is shown in Fig. 1. As can be seen, based on the structure of conventional switched reluctance machines, two pole pairs of PMs are installed in the stator yoke. The addition of PMs can enhance the air gap magnetic flux density and hence improve the torque density. Fractional-slot concentrated-winding configuration is usually adopted in doubly-salient PM machines. Besides the high torque density, doubly-salient PM machines inherit the advantages of simple structure and high reliability as switched reluctance machines do.

In doubly-salient PM machines, assuming that the fringing is negligible and the permeability of the iron core is infinite, the PM flux in each coil changes with the variation of rotor position is shown in Fig. 2. At the initial position, the stator pole with coil B is aligned with the rotor pole, and hence the flux of coil B is at its maximum value. When the rotor rotates counterclockwise, the flux of coil B decreases with the reduction of the pole alignment area. When the rotor turns at $\pi/12$, the stator pole with coil A is aligned with rotor pole, and the flux of coil A reaches the maximum value. When the rotor turns at $\pi/8$, the stator pole with coil B is completely out of alignment with the rotor pole, and hence the flux of coil B is at its minimum value. When the rotor turns at $\pi/4$, the stator pole with coil B is aligned with the rotor pole (another rotor pole) again, and hence the flux of coil B is at its maximum value again. So far, an electrical period has been completed. If the rotor rotates continuously, the aforementioned process will be repeated. The flux waveforms are as shown in Fig. 2b.

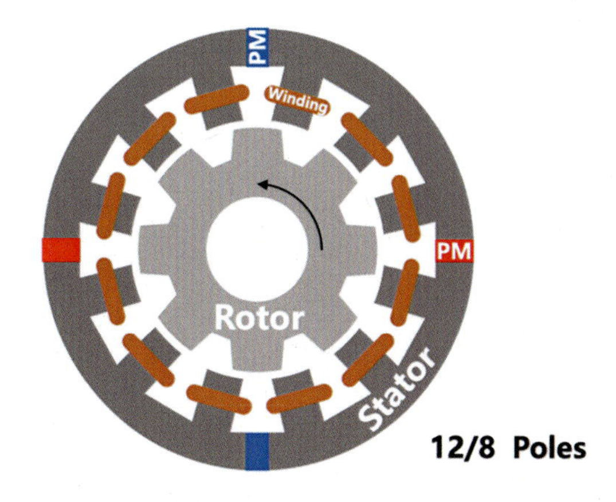

Fig. 1 Doubly-salient PM machine

Therefore, under no-load operation, a trapezoidal back-EMF is induced in each coil. As can be seen from Fig. 2, a PM excited torque can be produced by injecting a positive current when the PM flux increases, and injecting a negative current when the PM flux decreases. The torque of doubly-salient PM machines is produced by both the PM excited torque and the reluctance torque. Accordingly, compared to conventional switched reluctance machines, doubly-salient PM machines can provide higher torque due to the existence of PMs in the stator.

Doubly-salient PM machines have been gaining interest in the past two decades due to their improved torque density, simple structure, and high reliability. To further improve the torque density, many studies have been carried out. Conventional doubly-salient PM machines with 6/4 pole or 12/8 pole structure have a small variation of

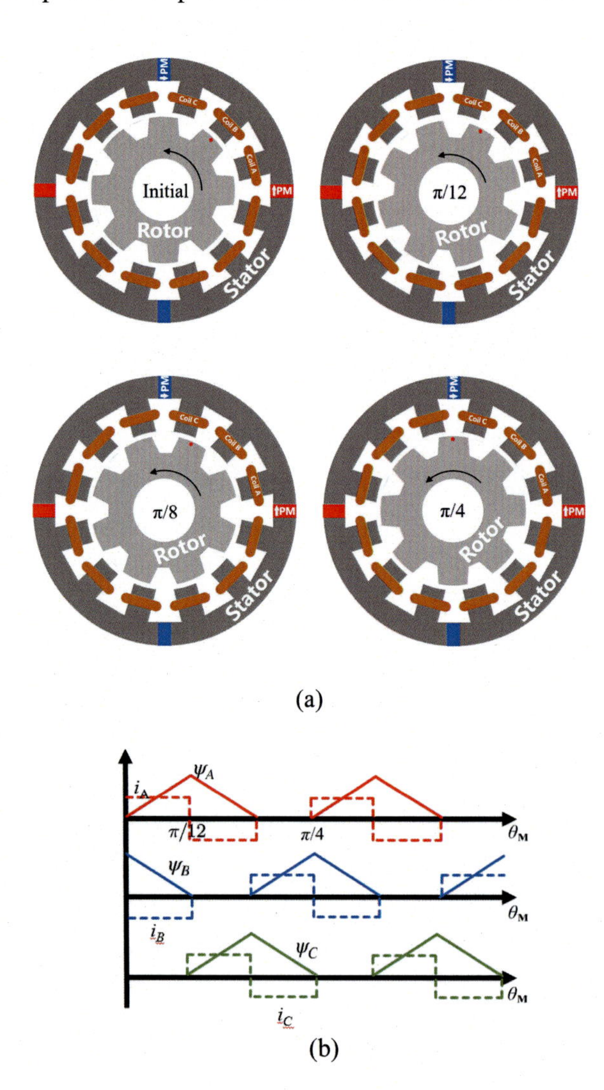

Fig. 2 Principle of doubly-salient PM machine. **a** doubly-salient PM machine at different rotor positions **b** current and magnet flux linkages

flux linkage, which is challenging for torque capability improvement. Accordingly, more doubly-salient PM machines with different stator/rotor pole combinations were proposed and compared in [4], including 6/4 pole, 6/5 pole, 6/7 pole, and 6/8 pole machines. It was shown that the 6/7 pole machine exhibits the highest torque density. For the conventional topology with round envelope as shown in Fig. 1, the PMs in the stator core result in the reduction of the slot area, which limits the electric loading and hence the torque density. By contrast, a doubly-salient PM machine with square envelope was proposed in [5], as shown in Fig. 3a. However, the stator back iron next to PMs is still thicker and the stator slot area is still small. Building upon the square envelope topoloty, an improved doubly-salient PM machine with square envelope was proposed in [6], as shown in Fig. 3b. In this machine, the thickness of stator back iron is roughly uniform and the PMs are inserted in the slots. Moreover, in order to achieve the balanced slot area, the tooth bodies of the side stator teeth are bended so that all sator tooth tips are kept uniformly distributed along the stator bore. It was shown that compared to the conventional square envelope machine, the improved version exhibits larger slot area under the same envelope and PM usage, and hence improved torque density, as well as improved the ratio of torque to PM volume.

On the other hand, for conventional doubly-salient PM machines, the PM flux linkage waveforms are different between winding A and B due to their inherit different magnetic circuit. This asymmetry between phases will lead to a high torque ripple when ideal three-phase trapezoidal currents are fed into armature windings. To overcome this issue, a conventional doubly-salient PM machine with E-shaped stator iron core and a new machine with Π-shaped stator iron core were investigated and compared in [7], as shown in Fig. 4a and b, respectively. It was shown that differing from the counterpart with E-shaped stator iron core, the asymmetric magnetic circuit phenomenon among different phases is avoided in the doubly-salient PM machine with Π-shaped stator iron core. This is due to the fact that in the machine with Π-shaped stator iron core, the three-phase magnetic circuits are in central symmetry, and

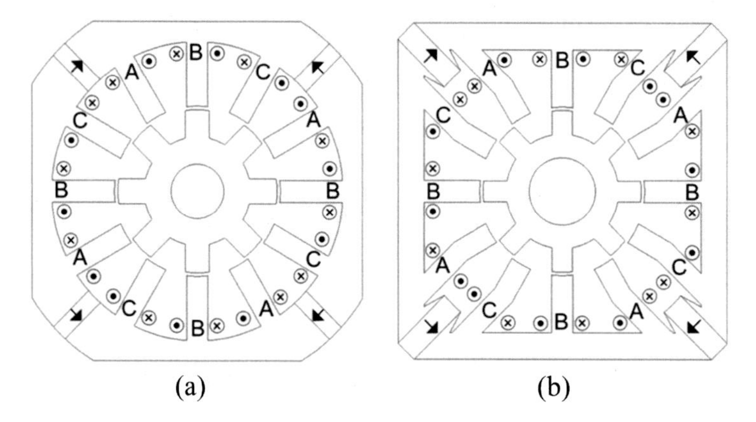

Fig. 3 **a** Square envelope doubly-salient PM machine [5]. **b** Improved square envelope doubly-salient PM machine [6]

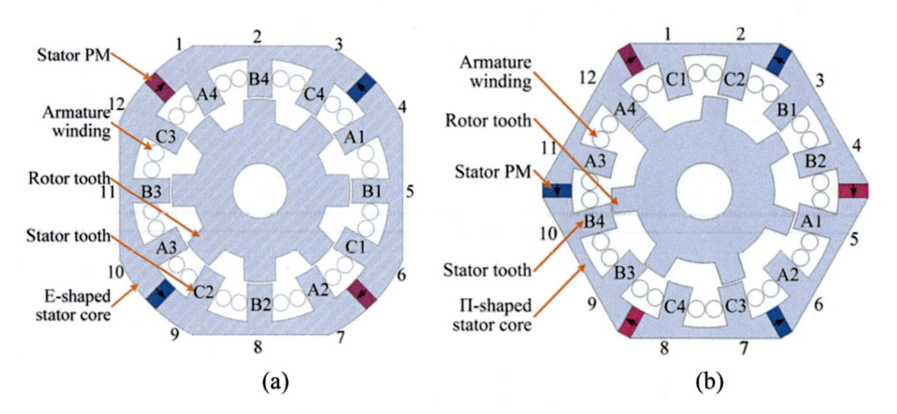

Fig. 4 E-core and Π-core doubly-salient PM machine. **a** 12/8 pole E-core doubly-salient PM machine [7]. **b** 12/7pole Π-core doubly-salient PM machine [7]

hence the PM flux linkages of all three phases are identical. As a result, the three-phase PM flux linkage and back-EMF waveforms of the Π-shaped core machine become balanced and sinusoidal. Moreover, the Π-shaped core machine exhibits higher average torque and lower torque ripple.

Although doubly-salient PM machines exhibit the advantages of simple structure and high reliability, there are still some shortcomings. The main challenges/difficulties of doubly-salient PM machines are as follows:

- Compared to other stator PM machines such as flux-switching PM machines and flux-reversal PM machines, the magnetic loading of doubly-salient machines is relatively low, which limits the ability to achieve higher torque density. This is due to the fact that compared to other stator PM machines, there is not enough space to accommodate PMs in doubly-salient PM machines.
- The asymmetric structure of doubly-salient PM machines will not only lead to the fluctuation of tangential force and hence output torque but also fluctuation of radial force, which results in high vibration and noise.
- The complex stator of doubly-salient PM machines consists of several core segments that are separated by PMs. The assembling of these segments is challenging for manufacturing in order to make all the stator units to be a perfect circle.
- The power factor of doubly-salient PM machines is typically low (around 0.7). This is because the inductance of doubly-salient PM machines is typically high due to the relatively high flux leakage in the stator. As a result, the flux-weakening capability of such machines is favorable and a wide constant power speed range could be achieved.

2.2 *Flux-Switching PM Machines*

The modern three-phase flux-switching PM machine was first introduced by Emmanuel Hoang in 1997 [8]. The topology of the flux-switching PM machine is shown in Fig. 5. As can be seen, the stator is relatively complex, which contains laminated "U"-shape segments, PMs, and armature windings. The PMs are circumferentially magnetized and sandwiched in the "U"-shape segments with alternative opposite polarity. the "U"-shape segments are wound by concentrated armature windings. Similar to the aforementioned doubly-salient PM machine, the rotor of this machine is also mechanically simple and hence robust since it is identical to that of switched reluctance machines. In addition, this topology is more favorable for PM thermal dissipation compared to conventional rotor-PM machines, due to the fact that the PMs are located in the stator instead of the rotor. Moreover, the risk of PM demagnetization of this machine is low, due to the fact that the armature reaction field is oriented perpendicular to the direction of the PM magnetization. On the other hand, flux-switching PM machines exhibit relatively high torque density due to their inherently "flux-focusing" effect. However, the PMs and the slots are competing to each other for the space, and hence, the slot area for the armature windings is relatively low. As a result, the electric loading is relatively low, which limits the further improvement of the torque density.

An arbitrary coil X depicted in Fig. 6 is used as an example to illustrate how the electrical power of a flux-switching PM machine is produced based on the fundamental principle of Faraday's law. Figure 6a demonstrates that when the rotor is situated at position A, the teeth of the rotor and stator are aligned, resulting in the minimum magnetic circuit reluctance. This, in turn, leads to a maximum value of flux-linkage in coil X. Once the rotor advances to position B, the teeth of the rotor align with the stator slots, resulting in the maximum magnetic circuit reluctance and no magnetic flux passing through coil X. As a result, the flux linkage of coil X reduces to its minimum value of 0. Following that, as the rotor rotates to position C, the teeth of the rotor once again align with the stator teeth, resulting in a minimum magnetic circuit reluctance. The flux-linkage of coil X increases to its maximum value, albeit

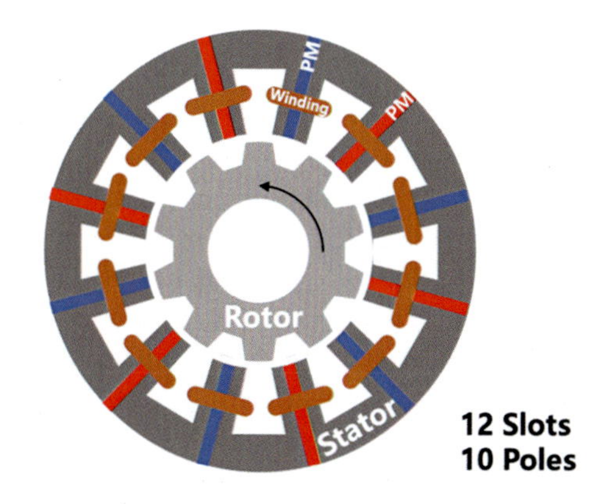

Fig. 5 Flux-switching PM machine

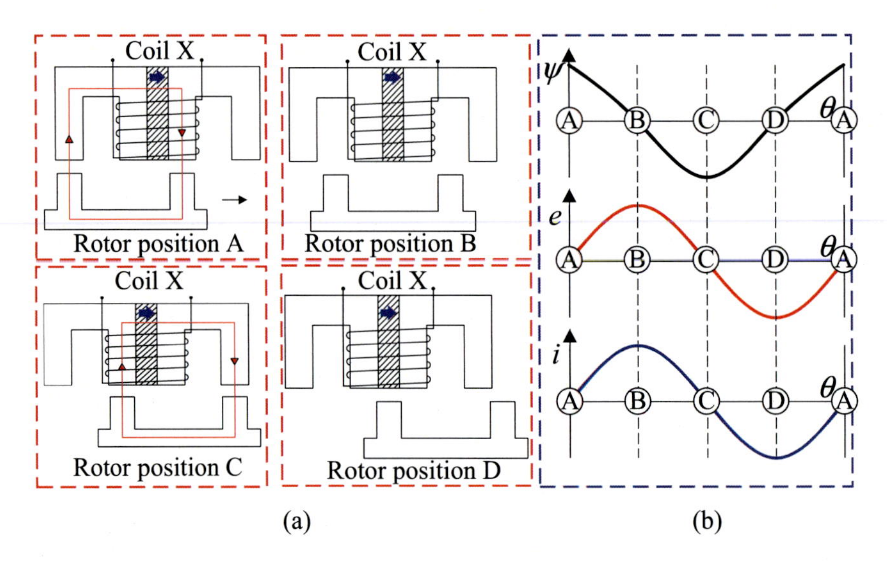

Fig. 6 Operating principle of flux-switching PM machines [9]. **a** Rotor at four typical positions. **b** Ideal flux-linkage, back-EMF, and phase current waveforms with respect to rotor angular position

with a reversed polarity. Lastly, as the rotor rotates to position D, the teeth of the rotor align with the PMs, resulting in maximum magnetic circuit reluctance and no magnetic flux passing through coil X. As a result, the flux-linkage of coil X reduces again to its minimum value of 0. Up to this point, one electrical period has been accomplished. If the rotor continues to rotate, the above-mentioned process will repeat. The idealized back-EMF waveform, as per Faraday's law of electromagnetic induction, is depicted in Fig. 6b. If a purely resistive load is connected, the current depicted at the bottom of Fig. 6b will be generated [9, 10].

Over the past few decades, flux-switching PM machines are gaining interest due to their high torque/power density, favorable thermal dissipation of PMs, and essentially sinusoidal back-EMF waveform. Several novel topologies of flux-switching PM machines have been proposed based on the typical topology shown in Fig. 5. One such machine is the C-core flux-switching PM machine, developed by removing half of the stator teeth of conventional machines to enlarge the slot area for armature winding, as shown in Fig. 7a. Compared to its conventional counterpart, the C-core machine exhibits improved torque density while using only half the volume and number of PMs [11]. Building upon this C-core machine, an E-core machine was developed by equipping fault-tolerant teeth between regular wound teeth, as shown in Fig. 7b. The fault-tolerance teeth provide physical isolation between the phase coils, resulting in a much lower mutual inductance than the C-core counterpart and significantly higher self-inductance, thereby reducing short-circuit current and improving fault-tolerance capability [12]. Another machine is the modular flux-switching PM machine, developed by removing the PMs in alternative stator teeth, as shown in Fig. 7c. While exhibiting lower average torque than conventional counterparts, the

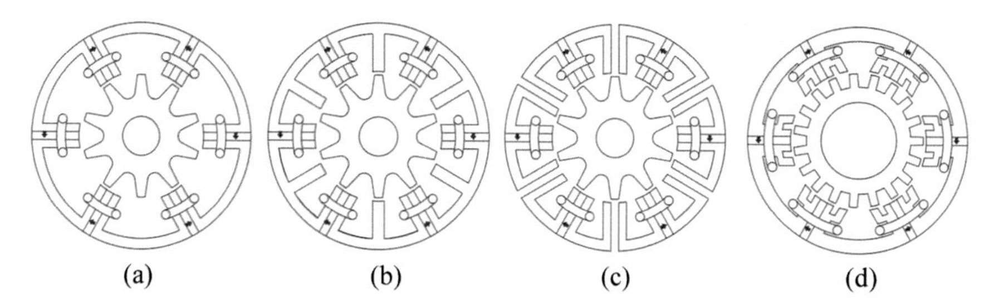

Fig. 7 Novel topologies of flux-switching PM machines. **a** C-core [11]. **b** E-core [12]. **c** Modular stator [13]. **d** Multi-tooth [14]

modular machine provides the best magnetic, thermal, and electrical decoupling between phases compared to other counterparts [13]. Furthermore, a multi-tooth flux-switching PM machine was developed, as shown in Fig. 7d, in which there are effectively two stator teeth on each side of the stator PMs. Compared to the conventional counterpart, the multi-tooth machine exhibits a higher torque density at low excitation current, even though it requires significantly less PM material. However, at very high electric loadings, its torque density is lower due to quicker saturation as the current is increased, caused by the higher armature reaction [14]. Table 1 provides a qualitative comparison of the pros and cons of these different flux-switching PM machines for better understanding.

Despite flux-switching PM machines having many merits mentioned above, they are not without their problems. The main challenges/difficulties of flux-switching PM machines are as follows:

- The PMs are surrounded by the armature windings, generally designed to operate at a higher temperature.
- The cost of flux-switching PM machines is usually high due to the typically high number and amount of PMs used.
- End-effect in flux-switching PM machines can be more severe than conventional rotor-PM machines due to the short stack length and the presence of PMs in the stator, resulting in a more significant flux leakage at the stator end and outer surface.
- The unique structure and flux-focusing effects of flux-switching PM machines can lead to local flux density in stator and rotor teeth exceeding 2 T, causing oversaturation and making accurate analysis of electromagnetic characteristics challenging. This can reduce efficiency, output torque capability, and overload capability, and potentially cause irreversible demagnetization.
- Manufacturing and assembling the complex stator of flux-switching PM machines, which consists of several core segments separated by PMs, can be challenging to achieve a perfect circle for all stator units.
- Like all other stator PM machines, the power factor of flux-switching PM machines is typically low (around 0.7). This is because the inductance of flux-switching PM machines is typically high due to the relatively high flux leakage in

Table 1 Comparison of different flux-switching PM machines

	Conventional machine	C-core machine	E-core machine	Modular stator machine	Multi-tooth machine
Saturation level	High due to the flux-focusing effects	Higher than that of the conventional machine	Relatively low due to the middle magnet-less tooth	Maximum	Relatively low
Torque ripple	Relatively low	Maximum	High	High	Minimum
Torque density	Minimum	High	Maximum	Higher	Low
Power density (with the same speed)	Minimum	High	Maximum	Higher	Low
Over-load capability	Low due to high saturation	Lower due to higher saturation	Slightly lower	Minimum due to the maximum saturation	Maximum
Efficiency	Maximum	Higher	High	Higher	Minimum
Demagnetization withstand capability	Relatively low	Relatively low	High because some demagnetizing flux is diverted	Highest because much demagnetizing flux is diverted	Higher than E-core machine
Flux-weakening capability	Relatively low	High	Higher	High	Highest due to a very high d-axis inductance
Fault-tolerance capability	Relatively low	High	Higher due to the auxiliary fault-tolerance tooth	Highest due to the unique phase magnetic separation	High

the stator. As a result, the flux-weakening capability of such machines is favorable and a wide constant power speed range could be achieved.

2.3 *Flux-Reversal PM Machines*

In recent years, there has been a growing interest in flux-reversal PM machines due to their advantages, including a simple rotor configuration, fast dynamic response, and high power density. Figure 8 shows a conventional flux-reversal PM machine with a 12-slot and 10-pole configuration. The rotor of the machine is as mechanically robust as those of doubly-salient PM machine and flux-switching PM machines. Each stator tooth has a pair of PMs with alternating polarities mounted on its inner

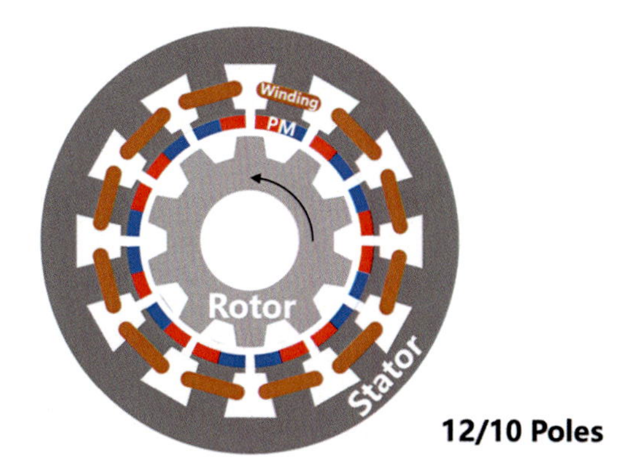

Fig. 8 Flux-reversal PM machine

surface. The stator teeth are typically wound by tooth windings, resulting in short end-windings. As the rotor rotates, the areas of the two PM segments facing the rotor teeth change with respect to the rotor position, creating bipolar flux linkage and back-EMF waveforms in the tooth windings for each stator tooth.

The operating principle of flux-reversal PM machines is illustrated in Figs. 9 and 10 [15]. In Fig. 9a, the machine is in an equilibrium position, and the flux generated by the PMs flows entirely within each stator pole, while there is no flux passing through the coils and the stator back-iron. However, when the rotor moves to position B, which is 30° displaced from position A, as shown in Fig. 9b, the rotor poles overlap with one or more of the PMs, resulting in flux passing through the coils and the stator back-iron. Meanwhile, the coil flux reaches its maximum at position B. As the rotor continues to move towards position C, which is 60° displaced from position A as illustrated in Fig. 9c, the flux linked by the coil and the flux in the stator back-iron return to zero since position C is another equilibrium position like position A. Upon reaching position D, which is 90° displaced from position A as shown in Fig. 9d, the coil flux reaches its maximum again but in the opposite direction to that of position B (as shown in Fig. 9b). This variation in the flux linkage linked by the coil results in the induction of back-EMF. Applying Faraday's law, the back-EMF reaches its maximum at position A, zero at position B, maximum but in the opposite direction at position C, and zero again at position D. When the current is energized in the same direction as the back-EMF waveform, a steady output torque is generated.

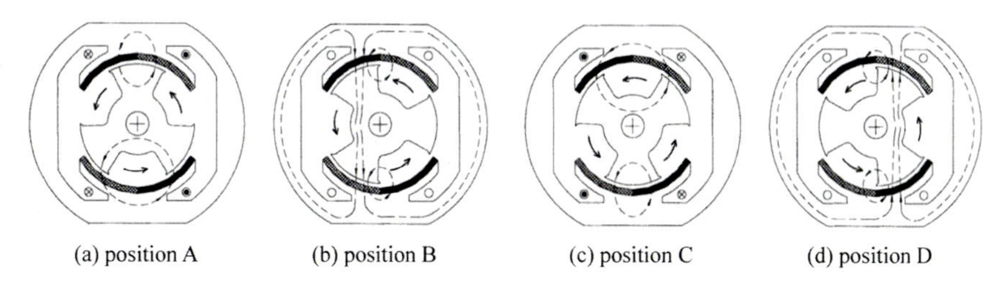

Fig. 9 The principle of operation of the flux-reversal machine [15]

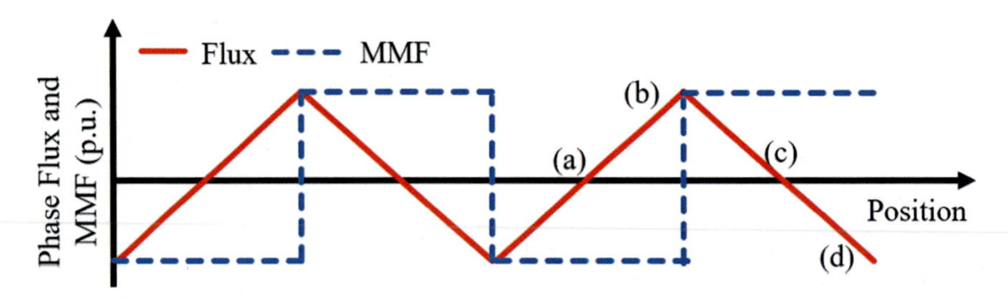

Fig. 10 The ideal variation of phase flux and MMF with the position for flux-reversal PM [15]

In addition to the conventional flux-reversal PM machine, several new topologies have been proposed. In [16], various flux-reversal PM machines with different PM arrangements were analyzed and compared, including NS-SN, NS-NS, NSNS-SNSN, and NSNS-NSNS types. These names correspond to the arrangement of PMs on two adjacent stator teeth, as depicted in Fig. 11. Upon comparing the flux-reversal PM machines with different pole numbers and PM arrangements, several conclusions can be drawn. Firstly, the torque produced by the flux-reversal PM machine is significantly influenced by the PM arrangement and is generated by several dominant harmonics. Secondly, in terms of torque density, for flux-reversal PM machines with 12 pairs of PMs, the optimal rotor pole number is approximately 14, and each PM arrangement performs better in a specific range of rotor pole numbers. Thirdly, the machines with four PM pieces on each stator tooth generate higher torque than those with two PM pieces. Finally, the flux-reversal PM machine with the NSNS-NSNS arrangement provides the highest torque density, as it produces the highest back-EMF fundamental wave. Different consequent-pole PM topologies were investigated and compared in [17], as shown in Fig. 12, including N/Fe–Fe/N, N/Fe–N/Fe, N/Fe/N/Fe–Fe/N/Fe/N, and N/Fe/N/Fe–N/Fe/N/Fe types of flux-reversal PM machines. The study revealed that the structure of N/Fe/N/Fe–N/Fe/N/Fe as shown in Fig. 12d exhibits the highest torque due to the utilization of additional magnetic field harmonic components, such as 2nd, 18th, and 26th. Furthermore, to reduce cogging torque, a flux-reversal PM machine with V-shape punches on the rotor pole was proposed as depicted in Fig. 13 [18]. It was demonstrated that the V-shape punched machine significantly reduced cogging torque and torque ripple by 84% and 31%, respectively, compared to the conventional counterpart.

As aforementioned, flux-reversal PM machines exhibit the advantages of robust rotor structure (suitable for high-speed applications), high fault-tolerance capability, quick dynamic response, etc. By contrast, the main challenges/difficulties of flux-reversal PM machines are as follows:

- PM flux leakage is a common issue in flux-reversal PM machines, leading to partial magnetic saturation and reduced power factor.
- The presence of stationary PMs on the stator teeth results in high eddy-current loss due to the time-varying armature magnetic fields.

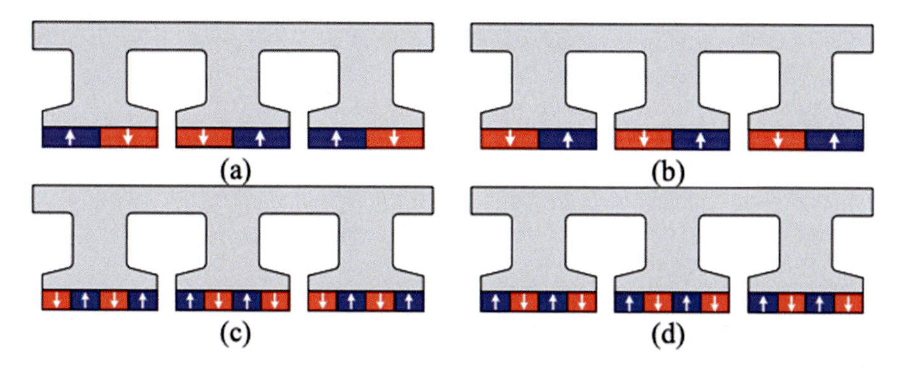

Fig. 11 Flux-reversal PM machines with different PM arrangements [16]. **a** NS-SN. **b** NS-NS. **c** NSNS-SNSN. **d** NSNS-NSNS

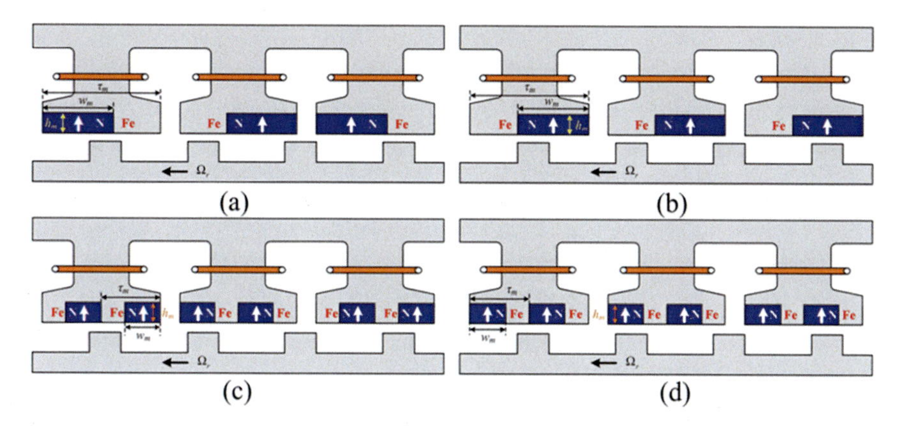

Fig. 12 Flux-reversal PM machines with different consequent-pole topologies [17]. **a** Type-1: N/Fe–Fe/N. **b** Type-2: N/Fe–N/Fe. **c** Type-3: N/Fe/N/Fe–Fe/N/Fe/N. **d** Type-4: N/Fe/N/Fe–N/Fe/N/Fe

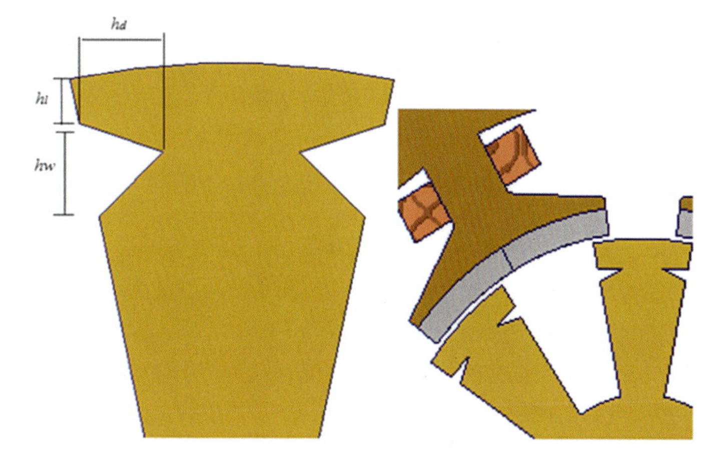

Fig. 13 Flux-reversal PM machine with V-shape punched rotor pole [18]

- The performance of flux-reversal PM machines is significantly influenced by the PM arrangement. Thus, further research is necessary to investigate the performance of these machines with different configurations.
- Like all other stator PM machines, the power factor of flux-reversal PM machines is typically low (around 0.7). This is because the inductance of flux-reversal PM machines is typically high due to the relatively high flux leakage in the stator. As a result, the flux-weakening capability of such machines is favorable and a wide constant power speed range could be achieved.

3 Magnetic Gear-Based Machines

High torque density is one of the most important features of electric machines used in EVs and HEVs. Magnetic gear-based machines are considered as one of the most promising candidates to achieve high torque density since the incorporated magnetic gearing effect is ideal for reducing the rotating speed and amplifying the output torque.

As for identifying the technical differences between conventional electric machines and magnetic gear-based machines, it is known that the electromagnetic torque in conventional electric machines is proportional to the electric loading and the magnetic loading, which can be expressed by:

$$T_e \propto M_g \times A \tag{1}$$

where T_e is the electromagnetic torque, M_g is the magnetic loading, and A is the electric loading. By contrast, magnetic gear-based machines are equivalent to a conventional electric machine with a virtual gear. The magnetic gearing effect plays an important role in boosting the electromagnetic torque. More specifically, the MMF is modulated from a high-speed low-pole pair number side to a low-speed high-pole pair number side, while both the frequency and power are the same. As a result, the rotor with high-pole pair number rotates at a low angular velocity. Since the power is the same between the input and output shafts, the torque is scaled up by the gear ratio, which can be expressed by:

$$T_e \propto M_g \times A \times G_r \tag{2}$$

where G_r is the gear ratio. That's why magnetic gear-based machines typically exhibit high torque density compared to conventional electric machines. Magnetic gear-based machines mainly include two types: vernier PM machines and magnetically-geared PM machines which will be introduced in detail in the following.

3.1 Vernier PM Machines

Vernier PM machines are becoming increasingly attractive due to their high torque density and simple mechanical structure. Figure 14 illustrates a typical vernier PM machine, which is structurally similar to conventional surface-mounted PMSMs. However, in conventional PMSMs, the number of stator poles matches the number of rotor poles, whereas vernier PM machines have different numbers of stator and rotor poles. Typically, vernier PM machines have a smaller number of stator poles and a greater number of rotor poles to highlight the "Vernier" effect. A small movement of the PM rotor generates a large flux-linkage movement in the stator armature windings. As the rotor speed steps down from the rotating magnetic field's speed, the torque increases [19].

The vernier PM machine typically features a toothed-pole/open-slot tooth structure on its stator, with overlapped windings wound around these open-slot teeth which serve as a flux modulator. The number of winding pole pairs, p_w, stator teeth, Q_S, and rotor PM pole pairs, p_r, must comply with the following rule:

$$p_w = Q_s - p_r \tag{3}$$

The relationship of the equivalent rotating speed of the magnetic field that the stator winding links, n_w, and the rotor speed, n_r, is governed by:

$$n_w p_w = n_r p_r \tag{4}$$

The frequency of stator winding, f_w, can be expressed as follows:

$$f_w = \frac{n_w p_w}{60} = \frac{n_r p_r}{60} \tag{5}$$

The torque relationship can be expressed as follows, based on the law of energy conservation:

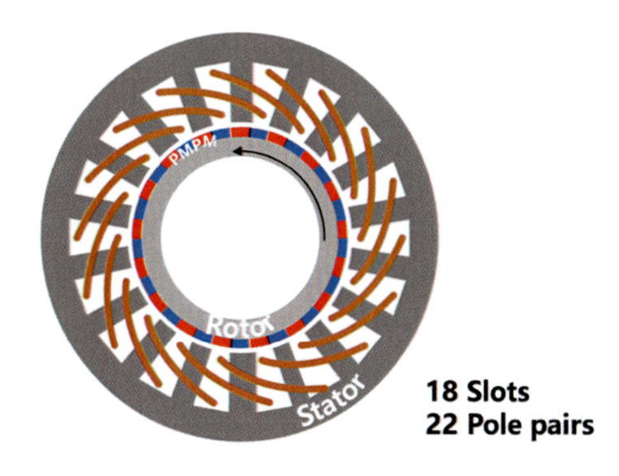

Fig. 14 Vernier PM machine

$$n_w T_{st} + n_r T_r = 0 \tag{6}$$

where T_{st} and T_r are the torques produced by the stator and the rotor, respectively. Substituting (4) into (6), the torque that is transferred from the stator to the rotor is determined by:

$$T_r = -(\frac{p_r}{p_w})T_{st} \tag{7}$$

$$G_r = -(\frac{p_r}{p_w}) \tag{8}$$

where G_r is the gear ratio between the rotor and the stator. As can be seen from Eq. (3), typically, p_r is greater than p_w. Hence, G_r is greater than 1, which means the torque transferred from the stator to the rotor is greater than the torque produced by the stator. This phenomenon is like a "reduction gear".

Vernier PM machines operate differently from conventional PMSMs, where the stator and rotor torques are typically equal. Instead, the vernier PM machine functions like a conventional PMSM with a "virtual reduction gear", resulting in an increased output torque and high torque density.

However, differing from conventional PMSMs, the power factor of vernier PM machines is typically low. Since the rotor saliency of vernier PM machines is negligible which is similar to surface-mounted PMSMs, $i_d = 0$ control mode is often used to drive vernier PM machines. If assuming the resistance is negligible, the power factor can be represented as follows:

$$P_f = \frac{1}{\sqrt{1 + (L_s I / \psi_m)}} \tag{9}$$

where L_s is the synchronous inductance, I is the phase current in RMS value, and y_m is the PM flux linkage. Due to the unique relationship between the winding pole pair number and the rotor PM pole pair number as aforementioned in Eq. (3), more specifically, the winding pole pair number is much lower than the rotor PM pole pair number. Hence, the pole pitch of the armature field is significantly larger than that of the rotor. Within one armature field pole, only half of the PMs are effectively contributing to the flux, while the other half mainly causes flux leakage. Accordingly, the winding inductance of vernier PM machines are typically high due to these flux leakage harmonic components, which results in a low fundamental flux density and a low power factor [20]. Vernier PM machines typically exhibit a power factor value of 0.66 or even lower. Consequently, these electric machines demand a high-capacity converter, leading to increased costs.

Various techniques have been proposed to enhance the power factor of vernier PM machines. For example: (1) a double-stator spoke-type PM vernier machine is presented in [20], as depicted in Fig. 15a. This machine has a spoke-type PM rotor and two unaligned stators which are shifted from each other by half tooth pitch. This structure improves magnet utilization and guides the flux path, resulting in a power factor of 0.83 measured in experiments. (2) A vernier PM machine with two-slot-pitch coil windings is presented in [21], as shown in Fig. 15b. This machine has twice the number of slots as conventional vernier PM machines, and the phase winding connection is adjusted to decrease the space harmonic content. Consequently, the phase inductance is decreased, and the power factor is enhanced, with a power factor of 0.92 measured in experiments.

Numerous novel topologies have been proposed to enhance various aspects of performance such as torque density and cost. For example, a PM vernier machine with non-uniform flux-modulation pole was proposed in [22], as illustrated in Fig. 15c. By setting the flux-modulation pole pitch different from the average slot pitch, additional working harmonics are introduced. The proposed machine was found to have a 20% increment in torque density compared to conventional non-overlapping winding PM vernier machines. In order to reduce the usage of rare-earth PM material, a PM vernier machine with toroidal winding and consequent-pole structure was proposed in [23], as depicted in Fig. 15d. The study demonstrated that the proposed machine achieved approximately 20% higher torque density while using only about 60% of the magnet material compared to a conventional PM vernier machine. In addition, a spoke-type ferrite PM machine with alternating flux barrier was presented in [24], as depicted in Fig. 15e. This machine exhibits over 50% increment in output torque compared to a conventional counterpart. Moreover, a PM vernier machine with Halbach array magnets in stator slot opening was introduced in [25], as depicted in Fig. 15f. The torque density of the proposed machine is as high as 23.1 kNm/m^3, and its power factor reaches up to 0.89.

Vernier PM machines can generate large output torque at low speed through the magnetic field modulation effect. Hence, direct-drive system with vernier PM machines used for EV/HEV becomes possible. The challenges/difficulties of vernier PM machines are as follows:

- Although many researchers are devoted to improving the power factor of vernier PM machines, the power factor of the vernier machine is still low compared to conventional electric machines such as PMSMs. However, the low power factor is caused by the relatively high inductance, which results in a good flux-weakening capability and a wide constant power speed range can be achieved.
- Vernier PM machines usually have more poles than conventional PM machines, which leads to high operating frequency and hence high eddy-current loss and high core loss. Therefore, the utilization of vernier PM machines in high-speed operation is limited.
- The overload capacity of vernier PM machines is limited and they have risk of irreversible demagnetization.

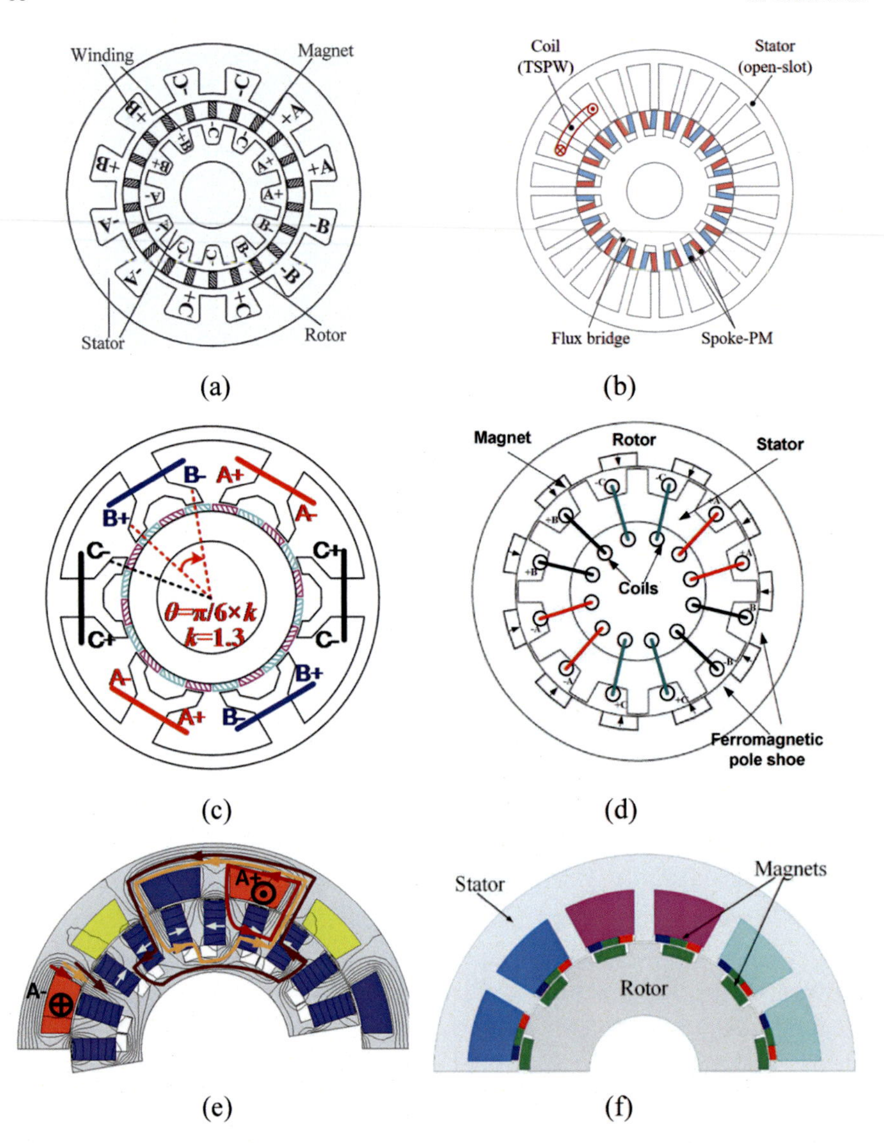

Fig. 15 New topologies vernier PM machines. **a** Double-stator spoke-type PM vernier machine [20]. **b** Vernier PM machine with two-slot-pitch coil windings [21]. **c** A PM vernier machine with non-uniform flux-modulation pole [22]. **d** A consequent-pole PM vernier machine with toroidal-winding [23]. **e** A spoke-type ferrite PM machine with alternating flux barrier [24]. **f** A PM vernier machine with Halbach array magnets in stator slot opening [25]

3.2 Magnetically-Geared PM Machines

Magnetic-geared PM machines are based on the concept of magnetic gears as illustrated in Fig. 16a, which can transfer torque between input and output shafts without mechanical contact. The operating principle of a magnetic gear is based on the modulation of magnetic fields produced by two PM rotors via ferromagnetic pole pieces [26]. By artfully incorporating a magnetic gear into a surface-mounted PM machine, a magnetic-geared PM machine can be achieved, as shown in Fig. 16b. This machine has armature windings in the stator and PMs on the rotor with different pole numbers and speeds. The rotating ferromagnetic pole pieces act as a flux modulator to match the two magnetic fields from the stator and rotor to have the same pole number and speed, thereby producing a steady torque like a reduction gear. This design allows for achieving high-torque direct-drive function.

The pole pieces of the outer rotor of the magnetic-geared PM machines work as the flux modulator. The number of winding pole pairs, p_w, the number of poles on the outer rotor, p_{or}, and the number of inner rotor PM pole pairs, p_{ir}, must comply with the following rule:

$$p_w = p_{or} - p_{ir} \tag{10}$$

The relationship of the equivalent rotating speed of the magnetic field that the stator winding links, n_w, the outer rotor speed, n_{or}, and the inner rotor speed, n_{ir}, is governed by:

$$n_w p_w = n_{or} p_{or} - n_{ir} p_{ir} \tag{11}$$

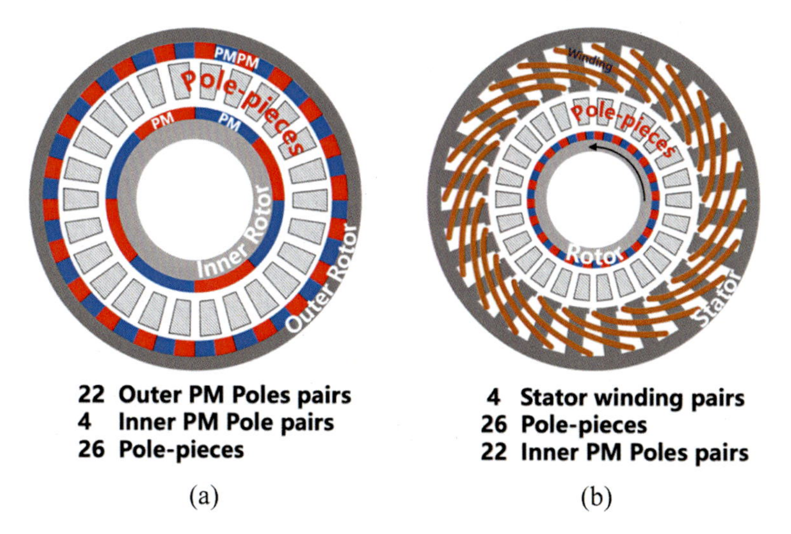

Fig. 16 Magnetic-geared PM machine. **a** Magnetic gear [26]. **b** Classical magnetic-geared PM machine

The frequency of stator winding, f_w, can be expressed by:

$$f_w = n_w p_w / 60 = (n_{or} p_{or} - n_{ir} p_{ir}) / 60 \tag{12}$$

The torque relationship can be expressed as follows, based on the law of energy conservation:

$$n_w T_{st} + n_{or} T_{or} + n_{ir} T_{ir} = 0 \tag{13}$$

where T_{st}, T_{or}, and T_{ir} are the torques generated from the stator, the outer rotor, and the inner rotor, respectively. Substituting (11) into (13), the torque transferred from the stator to the outer rotor to the inner rotor is governed by:

$$\left.\begin{aligned} T_{or} &= -(p_{or} / p_w) \cdot T_{st} \\ T_{ir} &= -(p_{ir} / p_{or}) \cdot T_{or} = (p_{ir} / p_w) \cdot T_{st} \end{aligned}\right\} \tag{14}$$

Hence, the gear ratios between the outer rotor and the stator, G_{or_st}, the inner rotor and the outer rotor, G_{ir_or}, as well as the inner rotor and the stator, G_{ir_st}, are as follows:

$$G_{or\_st} = -p_{or} / p_w; \; G_{ir\_or} = -p_{ir} / p_{or}; \; G_{ir\_st} = p_{ir} / p_w \tag{15}$$

Magnetic-geared PM machines primarily originate from a magnetic gear, and they can be categorized into four main types based on their topologies. These types include mechanically coupled machines (Type 1), pseudo machines (Type 2), mechanically and magnetically coupled machines (Type 3), and partitioned-stator machines (Type 4), as illustrated in Fig. 17a-d, respectively [27]. Figure 17a shows that the mechanically coupled machine is created by combining a magnetic gear and a conventional PMSM, which are linked together by a shared rotor. When the machine is in motor mode and its armature windings are energized, the high-speed PM rotor with a smaller number of poles from the magnetic gear, is driven by the armature magnetic field. The outer PM rotor of the magnetic gear or the flux modulation ring with pole pieces can be utilized as the output rotor. Consequently, the torque/speed can be converted/transmitted from the high-speed low-torque of the magnetic gear inner rotor to the low-speed high-torque of the outer rotor or the flux modulation ring of the magnetic gear. Figure 17b demonstrates the pseudo machine that is created by incorporating an additional set of armature windings on the low-speed element of the magnetic gear, while the high-speed rotor of the magnetic gear is driven to rotate. The PMs are mounted on the inside surface of the low-speed element, causing the pole piece (modulation ring) and high-speed rotor to rotate in the same direction at a fixed speed ratio. Figure 17c displays the mechanically and magnetically coupled machine, which is formed by utilizing a stator with armature windings to generate a magnetic field instead of the magnetic field produced by the PM rotor of the magnetic gear. In this case, the stator with armature windings replaces the inner PM rotor of the magnetic gear, while the flux modulation ring and/or the low-speed PM rotor rotate. Figure 17d illustrates the partitioned-stator machine, which results from combining a magnetic gear and a stator-PM machine (in this case, a flux-reversal PM machine). The windings and PMs of the stator-PM machine are separated, and the outer stator and inner PM structure constitute the machine stator. By removing the space conflict between the PM and the windings, the partitioned-stator machine offers improved torque density compared to the original stator-PM machine, i.e., a flux-reversal PM machine.

As a magnetic transmission device, magnetically-geared PM machines exhibit the advantages of non-contact transmission, no friction loss, low vibration, low noise, inherent overload protection, maintenance-free, oil-free, and high reliability. The challenges/difficulties of magnetically-geared PM machines are as follows:

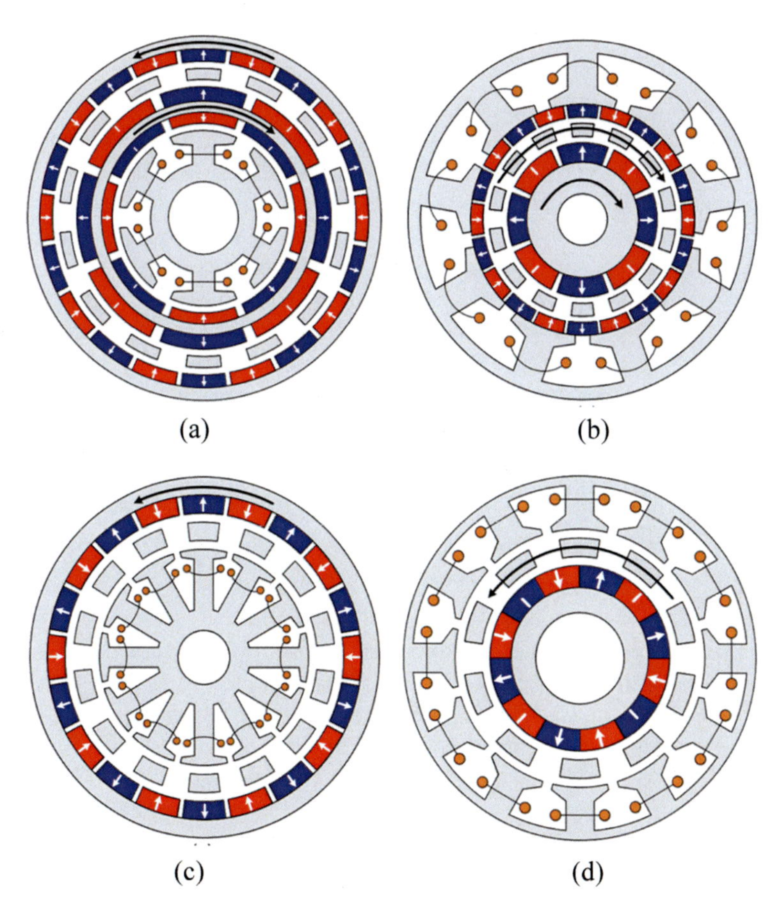

Fig. 17 Typical types of magnetic-geared PM machines. **a** Mechanically coupled machine [27]. **b** Pseudo machine [27]. **c** Mechanically and magnetically coupled machine [27]. **d** Partitioned-stator machine [27]

- Magnetically-geared PM machines usually have a large volume of PM material especially mechanically coupled machines, which leads to an increase in the material cost.
- Compared to conventional machines, magnetically-geared PM machines have a complex mechanical structure since they have at least two airgaps.
- The heat dissipation of the magnetically-geared PM machines especially the inner rotor is limited by the complex structure.
- In the magnetically-geared PM machines, there are two mechanical ports, i.e., the inner rotor and the outer rotor. By contrast, the armature windings represent an electrical power port. Therefore, how the magnetically-geared PM machine operates as a multi-power port motor needs further analysis and investigation.

4 Flux-Controllable Machines

Unlike conventional electric machines that usually operate under a single operating point, such as fans or pumps, the electric machines used in EV/HEV applications operate at varying torque-speed combinations. This means that the EV/HEV drive machines need to meet high torque requirements at low speeds, high-speed requirements, and have good flux controllability to ensure a wide constant power operating range. Flux-controllable machines are becoming increasingly attractive in EV/HEV applications because they can provide a wide constant power speed range. The main types of flux-controllable machines are hybrid-excited machines and memory machines. In hybrid-excited machines, both PM excitation and field coil excitation sources are used, while in memory machines, the hysteresis loop of the permanent magnet material can be changed through a pulse current to vary the excitation state of the machine [28].

4.1 Hybrid-Excited Machines

Conventional PM machines exhibit high torque/power density and high efficiency due to the use of high-energy PM materials. However, the PM flux in airgap cannot be directly adjusted which limits the speed range of the machine, especially for non-salient PM machines, e.g., surface-mounted PM synchronous machines. In order to achieve a wide speed range, the flux-weakening control mode should be adopted in which a negative d-axis current needs to be utilized to oppose/weaken the PM excitation. As a result, the efficiency of the machine in high-speed region and especially at partial load is reduced. By contrast, for conventional electrically-excited machines, the air-gap flux can be feasibly adjusted directly through controlling the field winding current in the rotor. However, the torque density and efficiency of the electrically-excited machines are lower than those of the PM machines in the low-speed region. Hybrid-excited machines exhibit the combined advantages of both PM machines and

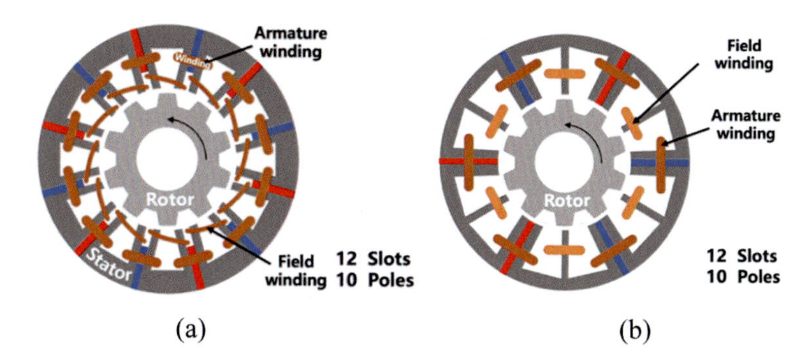

Fig. 18 **a** Series hybrid-excited flux-switching PM machine. **b** Parallel hybrid-excited flux-switching PM machine with E-core structure

electrically-excited machines. Therefore, hybrid-excited machines show the advantages of high torque density and relatively high efficiency in the low-speed region, as well as high efficiency in the high-speed region and wide speed range.

Hybrid-excited machines can be divided into two categories based on their magnetic circuits: series hybrid-excited machines, where the flux lines excited by the field coils pass through the PMs, and parallel hybrid-excited machines, where the flux lines of field coils do not pass through the PMs. For example, Fig. 18a shows a series hybrid-excited flux-switching machine. Compared to conventional flux-switching PM machines, the stator flux of this machine can be partially adjusted by controlling the field winding current, resulting in improved flux-enhancing and flux-weakening capabilities. By contrast, Fig. 18b shows a parallel hybrid-excited flux-switching machine. Unlike U-shaped core flux-switching PM machines, this machine has an E-shaped core stator with excitation windings on the middle tooth of the stator E-core. In this E-core structure, the flux produced by the PM and field windings are parallel.

According to the location of the excitation source, hybrid-excited machines can be classified into several categories [29] including Type 1: both the PMs and DC field coils are in the rotor, such as a surface-mounted PM hybrid-excited machine as shown in Fig. 19a, and an interior PM hybrid-excited machine as shown in Fig. 19b, respectively; Type 2: both the PMs and DC field coils are in the stator, such as a series hybrid doubly-salient PM machine as shown in Fig. 19c, and a parallel hybrid doubly-salient PM machine as shown in Fig. 19d, respectively; Type 3: PMs are in the rotor and DC field coils are in the stator, such as a consequent-pole rotor PM hybrid-excited machine with equal PM pole and iron pole as shown in Fig. 19e, and a consequent-pole rotor PM hybrid-excited machine with enlarged axial PM pole as shown in Fig. 19f, respectively.

Type 1 hybrid-excited machines, which have both PMs and DC field coils in the rotor, can be viewed as the integration of a rotor-PM machine and a rotor-wound field machine. However, since DC field coils are located in the rotor, the use of brushes/sliprings or exciter and rotating diodes is necessary, making this type of

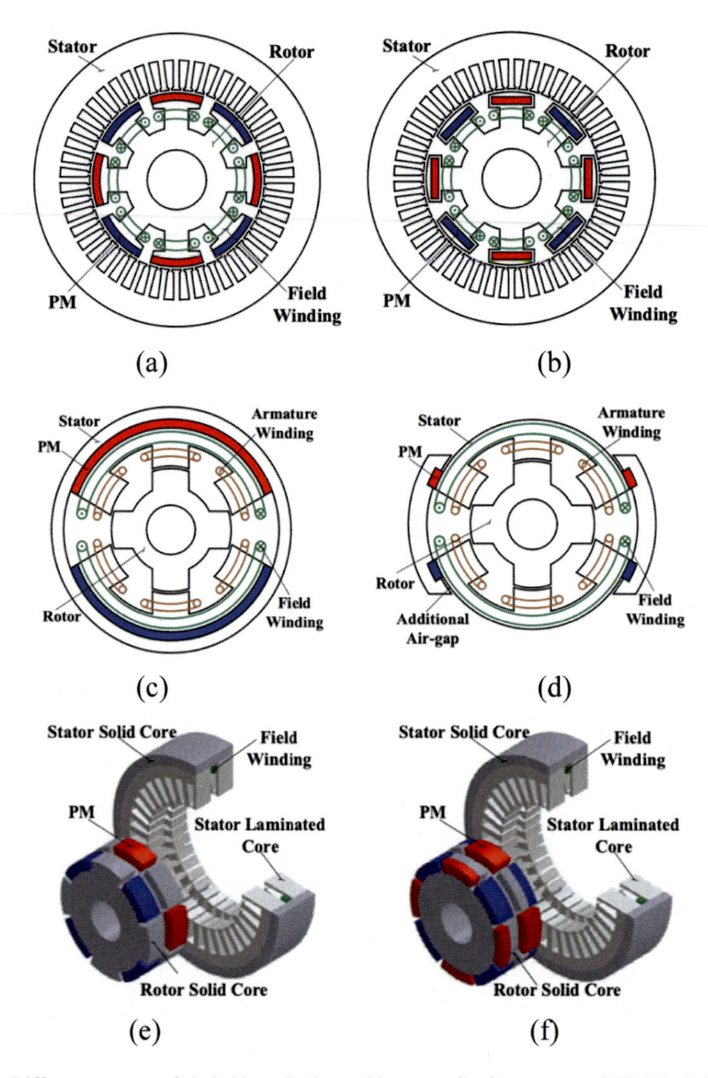

Fig. 19 Different types of hybrid-excited machines. **a** Surface-mounted PM hybrid-excited machine [29]. **b** Interior PM hybrid-excited machine [29]. **c** Series hybrid doubly-salient PM machine [29]. **d** Parallel hybrid doubly-salient PM machine [29]. **e** Consequent pole rotor with equal PM pole and iron pole [29]. **f** Consequent pole rotor with enlarged axial PM pole [29]

hybrid-excited machine less appealing than others. By contrast, Type 2 hybrid-excited machines with both PMs and DC field coils in the stator are more suitable since both components are stationary, eliminating the need for brushes. Type 3 hybrid-excited machines, which have PMs in the rotor and DC field coils in the stator, have the potential for achieving high torque density by using PMs in the rotor. Additionally, slip rings or brushes are not required since the DC field coils are in the stator. An example of a hybrid-excited consequent-pole PM machine in this category is shown in Fig. 19e, as reported in [30]. This machine has a circumferential DC field winding placed in the middle of the stator to regulate the magnetic flux, and a wide speed range can be achieved by controlling the excitation DC field winding current without the risk of demagnetization for the PM pieces.

Hybrid-excited machines combine the advantages of electrically-excited machines and PM machines, and hence high toque density and wide speed range can be achieved simultaneously. The challenges/difficulties of hybrid-excited machines are as follows:

- The additional excitation source results in a complicated structure.
- The use of brushes/sliprings or exciter and rotating diodes in the hybrid-excited machines with both PMs and DC field coils in rotor is inevitable. Moreover, thermal dissipation issues exist for the DC field coils in the rotor.
- Hybrid-excited machines with both PMs and DC field coils in stator have complicated stator structure due to the fact that PMs, DC field coils, and armature windings are competing for space in the stator.
- The DC field excitation MMF can be high, which affects the machine power density and efficiency.

4.2 Memory PM Machines

Memory PM machine was first proposed by V. Ostovic in 2003 [31]. A typical memory machine using tangential magnetized low coercive force (LCF) PMs is shown in Fig. 20. The stator is identical to that of a conventional PMSM, while the rotor has a layered structure consisting of PMs (red and blue), soft iron (grey), and nonmagnetic material (yellow). All of these components are mechanically fixed to a non-magnetic shaft. Trapezoidal AlNiCo magnets are used as LCF PMs, which can be magnetized differently by applying a pulse of d-axis current in the stator windings. As a result, the air-gap flux of each rotor pole can be adjusted. This adjustable-flux feature sets memory PM machines apart from conventional PMSMs where the PM flux cannot be regulated without flux-weakening.

The variable flux operating principle of memory PM machines can be explained by a simplified illustration of the hysteresis model of PM materials as shown in Fig. 21. The residual magnetism of the PM moves along the hysteresis loop, which is determined by both the PM material and the magnetic field. This loop is composed of reversible and irreversible regions. In the reversible region, the PM operates above the knee point, whereas in the irreversible region, the PM is irreversibly demagnetized when the operating point is below the knee point.

For conventional PMSMs, the PM working points should be along the reversible magnetization line B-F, in which F is the cross point of the reversible magnetization line and demagnetizing curve. The working points should be kept above point F and avoid demagnetization. By contrast, for memory PM machines, the PM working point can move to below the cross point, F, by applying a negative d-axis current pulse in the stator windings. When the PM is demagnetized from point F to point G,

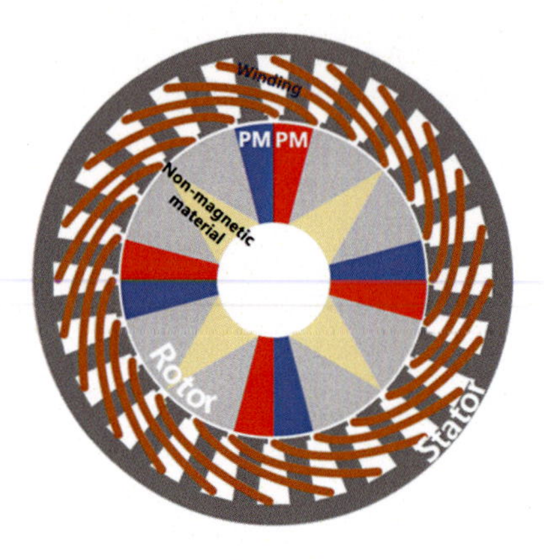

Fig. 20 Topology of the original variable flux memory machine

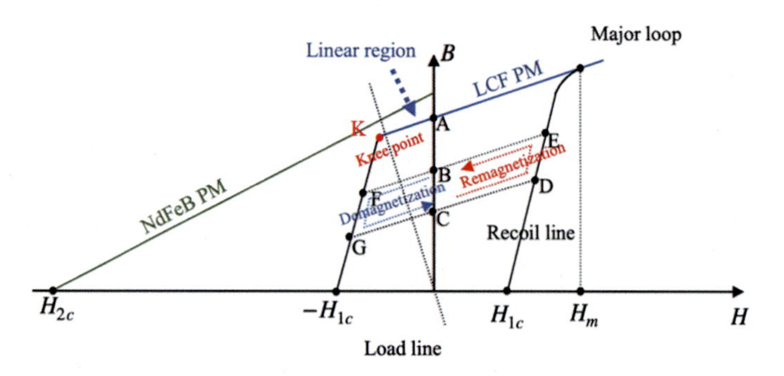

Fig. 21 Simplified illustration of hysteresis model of LCF PM

the PM will work along a new reversible magnetization line G-D. On the other hand, when the PM working point is initially at the right of point D, by giving a certain positive *d*-axis current pulse in the stator windings, the PM will be magnetized, and the recoil line will move upwards. Accordingly, in memory PM machines, the PMs are able to work along different recoil lines by giving a certain *d*-axis excitation to change the flux density, *B*, of the PM material. The PM could be maintained at the last magnetizing state. Hence, these machines are named as memory machines. Since the PM flux of the memory PM machine is adjustable, the speed range can be extended by changing the PM working point without adopting flux weakening control method by applying a negative *d*-axis current anymore.

To improve torque density and speed range, several memory machines with different structures have been developed. The memory machine concept can also be combined with the stator-PM machine, e.g., a stator-PM doubly-salient flux memory motor was presented in [32] as shown in Fig. 22. This machine features a three-phase 24/16-pole doubly-layer-stator outer-rotor topology. In the stator, two types of PMs, i.e., AlNiCo and NdFeB, and the magnetizing windings are in the inner layer, while the armature windings are in the outer layer. It was found that this machine

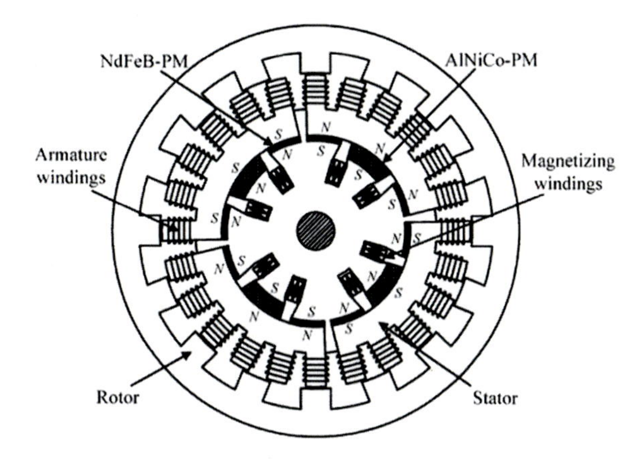

Fig. 22 Stator-PM doubly-salient flux memory machine [32]

achieved both high torque/power density and high efficiency through the use of high-energy NdFeB magnets in conjunction with AlNiCo magnets, and the application of temporary current pulses for online PM magnetization to efficiently control air-gap flux.

Another novel machine design is the magnetic-geared memory PM machine proposed in [33], as shown in Fig. 23. This machine comprises an inner stator, a static modulation ring, and an outer rotor. The stator consists of two sets of windings and three pole pair of AlNiCo magnets, while the static modulation ring features 18 pole/iron pieces, and the outer rotor is a surface-mounted PM rotor with 16 pole pair of NdFeB magnets. The AlNiCo magnets can be properly magnetized or demagnetized. Hence, this hybrid-structure machine artfully integrates the magnetic-gearing effect and flux-mnemonic capability together, which provide the high torque/power density and the wide speed range operation. These features are highly desirable for EV/HEV applications.

Due to the adjustable PM flux, the memory PM machines can not only retain high output torque but also achieve a wide speed range operation. The challenges/difficulties of the memory PM machines are as follows:

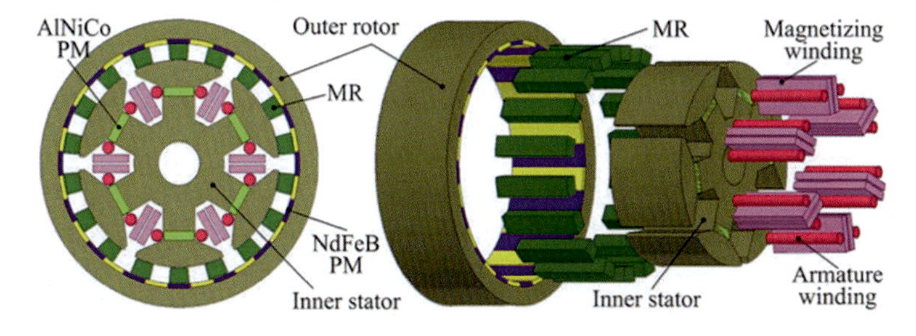

Fig. 23 Magnetic-geared memory PM machine [33]

- The remanence of the memory magnets is relatively lower than conventional magnets, which leads to the relatively low torque density of memory PM machines.
- The magnetizing and demagnetizing processes of the magnets are nonlinear. Hence, it is challenging to accurately control the working point of the magnets.
- When magnetizing or demagnetizing, a large current pulse is needed to change the working point of PMs which is challenging for the drive system.

5 Performance Comparison

For a better understanding of the pros and cons of these advanced machines, a qualitative comparison is conducted and listed in Table 2. In terms of structure complexity, the stator PM machines usually have simple structures, while the flux-controllable machines exhibit more complex structure since additional excitation windings (field windings for the hybrid-excited machines and magnetizing windings for the memory PM machines) are required. In addition, magnetically-geared PM machines exhibit the most complicated structure since two air gaps are included in such machines, which is challenging for manufacturing. In terms of torque/power density, magnetic gear-based machines exhibit the highest torque density due to the magnetic-gearing effect. In terms of speed range, flux-controllable machines outperform others due to their unique variable-flux capability. In terms of reliability, the stator-PM machines outperform others since they have simple and robust rotor structure as well as more favorable PM thermal dissipation capability.

In general, even though these advanced electric machines exhibit advantages in some aspects compared to conventional electric machines in EV/HEV applications, there are still a lot of technical challenges to be overcome. At present, all these advanced electric machines have not been widely used in EV and HEV applications. Therefore, further research is required to iteratively improve the all-round performance of these electric machines.

Table 2 Qualitative comparison of advanced electric machines

		Structure complexity	Torque/power density	Speed range	Reliability
		★ → ★★★★★ Simple → Complex	★ → ★★★★★ Low → High	★ → ★★★★★ Narrow → Wide	★ → ★★★★★ Low → High
Stator-PM machines	Double-salient PM machine	★	★★	★★★	★★★★★
	Flux-switching PM machine	★★	★★★	★★★	★★★★★
	Flux-reversal machine	★★★	★★★	★★★	★★★★★
Magnetic gear-based machines	Vernier PM machine	★	★★★★★	★★★★	★★★★
	Magnetically-geared PM machine	★★★★★	★★★★★	★★★★	★★★
Flux-controllable machines	Hybrid-excited machine	★★★★	★★★	★★★★★	★★
	Memory PM machine	★★★★	★★★	★★★★★	★★

References

1. Haddoun A, Benbouzid MEH, Diallo D, Abdessemed R, Ghouili J, Srairi K (2008) Modeling, analysis, and neural network control of an EV electrical differential. IEEE Trans Industr Electron 55(6):2286–2294
2. Chau KT, Li W (2014) Overview of electric machines for electric and hybrid vehicles. IJVD 64(1):46. https://doi.org/10.1504/IJVD.2014.057775
3. Liao Y, Liang F, Lipo TA (1992) A novel permanent magnet motor with doubly salient structure. In: Conference record of the 1992 IEEE industry applications society annual meeting, vol 1. pp 308–314. https://doi.org/10.1109/IAS.1992.244279
4. Shi JT, Zhu ZQ, Wu D, Liu X (2014) Comparative study of novel synchronous machines having permanent magnets in stator poles. In: 2014 international conference on electrical machines (ICEM). pp 429–435. https://doi.org/10.1109/ICELMACH.2014.6960216
5. Mingyao L, Ming C, Z E (2002) Design and performance analysis of new 12/8-pole doubly salient permanent-magnet motor. In: Sixth international conference on electrical machines and systems, ICEMS 2003, vol 1. pp 21–25
6. Zhang L, Wu LJ, Huang X, Fang Y, Lu Q (2019) A novel structure of doubly salient permanent magnet machine in square envelope. IEEE Trans Magn 55(6):1–5. https://doi.org/10.1109/TMAG.2019.2906772
7. Du Y et al (2019) Principle and analysis of doubly salient PM motor with Π-shaped stator iron core segments. IEEE Trans Industr Electron 66(3):1962–1972. https://doi.org/10.1109/TIE.2018.2838060
8. Hoang E, Ahmed HB, Lucidarme J (2019) Switching flux permanent magnet polyphased synchronous machines. Presented at the EPE 97. [Online]. https://hal.archives-ouvertes.fr/hal-00533004/document. Accessed 27 Feb 2019
9. Chen H, EL-Refaie AM, Demerdash NAO (2020) Flux-switching permanent magnet machines: a review of opportunities and challenges-part I: fundamentals and topologies. IEEE Trans Energy Convers 35(2): 684–698. https://doi.org/10.1109/TEC.2019.2956600
10. Shi Y, Jian L, Wei J, Shao Z, Li W, Chan CC (2016) A new perspective on the operating principle of flux-switching permanent-magnet machines. IEEE Trans Industr Electron 63(3):1425–1437. https://doi.org/10.1109/TIE.2015.2492940
11. Zhu ZQ, Chen JT (2010) Advanced flux-switching permanent magnet brushless machines. IEEE Trans Magn 46(6):1447–1453. https://doi.org/10.1109/TMAG.2010.2044481
12. Raminosoa T, Gerada C, Galea M (2011) Design considerations for a fault-tolerant flux-switching permanent-magnet machine. IEEE Trans Industr Electron 58(7):2818–2825. https://doi.org/10.1109/TIE.2010.2070782
13. Taras P, Li GJ, Zhu ZQ (2017) Comparative study of fault-tolerant switched-flux permanent-magnet machines. IEEE Trans Industr Electron 64(3):1939–1948. https://doi.org/10.1109/TIE.2016.2627022
14. Zhu ZQ, Chen JT, Pang Y, Howe D, Iwasaki S, Deodhar R (2008) Analysis of a novel multi-tooth flux-switching PM brushless AC machine for high torque direct-drive applications. IEEE Trans Magn 44(11):4313–4316. https://doi.org/10.1109/TMAG.2008.2001525
15. Deodhar RP, Miller TJE (1997) The flux-reversal machine: a new brushless doubly-salient permanent-magnet machine. IEEE Trans Ind Appl 33(4):10
16. Li H, Zhu ZQ (2019) Influence of adjacent teeth magnet polarities on the performance of flux reversal permanent magnet machine. IEEE Trans Ind Appl 55(1):354–365. https://doi.org/10.1109/TIA.2018.2867818
17. Li HY, Zhu ZQ (2018) Analysis of flux-reversal permanent-magnet machines with different consequent-pole PM topologies. IEEE Trans Magn 54(11):1–5. https://doi.org/10.1109/TMAG.2018.2839708
18. Vidhya B, Srinivas KN (2017) Effect of stator permanent magnet thickness and rotor geometry modifications on the minimization of cogging torque of a flux reversal machine. Turk J Elec Eng & Comp Sci 25:4907–4922. https://doi.org/10.3906/elk-1703-33

19. Wu F, El-Refaie AM (2019) Permanent magnet vernier machine: a review. IET Electr Power Appl 13(2):127–137
20. Li D, Qu R, Lipo TA (2014) High-power-factor vernier permanent-magnet machines. IEEE Trans Ind Applicat 50(6):3664–3674
21. Xie S et al (2022) Investigation on stator shifting technique for permanent magnet vernier machines with two-slot pitch winding. IEEE Trans Ind Electron 1–11. https://doi.org/10.1109/TIE.2022.3201308
22. Zou T, Li D, Qu R, Jiang D, Li J (2017) Advanced high torque density PM vernier machine with multiple working harmonics. IEEE Trans Ind Appl 53(6):5295–5304. https://doi.org/10.1109/TIA.2017.2724505
23. Li D, Qu R, Li J, Xu W (2015) Consequent-pole toroidal-winding outer-rotor vernier permanent-magnet machines. IEEE Trans Ind Applicat 51(6):4470–4481
24. Liu W, Lipo TA (2018) Analysis of consequent pole spoke type vernier permanent magnet machine with alternating flux barrier design. IEEE Trans Ind Appl 54(6):5918–5929. https://doi.org/10.1109/TIA.2018.2856579
25. Xie K, Li D, Qu R, Gao Y (2017) A novel permanent magnet vernier machine with halbach array magnets in stator slot opening. IEEE Trans Magn 53(6):1–5. https://doi.org/10.1109/TMAG.2017.2658634
26. Atallah K, Howe D (2001) A novel high-performance magnetic gear. IEEE Trans Magn 37(4):2844–2846
27. Zhu ZQ, Li HY, Deodhar R, Pride A, Sasaki T (2018) Recent developments and comparative study of magnetically geared machines. CES Trans Electr Mach Syst 2(1):13–22. https://doi.org/10.23919/TEMS.2018.8326448
28. Overview of flux-controllable machines_Electrically excited machines, hybrid excited machines and memory machines. https://reader.elsevier.com/reader/sd/pii/S1364032116305901?token=DAA4939C76AB72C6CEF0BD1E2FE32AC8DDF3306540974A8A09ADFBE334EFAA8114F24C0E817D40E8E7484A88C3459F94&originRegion=eu-west-1&originCreation=20221023135434. Accessed 23 Oct 2022
29. Zhu ZQ, Cai S (2019) Overview of hybrid excited machines for electric vehicles. In: 2019 fourteenth international conference on ecological vehicles and renewable energies (EVER). pp 1–14. https://doi.org/10.1109/EVER.2019.8813587.
30. Tapia JA, Leonardi F, Lipo TA (2003) Consequent-pole permanent-magnet machine with extended field-weakening capability. IEEE Trans Ind Appl 39(6):1704–1709. https://doi.org/10.1109/TIA.2003.818993
31. Ostovic V (2003) Memory motors. IEEE Ind Appl Mag 9(1):52–61. https://doi.org/10.1109/MIA.2003.1176459
32. Zhu X, Quan L, Chen D, Cheng M, Hua W, Sun X (2011) Electromagnetic performance analysis of a new stator-permanent-magnet doubly salient flux memory motor using a piecewise-linear hysteresis model. IEEE Trans Magn 47(5):1106–1109. https://doi.org/10.1109/TMAG.2010.2072986
33. Liu C, Chau KT, Qiu C (2014) Design and analysis of a new magnetic-geared memory machine. IEEE Trans Appl Supercond 24(3):1–5. https://doi.org/10.1109/TASC.2013.2295680

Conductive Common-Mode EMI Suppressing Methods in Inverter Fed Motor Drives

Yihua Hu and Xiao Chen

Abstract The impact of electromagnetic interference (EMI) is an increasingly important aspect of the performance of switching inverters. The challenges of managing EMI continue to grow with the emergence of wide bandgap (WBG) devices, the trend towards ever-greater integration and higher power rating. This paper reviews suppression methods for the conductive common-mode (CM) EMI in inverter fed motor drives. In order to span EMI suppression across the full system design process, the review considers both mitigation from the sources and suppression along the conduction paths. Furthermore, the shortcomings and merits of the reviewed publications are discussed, and their attenuation frequency range and attenuation level are compared. It is demonstrated that the CM EMI at low frequency is mainly determined by the PWM strategies and can be reduced or even theoretically eliminated through zero common-mode control. On the other hand, the CM EMI at high frequency is markedly influenced by the switching transients of the power devices. Thus, various drive circuits are reviewed, which improve the switching behaviour. Finally, the deployment of passive and active filters to suppress or compensate for the EMI is discussed.

1 Background

Electric drives have been increasingly adopted in many industrial sectors and defence applications due to their ability to yield highly controllable, fast response and high-power density systems solutions. However, as the switching frequency and switching speed of power devices and the inverter power continue to increase, the problems posed by Electromagnetic Interference (EMI) in electric drives are exacerbated. EMI has the potential to cause deterioration in the performance of electrical machines and

Y. Hu (✉)
University of York, York, UK
e-mail: Yihua.hu@york.ac.uk

X. Chen
University of Sheffield, York, UK
e-mail: xiao.chen@sheffield.ac.uk

C. H. T. Lee (ed.), *Emerging Technologies for Electric and Hybrid Vehicles*,
Green Energy and Technology, https://doi.org/10.1007/978-981-99-3060-9_4

its drive system through a variety of failure mechanisms, including, inter-alia, bearings damage, unexpected actions of power devices, interference on analogue feedback signals, etc. These problems eventually lead to system performance degradation or even shutdown.

EMI in electric drives can be classified as conductive EMI and radiated EMI. The conductive EMI is defined with the frequency range from 150 kHz to 30 MHz [1]. Accounting for different conductive paths, conductive EMI can be divided into differential-mode (DM) EMI and common-mode (CM) EMI. DM EMI is caused by the voltage differences between phases and results in conduction between the three phases through the load circuit. CM EMI is caused by the voltage differences between the neutral points of three phases and the ground, which is referred to as the common-mode voltage (CMV). In a traditional three-phase inverter, the CMV is expressed as $V_{cm} = (V_{AO}+V_{BO}+V_{CO})/3$. It can be seen that the CMV in a traditional switching inverter (in which the outputs of each phase at any operating switching state are either $\pm$ $V_{DC}/2$) cannot be eliminated.

The parasitic *LC* components in drives form the various CM conductive paths. The main CM path in a typical three-phase, six-switch inverter is shown in Fig. 1. The CMV is coupled to the ground through parasitic capacitances and is conducted back to the DC side. The parasitic capacitances mainly exist between the inverter heatsink to ground, the motor windings, and the motor windings to the motor frame. The magnitude of the capacitances encountered between various components in electrical machines and its drive system has been modelled and measured in several publications, including the machine winding to frame capacitances in the context of bearing currents [2–4], power cable to cable capacitances [5, 6], power module internal capacitances [7, 8], and power module to heat sink capacitances [9–11]. These capacitances tend to have values in a range from 10 to 1000pF.

The EMI sources in inverter fed motor drives predominantly comprise two elements, viz. the CMV caused by the discrete voltages generated by the inverter

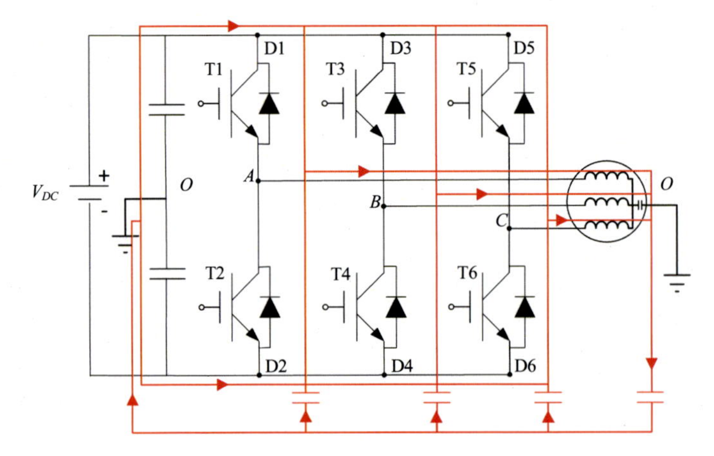

Fig. 1 The main conduction paths of common-mode EMI in motor drives

and the PWM; the high *dV/dt* and *di/dt* generated during the switching transient of power devices. The spectral distributions of EMI caused by the PWM sequences and switching transient of power devices are analysed in detail in [12]. The result therein indicates that the EMI between 150 kHz and 5 MHz is determined mainly by the PWM sequences, while the switching transient influences the EMI between 5 and 30 MHz. On the one hand, the harmonics of discrete output voltage induces ripple current at low frequency. On the other hand, due to the rapid switching actions of the power devices, the high *dV/dt* induces large charging/discharging currents in parasitic capacitances at high frequency. The higher the switching frequency and DC bus voltage of the inverter, the more problematic the EMI caused by the inverter is likely to be.

There are two main research topics on conductive EMI, viz., conductive EMI suppression and the development of high-frequency models of conduction paths of EMI. Establishing high-frequency models aims to intrinsically suppress conducted EMI better, for example, optimizing the design of EMI filters based on high-frequency models. Several review papers have been published on these two topics. In [13], Wang et al. review the advances in EMI modelling techniques for adjustable speed drives. The equivalent circuits for high-frequency models of induction machines and rectifier inverters are presented. Though EMI suppression is discussed in [13], only EMI filters and reduced CMV PWM techniques are explored. The EMI characteristics and EMI reduction methods for wide-bandgap (WBG) devices are reviewed in [14]. However, [14] mainly focuses on the reduction methods based on the power devices, such as advanced gate drive and packaging optimization. Furthermore, In [15], Amin and Choi also pay their attention to WBG devices and emphasize the role of passive filters and active filters in EMI reduction. In [16], EMI reduction methods for DC-DC converters are discussed, but zero CMV techniques are not included. Collectively, these review papers are somewhat lacking in quantitative analysis and arguably do not comprehensively review the full spectrum of EMI reduction methods for inverter fed drives.

This paper aims to review the existing CM EMI reduction methods that can be adopted at either the early stage or the later stages of the machine drive design process. At the early stage, the EMI can be suppressed from its sources while at the later stages, filters can be applied to attenuate the EMI along conductive paths. Hence, the reviewed publications are categorized into two classes as shown schematically in Fig. 2. For each publication reviewed, the effective frequency range and reduction effects are summarized. In section II, methods employed for mitigating EMI from sources are presented. The methods that are able to suppress EMI along conductive paths are presented in section III. In section II and section III, the principles and status of each method are presented. In section IV, the attenuation frequency range and attenuation level provided by each method are compared. The merits and drawbacks and likely future trends in the development of EMI suppression methods are discussed in section IV. Section V draws key conclusions from the review.

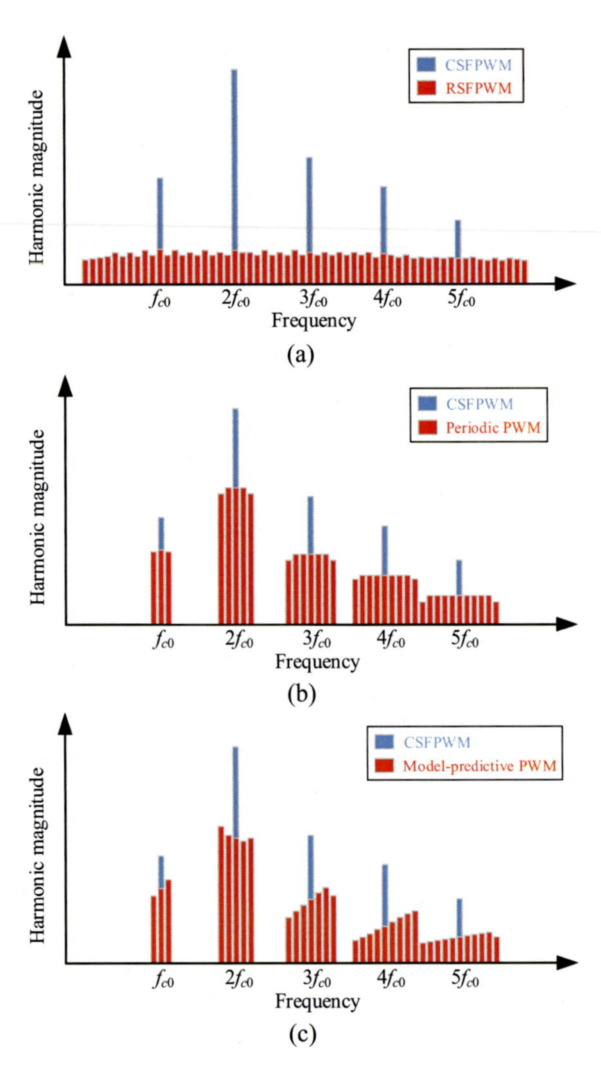

Fig. 2 Comparison of representative current harmonic magnitude patterns of variable switching frequency PWM and constant switching frequency PWM. **a** RSFPWM versus CSFPWM. **b** Periodic PWM versus CSFPWM. c Model-predictive PWM versus CSFPWM

2 EMI Mitigation from EMI Sources

From the perspective of EMI sources, EMI mitigation can be achieved by applying variable switching frequency PWM (VSFPWM) and improving the switching transient of power devices. Besides, although the CMV of three-phase two-level switching inverters cannot be eliminated, complex topologies of inverters and motors developed in recent years provide options for theoretically eliminating CMV.

A. Variable switching frequency modulation strategies

Constant switching frequency PWM(CSFPWM), such as space-vector PWM (SVPWM), has been widely applied in conventional inverter control strategies.

The harmonics of voltage and current are concentrated at integer multiples of the switching frequency under CSFPWM. These concentrated harmonics generate EMI at low frequency, i.e. tens to hundreds of kHz.

To reduce EMI peaks at the switching frequency and its integer multiples, spread spectrum modulation strategies have been introduced to achieve VSFPWM. VSFPWM spreads the concentrated harmonics over a wider frequency band, which has the effect of reducing EMI peaks. According to different means of varying the switching frequency, VSFPWM applied in motor drives can be classified as random switching frequency PWM (RSFPWM), periodic PWM, and model predictive PWM. By way of illustration, Fig. 2 compares representative current harmonic magnitude patterns of RSFPWM, periodic PWM and model-predictive PWM against CSFPWM. Fundamentals are excluded in Fig. 2, and f_{c0} denotes the carrier frequency.

RSFPWM controls the switching frequency varying randomly on both sides of the constant switching frequency. Assuming the constant switching frequency of the inverter is $f_c(t_0)$, Rand$\in(-1,1)$ is a random number, $\gamma\in(0,1)$ is a constant coefficient, the switching frequency f_c of RSFPWM can be expressed as $f_c = f_c(t_0) \times (Rand \times \gamma)$. The generation of pseudo-random signals and SVPWM was implemented on an FPGA-based electric drive in [17], which demonstrated that RSFPWM can spread the spikes of line voltage harmonics to nearby frequencies. For the purpose of enhancing the harmonic power spreading ability of RSFPWM, multiple carrier-based RSFPWM is proposed in [18] to improve the randomness of the switching frequency. In [19], due to the excellent performance of RSFPWM in reducing harmonic peaks, RSFPWM is introduced to interior permanent magnet synchronous machine (IPMSM) sensorless control based on high-frequency signal injection to spread out the additional harmonics, yielding a 20 dB attenuation in the power peak of full-spectrum. The voltage, current and acoustic noise spectral characteristics of five state-of-the-art RSFPWM strategies in adjustable-speed drives (ASD) are quantitatively evaluated in [20], and the experimental results demonstrate that RSFPWM can attenuate the acoustic noise peaks by more than 20 dB in an induction motor drive. These RSFPWM publications focus mainly on noise reduction rather than EMI suppression. In [21], the EMI spectrum of RSFPWM is analysed in-depth, indicating that RSFPWM can eliminate discrete common-mode voltage spikes in the 1-100 kHz band. The randomness inherent in the switching frequency raises the following concerns:

(a) The switching loss cannot be predicted as the switching frequency changes in a random manner, which makes efficiency calculation challenging. Hence, the thermal subsystem cannot be optimally designed [22];
(b) RSFPWM cannot arrange the spectral distribution of harmonic energies, and it is not effective in suppressing specific harmonics;
(c) In some applications, RPWM increases the requirements for EMI filters [21].

In contrast to RSFPWM, periodic PWM changes the switching frequency periodically within a certain range, and the switching frequency of periodic PWM can be expressed as: $f_c = f_c(t_0) + Func(t) \times \Delta f_c$, where $Func(t)$ indicates specific changing patterns and Δf_c is the varying width of the switching frequency. According to different frequency-changing patterns, periodic PWM can be classified as having a sinusoidal profile, exponential profile, triangular profile [23–25], or sawtooth pattern [26]. It is noted in [27] that the Δf_c is a key factor affecting the spreading effect of periodic PWM. EMI peaks decrease as Δf_c increases. However, when Δf_c increases to a certain degree, overlapping of harmonics will occur, which acts to counter the suppression of EMI. In [26, 28], Xu et al. conducted a comparative study on inverter loss and motor loss in PMSM drives using CSFPWM, sinusoidal periodic PWM, and sawtooth periodic PWM, demonstrating that periodic PWM does not increase inverter loss, but it will marginally increase motor loss compared to CSFPWM. Furthermore, in [29], it is demonstrated that in the frequency band between 1 to 100 kHz, sawtooth periodic PWM has a superior ability to suppress common-mode conducted EMI than sinusoidal periodic PWM. Since the switching frequency of periodic PWM varies periodically, inverter loss prediction can be performed to improve efficiency and optimize thermal design.

RSFPWM and periodic PWM achieve the effect of reducing the values of EMI spectrum peaks by changing the switching frequency within a certain range. However, these two modulation strategies are the direct application of spread spectrum modulation strategies in inverters, and the changes on the switching frequency of inverters are not calculated theoretically and lack an underpinning analytical basis.

To precisely control the current ripple caused by different switching strategies, in [30, 31], Jiang and Wang et al. presented two VSFPWM strategies based on current ripple prediction. An analytical model is developed to predict three-phase current ripple using the switching frequency and PWM duty ratio. The switching frequency is then adjusted to control the ripple to meet the desired ripple specification. Furthermore, in [32], the EMI mitigation effects of model predictive VSFPWM on AC motor drives are investigated.

It is can be seen that VSFPWM can realize the suppression of EMI at low-frequency without the need for additional hardware. Furthermore, the emergence and growing adoption of WBG devices bring the prospect of much higher switching frequencies, so that Δf_c can increase to further improve the spectral dispersion effect of VSFPWM on EMI spikes. The application cases, effective frequency and effects of variable switching frequency PWM are summarized in Table 1.

However, because of the relatively large duration of PWM sequences (typically tens to hundreds of milliseconds), VSFPWM tends to achieve limited suppression of EMI at high-frequency which is predominantly induced by the switching transients of power devices. To suppress conducted EMI at tens of MHz, it is necessary to ameliorate the switching transients, e.g. by adopting improved gate drive circuit.

B. Improved gate drive circuit for power devices

Compared with the tens to hundreds of milliseconds durations of typical PWM sequences, the time scale of switching transient of power devices lies in the range of tens to hundreds of nanoseconds depending on the devices and the gate driver. Therefore, during switching transients, the high *dV/dt* and accompanying voltage ringing caused by impedance mismatch tend to worsen the EMI performance at high frequencies up to hundreds of MHz.

Table 1 Summary of variable switching frequency PWM method key findings

Method	Refs.	Cases	Theme	Frequency range of interest	Contribution
RSF PWM	Pu et al. [17]	VSI fed AC motor	Implementation of conventional RSFPWM	Below 150 kHz	Less than 5 dB attenuation on line voltage harmonic peaks compared with SVPWM
	Sivarani et al. [18]	VSI fed induction motor	To enhance the randomness, the random carrier is synthesized by four carriers of different frequencies	Below 150 kHz	Compared with conventional RSFPWM, harmonic spread factor of the output voltage is reduced by 21% while THD and PSD remain the same
	Zhang et al. [19]	IPMSM sensorless control	Random frequencies for switching and HF signal are adopted to spread the power spectrum	Around the switching frequency	More than 20 dB/Hz attenuation on stator current PSD peaks
	Lee et al. [20]	VSI fed induction motor	A comparative study of 5 RPWM strategies under different fundamental frequencies	Below 50 kHz	Greater than 20 dB/Hz attenuation of PSD peaks of line voltage and phase current
	Lezynski et al. [21]	VSI fed induction motor	Frequency-domain and time-domain analysis of common-mode voltage	1 kHz–100 kHz	Up to 4 dBμV and 11 dBμV reductions on peak and average values of CMV harmonics
Periodic PWM	Santolaria et al. [23–25]	Buck converter	Study on the EMI spectrum when the carrier frequency varies in sinusoidal, exponential and triangle profiles	150 kHz–30 MHz	Approximately 20 dBμV reductions on the EMI peak values and these three modulation profiles show small differences

(continued)

Table 1 (continued)

Method	Refs.	Cases	Theme	Frequency range of interest	Contribution
	Yongxiang et al. [26, 28]	VSI fed PMSM	Study on inverter losses and motor losses in PMSM drives using periodic PWM methods	Around the switching frequency	The periodic PWM does not increase inverter losses, but it will slightly increase motor losses
	Yongxiang et al. [29]	VSI fed PMSM	A comparative study on EMI suppression abilities of periodic PWM modulated in sinusoidal and sawtooth profiles	Below 100 kHz	Sinusoidal and sawtooth profiles both bring more than 10 dBμV reductions on EMI peaks and have small differences
Model predictive PWM	Jiang et al. [30, 31]	Three-phase converter	Propose two model predictive PWM methods to reduce EMI and control the peak value and RMS value of current ripple	10 kHz–1 MHz	These two methods bring 10 dBμA attenuation on EMI current and more than 10% reduction on switching losses
	Jiang et al. [32]	Multi-level inverter fed PMSM	Apply model predictive PWM on AC motor drive to control the current ripple and EMI	10 kHz–1 MHz	More than 10 dBμA attenuation on EMI current and about 33% reduction on switching losses

It is shown in [33] that the corner frequencies of the spectrum of EMI generated by power devices are determined by the duty ratio of the PWM sequence and the switching time, indicating that the device switching time has a significant impact on the EMI at high frequencies. The EMI at high frequencies can be suppressed by slowing switching transients, albeit at the expense of increased switching losses. In [34], the switching losses and EMI of IGBTs are quantitatively compared for SiC devices and Si-SiC combination devices. In order to obtain the best trade-off between EMI suppression and switching loss, switching transient shaping based on active voltage control (AVC) and active gate driver (AGD) have attracted much research interest of late. Typical circuit schematics for implementing these two techniques are illustrated in Fig. 3.

AVC is a hardware-based solution for forcing the IGBT collector voltage (V_{ce}) transient to follow a predefined trajectory through feedback using quantities such as V_{ce}, V_{ge}, and *dV/dt*. Yang et al. proposed a method for controlling the switching transients of IGBT into Gaussian shapes in [35] so that the discontinuities of PWM edges are eliminated due to the infinitely differentiable characteristics of Gaussian function, resulting in a significant reduction of EMI in the range 1–100 MHz. Based on the Gaussian S-shaped IGBT switching transients, the analysis on the trade-off between switching losses and high-frequency EMI suppression is investigated in [36] using an accurate IGBT high-frequency model demonstrated in [37].

AGD is a gate-driving circuit that adjusts the switching speed by controlling *dV/dt* and *di/dt*. Compared to AVC, AGD tends to introduce additional switching losses as a consequence of longer switching times. In [38], the gate driving circuit varies the gate drive voltage as different stages of switching transient, so as to exercise control over the switching speed. To further optimize the switching losses of IGBT caused by AGD, a digital active gate driver (DAGD) based on FPGA is designed and reported in [39] and [40]. Through the condition monitoring of the load current, the most appropriate parameters of the drive circuit are determined by an FPGA. Compared with AGD, switching losses can be significantly reduced (reducing by

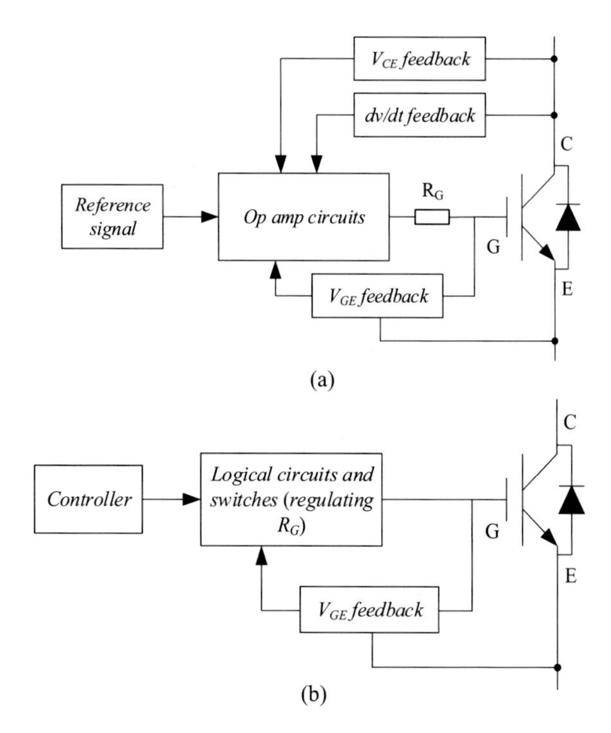

Fig. 3 Schematics of the improved gate driver techniques. **a** AVC. **b** AGD

42% in [39] and 64.7% in [40]) using DAGD, but at the expense of increased circuit complexity.

In comparison to more established Silicon-based switching device technologies, WBG devices can operate at higher switching frequency with low switching loss. However, WBG devices give rise to greater *di/dt* and *dV/dt*, which make WBG devices not only the source of more severe conductive EMI but also more likely to be affected by cross-talk between devices in the same leg. This cross-talk can lead to unexpected switching action and has the potential to cause the short-circuit failure or reverse breakdown. In [41], the mechanism and conduction path of crosstalk between two devices in the same leg is described, and a clamped circuit is proposed to eliminate any positive or negative spurious voltage caused by cross-talk. A summary of the requirements for the SiC drive circuits and the designed circuits to suppress crosstalk is presented in [42] and [43]. In [44] and [45], AGD is introduced into a SiC drive circuit to suppress overshoots by controlling the value of the gate resistor (R_g) during the Miller plateau effect. In [46] and [47], a logical circuit-based AGD is applied to dynamically adjust the effective value of R_g to achieve an improved trade-off between fast switching duration and suppressing voltage ringing. Instead of increasing the R_g to slow down the transient, in [48], the V_{gs} of the SiC MOSFET is adjusted using multiple power supplies. In addition to adjusting R_g and V_{gs}, a current proportional to *dV/dt* is subtracted from the gate current (I_g) to reduce the *dV/dt* and switching loss of GaN transistors in a most recent study [49]. Due to the faster transient of GaN transistors, the AGDs applied on GaN transistors are required to achieve fast response. In [50], the R_g update rate achieves sub-nanosecond resolution (6.7 GHz).

As the two gate drive circuits shown previously in Fig. 3, AGD can control the switching speed while it is unable to shape the switching transient. Hence, the optimal trade-off between switching loss and switching speed is more difficult to achieve in AGD compared with AVC, which is a voltage control system based on operational amplifiers. The features of reviewed papers on improved gate drive circuits are summarized in Table 2.

A. Complex topologies of inverter and machines

As previously discussed in Section I, at the CMV of conventional three-phase two-level inverters cannot be zero at any instant, and hence CM EMI cannot be eliminated in such inverters. To further reduce, or even eliminate, the CMV, a variety of modified, and in some cases complex, inverter and machine topologies has been developed to minimize the CMV. The topologies used to eliminate or mitigate EMI include three-phase four-leg inverters, three-level inverters, paralleled inverters and dual three-phase motors. Schematics of these various arrangements are shown in Fig. 4.

The concept of a four-leg inverter is introduced to motor drives in [51]. Four *LC* passive filters are mounted at the output points of four legs to connected the fourth leg to the three-phase system, which introduces a negligible load unbalance. To eliminate the CMV, special modulation strategies need to be designed in which zero vectors (i.e. 000 and 111) are not permitted.

Lee et al. first proposed the concept of three-dimensional space vector modulation (3-D SVM) in [52] and [53]. 3-D SVM introduces the r-axis into α-β coordinates,

Table 2 Summary of improved drive circuits' key findings

Method	Ref	Cases	Theme	Frequency range of interest	Contribution
Active voltage control	Oswald et al. [34]	Single power device	Investigation on EMI and losses caused by switching transient of IGBT, SiC and Si-SiC	Up to 100 MHz	Compared with IGBTs, the high *dV/dt* of SiC incurs EMI increasing by maximum 30 dBμV
	Oswald et al. [33]	IGBT	Compare S-shaped (twice differentiable) switching transient and trapezoidal (thrice differentiable) transient	Up to 100 MHz	S-shaped transient introduces an additional 20 dB/dec steeper gradient at high frequencies
	Yang et al. [35]	IGBT modules	Apply Gaussian switching transient (infinitely differentiable) on IGBT modules	1–100 MHz	The maximum EMI reduction goes to 40 dB and occurs between 10 and 20 MHz
Advanced gate drive	Idir et al. [38]	IGBT	Control the *dV/dt* and *di/dt* by setting different voltage levels of driving signal	–	The switching time is increased depend on different signal voltages so that the overshoot is suppressed
	Paredes et al. [44, 45]	SiC MOSFET	Change the gate resistance during the Miller plateau effect to suppress the overshoot and ringing	Up to 100 MHz	Voltage overshoots is reduced by 28% and the EMI noise with a resonant frequency of 5.4 MHz is eliminated
	Nayak et al. [46, 47]	SiC-based induction motor drive	In a switching duration, reduce the gate resistance at the start and end stages and increase the resistance at the middle stage	–	Voltage overshoot is reduced from 45% of the DC voltage to 5%
	Yang et al. [48]	SiC MOSFET	The AGD is fed by multiple power supplies, and these power supplies are selected by switches controlled by CPLD	Below 1 GHz	The AGD eliminates the noise in V_{ds} with resonant frequency of 9.2 MHz

(continued)

Table 2 (continued)

Method	Ref	Cases	Theme	Frequency range of interest	Contribution
	Bau et al. [49]	GaN MOSFET	A current mirror amplifying the feedback current are integrated with a push–pull buffer into an ASIC	Below 1 GHz	The noise at the frequency band of 20 MHz–200 MHz is reduced by 5 dB on average
	Dymond et al. [50]	GaN MOSFET	A programmable IC driver is designed to realize variable output resistance (R_g) with a ultra-high timing resolution	Below 2 GHz	The load voltage noise at the frequency band of 200 MHz–2 GHz is reduced by more than 5 dB on average

so as to demonstrate the 16 switching states in 3-D α-β-r coordinate. The CMV can be eliminated when the inverter is operated with six switching states from among the total of 16 states. Hence, the SVPWM used in conventional inverters can be performed using the 6 zero CMV (ZCMV) space vectors. Although the CMV is eliminated, the number of optional space vectors decreases from 8 to 6, which reduces the utilization of the DC bus voltage.

A three-level inverter is combined with a four-leg inverter in [54]. The three-level inverter offers more options for avoiding zero vectors and does not reduce the utilization of DC voltage. The fourth leg can be used not only to eliminate the CMV but also for fault-tolerant design of electric drives. In [55], the options provided by the fourth leg are fully utilized: the fourth leg is used to eliminate the CMV when the inverter is healthy, to operate fault-tolerant control when IGBT faults occur and to be operated as a boost converter during voltage sags.

In three-level inverters, the voltage between the neutral point of each phase and ground can take one of 3 values, viz. $\pm V_{DC}/2$ or 0. With the extra 0 level, 7 space voltage vectors can achieve ZCMV output among the total 27 space vectors of three-level inverters. Zhang et al. first realized the ZCMV PWM in [56]. However, implementing ZCMV modulation directly in three-level inverters deteriorates the performance of three-level inverters due to the constraint of avoiding zero vectors. Therefore, following the initial proposal of the concept of ZCMV PWM, many subsequent studies have focused on reducing the side effects caused by ZCMV PWM, such as increases in output voltage THD, current ripple, and switching loss. To reduce the switching loss, a three-segment switching sequences for zero CMV modulation is realized in [57]. To reduce the current ripple and THD, a four-state ZCMV PWM is presented in [58] with its four PWM sequences. According to synthesizing reference vectors with different space vectors, three types of ZCMV PWM for three-level inverters are listed and comparatively studied in [59]. In [60] and [61], VSFPWM

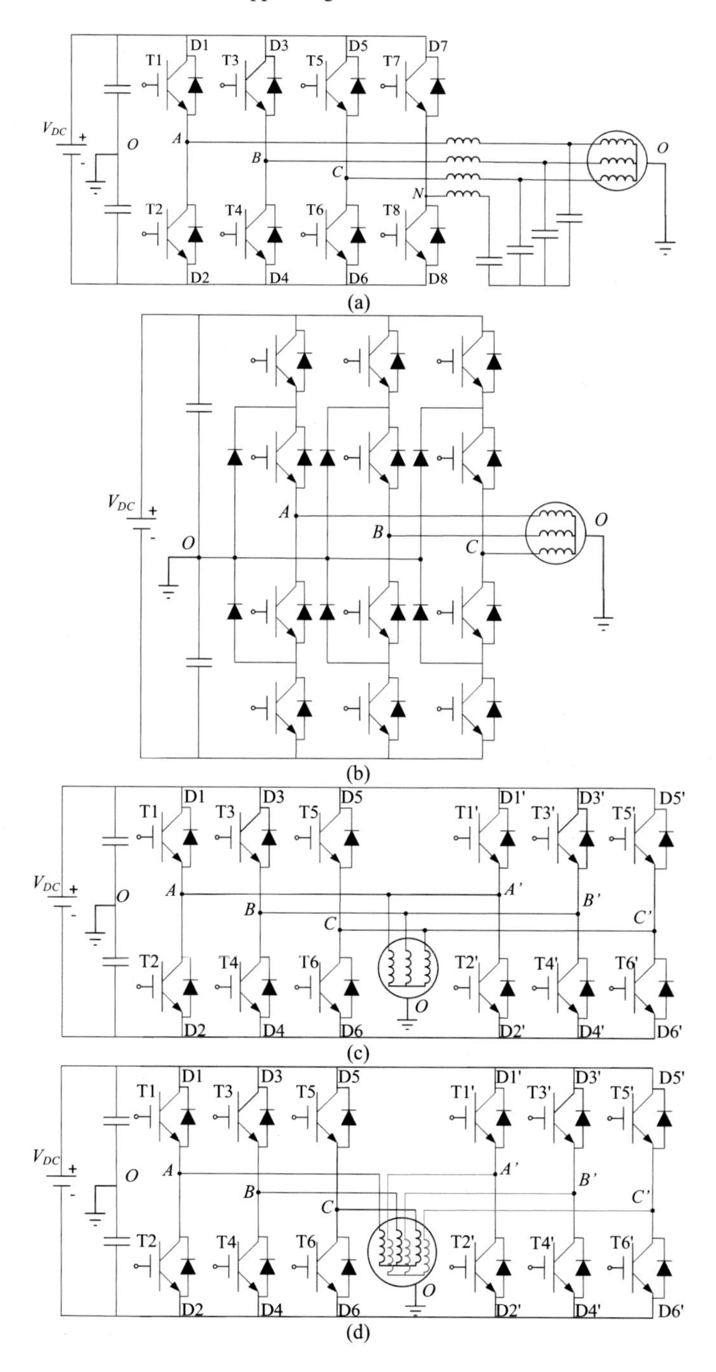

Fig. 4 Schematics of modified topologies of inverter and machines. **a** Three-phase four-leg inverters. **b** Three-level inverters. **c** Paralleled inverters. **d** Dual three-phase motors

is applied in three-level inverters to obtain an improved trade-off between EMI suppression and the performance of three-level inverters, albeit that CMV cannot be eliminated in this way.

A paralleled inverter configuration has more options in terms of switching patterns compared with three-level inverters. Integrated inter-phase inductors are adopted to connect two inverters in [62]. The inter-phase inductors are used for limiting the circulating current between the two inverters and are improved in [63] by optimizing the magnetic circuit to reduce its weight and volume.

Using the paralleled inverter arrangement in [62], the elimination of specific harmonics is realized through carrier phase shifting. Furthermore, the effect of the interleaving angle between the carriers of two inverters on harmonic elimination is analysed. In [64], carrier phase shifting combined with discontinuous SVPWM is applied on a paralleled inverter to limit the CM circulating currents. Since paralleled inverters have even numbers of legs, ZCMV can be achieved. The concept of paralleled space vectors is proposed in [65] and [66], with 6 paralleled space vectors explored to realize ZCMV modulation. In [67], a novel dual-segment three-phase PM machine is introduced to eliminate the need for the inter-phase inductors between two inverters, with consequent benefits in power density.

Dual three-phase arrangements proposed by Zhang achieves CMV cancellation utilizing two sets of symmetrical winding [68, 69]. The two sets of windings are directly paralleled without inter-phase inductors and are supplied by two inverters on a common DC bus. The CMVs of the two inverters have the same magnitude but opposite phases, so that the two voltages cancel each other, thereby achieving near ZCMV. In [70], ZCMV is achieved in an asymmetrical six-phase motor with interleaved windings by PWM signal shifting arrangement. Also, Carrier phase shifting [71] and model predictive PWM [72] are applied on dual three-phase motors to improve the performance.

Although the use of complex topologies can, at least in principle, eliminate common-mode voltages, ZCMPWM will increase output ripple and harmonics, increase switching losses, and reduce bus voltage utilization. Therefore, in addition to ZCMPWM, the combination of advanced modulation strategies and complex topologies has become a research hotspot in recent years, in large part since this combination of approaches has the potential to achieve a good balance between EMI suppression and inverter performance. Table 3 summaries the reviewed publications on complex topologies of machines and inverters.

3 EMI Suppression Along Propagation Paths

The methods of suppressing EMI from sources are mainly adopted at the early stage of system design. However, for drives in which the EMI has not been fully mitigated at source or drives that have already been designed and manufactured, EMI filters are often required to meet the often stringent EMI requirement of the relevant standards. It is also worth noting that in some case, the environment into which the drive

Table 3 Summary of complex topologies' key findings

Method	Refs.	Cases	Theme	Frequency range of interest	Contribution
Three-phase four-leg inverter	Zhang et al. [52, 53]	Four-leg inverter with *LC* load	Proposes the concept of 3-D SVPWM to adapt the space vector modulation to three-phase four-leg inverter	–	The 3-D SVPWM provides six ZCMV vectors to realize ZCMV modulation
	Julian et al. [51]	Four-leg inverter fed Induction motor	Compares the CMV generated by the three-phase inverter, three-phase inverter with non-zero state modulation, and three-phase four-leg inverter	Up to 100 kHz	The fourth leg reduces CMV by 30 dB compared with three-phase inverter and reduces extra 20 dB above 60 kHz compared with three-phase inverter with non-zero state modulation
	Chen et al. [55]	Three-level four-leg inverter motor drive	Introduces the fourth leg to a three-level inverter, so as to obtain more ZCMV vectors	150 kHz to 10 MHz	Attenuate the EMI by 20 dBμA across the all frequency range
	Garg et al. [54]	Four-leg inverter fed induction motor	Active zero state PWM is applied to eliminate the CMV generated by zero vectors	Up to 150 kHz	The RMS value of CMV and ground current is reduced by 60% and 80% respectively
Three-level inverter	Zhang et al. [56]	Three-level inverter fed synchronous motor	First utilizing the 7 ZCMV vectors among the 27 vectors of three-level inverter to realize CMV control	Up to 20 kHz	The CMV is almost eliminated, but the THD of output line-to-line voltage is increased due to the ZCMV control

(continued)

Table 3 (continued)

Method	Refs.	Cases	Theme	Frequency range of interest	Contribution
	Duang-upra et al. [57]	Three-level inverter with *RL* load	Proposes three-segment ZCMV switching sequences to reduce the switching commutation numbers	–	The switching commutation number is reduced by 50% without increases on the THD of output line-to-line voltage
	Nguyen et al. [58]	Three-level inverter fed induction motor	Proposes a four-state ZCMV PWM with its four sequences to reduce current ripple and THD	Integral multiples of switching frequency	The RMS value of current ripple is reduced from 0.131 A to 0.075 A, and the current THD is reduced by half
	Kai et al. [59]	Three-level inverter with *RLC* load	Compares three ZCMV PWM methods in terms of DC utilization, DC current ripple, THD and power losses	–	The ZCMV PWM method that synthesizes reference vectors with 2 medium vectors and 1 zero vector performs better on all aspects
Paralleled inverter	Zhang et al. [62]	Paralleled inverter with *RL* load	Discusses the effects of interleaving angle on reducing the output current ripple and the size of interphase inductors	Integral multiples of switching frequency	With 180° interleaving, the THD of ripple current and the weight of inductors can be reduced by 70% and 60%, respectively
	Zhang et al. [64]	Paralleled inverter with *RL* load	Eliminates the low-frequency CM circulating current by avoiding the coexistence of different zero vectors	–	The proposed PWM method can eliminate low-frequency CM current under symmetric or asymmetric interleaving

(continued)

Table 3 (continued)

Method	Refs.	Cases	Theme	Frequency range of interest	Contribution
	Jiang et al. [65, 66]	Paralleled inverter with *RL* load	Propose paralleled voltage vectors to achieve ZCMV control	Up to 2 MHz	Greater than 20 dBμV EMI reduction compared with conventional SVPWM
	Shen et al. [67]	Paralleled inverter fed dual-segment PMSM	Designs a special dual-segment PMSM to cancel the interphase inductors of paralleled inverter	150 kHz to 1 MHz	ZCMV PWM brings 10 dBμV reduction on peak EMI compared with conventional SVPWM
Dual three-phase motor	von Jauanne et al. [68, 69]	Dual-inverter fed dual three-phase motor	Two inverters generate reverse CMV to achieve ZCMV on the two sets of windings	Integral multiples of switching frequency	The leakage current to ground is reduced by more than 20 dB
	Shen et al. [70]	Dual-inverter fed dual three-phase motor	Proposes a ZCMV PWM scheme on asymmetrical six-phase PMSM based on PWM shifting	150 kHz to 2 MHz	The proposed PWM shifting ZCMV scheme brings 20 dB average peak EMI reduction comparing to SVPWM
	Ye et al. [72]	Dual-inverter fed dual three-phase PMSM	Propose a variable switching sequence PWM based on model predictive method to suppress current harmonics	Up to 10 MHz	Maximum 40 dBμA reduction on DC current EMI below 1 MHz

is integrated may require the adoption of additional filtering. EMI filters can be categorized into two types: passive filters and active filters. The advances in these two types of filters are reviewed in this section.

A. Passive EMI filters

Passive EMI filter is one of the most important and effective solutions to suppress conductive EMI. Inserting a filter at the input of inverters can improve the qualities of feedback signals while inserting a filter at the inverter output can reduce the three-phase current ripple, hence improving the overall performance of the motor drive.

Passive filters form circuits in EMI propagation paths to reflect or shunt EMI currents. In practice, several filters can be cascaded to suppress EMI at different frequencies and enhance the insertion loss. There are several common filter configurations, including *LC*, *CL*, *CLC* and *LCL*. For different source impedances and load impedances, suitable configurations should be selected according to the principle of impedance mismatch [73]. The preferred filter configurations involve connecting the inductances to the side with the lower series impedance and the capacitors to the side with higher parallel impedance, as shown in Fig. 5. In this way, the reflection coefficients of the transmission line are a maximum to reflect high frequency EMI as much as possible.

Akagi et al. designed passive filters to suppress the leakage currents between inverter to the ground [74, 75], machine frame to the ground [74, 76] and the bearing current [77], so that the leakage currents and the bearing current were reduced by 90%. In [77], a line filter (DC side of the inverter) and machine filter (AC side of the inverter) are combined to attenuate the CM EMI. To explore the influence of different filter mounting positions, Xue and Wang compare the differences among EMI filters installed at the inverter output, motor input, and motor chassis in [78], discussing the application scenarios of the three installation methods. In [79] and [80], comprehensive design procedures for the inverter output filter in electric drives are described. In addition, these papers note that the deriving parasitic parameters on the conductive path have an important influence on the filter design and performance.

To achieve an overall optimal filter design, the accurate extraction of parasitic parameters and high-frequency models of CM impedance are key considerations. In [81], Chen et al. proposed a method for extracting the CM impedance of induction motors based on the measured values of CM voltage and current. An equivalent circuit of the CMV conductive loop was established, and the circuit parameters obtained by fitting the impedance curve to the measured values of CM voltage and current. Based on [81], nonlinear programming and a Monte Carlo algorithm were applied in [82] to fit the CM impedance curve of induction motors. The accuracy of the

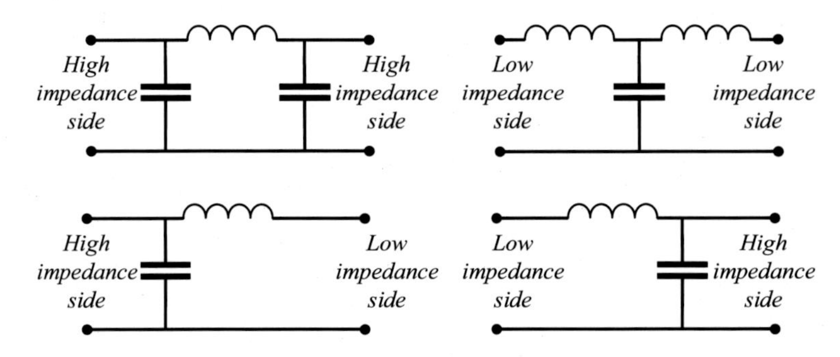

Fig. 5 Illustration of impedance mismatch principle

proposed extraction method was verified by comparing with the measured values of CM impedance.

B. Active EMI filters

Active EMI filters, also referred to as active common noise canceller (ACC), is a CM voltage/current compensator used to reduce or even eliminate the CM EMI.

ACC first detects the CM voltage or current from the conductive path, then generates a reverse voltage or current using amplifier circuits, finally injects the reverse signals into the power circuit to realize the cancellation of CM interference. S. Ogasawara and H. Akagi first introduced ACC to motor drives in [83, 84].

A typical ACC, as depicted in Fig. 6, consists of a noise-sensing circuit, a noise-processing circuit, and a noise-compensating circuit [85]. In [83] and [84], three capacitors in parallel are used to detect the CMV of the inverter, then the amplifier circuit, fed from the DC bus, injects current into the primary side of the CM transformer, and hence inducing a compensation voltage opposing the CMV on the secondary side of the CM transformer, thereby cancelling the CMV. Experimental testing in [83] and [84] has verified that ACC can attenuate EMI in the 10 kHz–3 MHz band by 20 dB.

The performance of ACC on CM EMI cancellation depends on the measurement accuracy of the CM signals. In [86] and [87], Blaabjerg et al. surveyed the commonly used methods for CM signals detection in ACC. M. Piazza improved the CM signal detection circuit, the amplifier circuit, and the CM transformer of ACC in [88], and used a dedicated DC power supply for the amplifier circuit, improving the compensation performance of ACC. The CM transformer in ACC increases the cost and volume of ACC. By changing the injection location of the compensation voltage or current, the elimination of the CM transformer can be achieved. Akagi et al. designed a hybrid active filter which directly connects the ACC to the AC side through a passive RC filter in [89] and [90]. The passive filter has an additional function of power-factor correction for other inductive loads connected on the same 6.6 kV industrial power distribution system. Instead of compensating the voltage

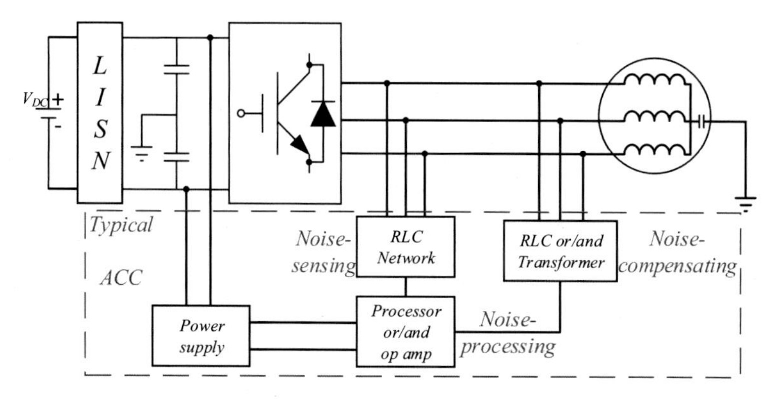

Fig. 6 Schematic of the typical active EMI filter

or current to the DC side or AC side, in [91], Zhu et al. injected compensations to the chassis relative to the inverter ground, so the CM transformer can be eliminated. In **Error! Reference source not found.**, Zhang et al. build a CM impedance network on the AC side which has same CM impedance as the motor and injects the ACC compensation voltage into the AC side CM impedance network to generate a compensation current, with no transformer required. However, the effectiveness of the compensation for CM EMI depends on the matching of the AC-side impedance network and the load-side CM impedance. However, this match is difficult to achieve in practice because the motor on-load impedance is difficult to measure accurately.

An active filter is a compensation system that exhibits good CM EMI suppression in the low frequency range. However, the performance of ACC is limited by many factors, such as the accurate measurement of system CM signals and impedance, and the reliability of the compensation circuit. Table 4 summarizes the reviewed papers on passive and active EMI filters.

4 Bearing Current Due to Common Mode Voltage

Electrical machines operating in tandem with inverters provide high performance electromechanical energy conversion solutions for many industrial sectors, including, inter-alia, automotive, aviation, marine, wind energy. For applications spanning powers from mW to MW such drivetrains are able to provide both high efficiency and precise controllability. However, the high frequency switching in the inverters can cause high frequency common mode voltage between the machine neutral point and the ground. Their high frequency effects are usually regarded as second-order and parasitic effect in terms of primary performance features but can cause significant in-service deterioration and sometime laminations is also due to the skin effect which forces $i_{B,circ}$ to flow on the lamination surfaces. Meanwhile, the common mode voltage can lead to a frame-rotor voltage via the capacitive coupling among the windings, the stator lamination/frame, and the rotor. Then, the frame-rotor voltage generates the electrical discharge machining (EDM) bearing current, as denoted with the $i_{B,EDM}$ path in Fig. 7.

These high frequency voltages and currents can bring a set of challenges to the reliability of electrical machines, including: (a) bearing life reduction due to excessive bearing currents; (b) winding insulation life reduction as a consequence of high voltage stress in inter-turn insulations; (c) electromagnetic interference emergence due to high dV/dt. These problems will be exacerbated as WBG semiconductor devices start to feature in the inverters with faster switching rise times.

To accurately model the bearing currents, Muetze and Binder presented an analytic approach to the calculation of leading capacitances in an electrical machine, thus predicting the bearing voltage ratio (BVR) to estimate the bearing failure risk due to the high frequency circulating bearing currents [2]. They also proposed a circulating bearing current model based on eddy-current calculations, investigated the relationship between the circulating bearing current and the common-mode current,

Table 4 Summary of EMI filters key findings

Method	Ref	Cases	Theme	Frequency range of interest	Contribution
Passive EMI filters	Akagi et al. [74–76]	Inverter fed induction motor	Designs passive filters to eliminate inverter-ground and motor frame-ground leakage currents	–	The RMS value of the leakage currents and the bearing current is reduced by more than 90%
	Akagi et al. [77]	Inverter fed induction motor	Designs two filters inserted into both sides of the inverter to attenuate the CM EMI	150 kHz to 30 MHz	With the designed filters, greater than 40 dB reduction at low-frequency EMI (below 10 MHz) and approximately 20 dB reduction at high frequency (up to 30 MHz)
	Xue et al. [78]	Inverter fed AC motor	Compares the suppressing effects of inverter-end filter, motor-front-end filter and motor-chassis-end filter	Up to 10 MHz	Below 1 MHz, the filters show slight differences; From 1 to 10 MHz, motor-front-end filter shows better performance
	Chen et al. [81, 82]	Inverter fed induction motor	Extracts accurate CM impedance through equivalent CM circuit measuring CM voltage/current	Up to 10 MHz	Compared to two traditional methods, the estimated error of the proposed method is reduced by 10 dB
Active EMI filters	Ogasawara et al. [83, 84]	Inverter fed induction motor	Proposes the scheme of voltage-sensing voltage compensating active EMI filter	10 kHz to 3 MHz	Approximately 20 dBμV CMV reduction in low frequency (10 kHz to 3 MHz), but little or no reduction at higher frequency
	Piazza et al. [88]	Inverter fed induction motor	Improves the detecting circuit, amplifier circuit and power supply of active filter to enhance the compensating ability	150 kHz to 30 MHz	About 10 dBμV CMV reduction in low frequency (below 1.5 MHz), but passive filter performs better at high frequency

(continued)

Table 4 (continued)

Method	Ref	Cases	Theme	Frequency range of interest	Contribution
	Akagi et al. [89, 90]	Three-level inverter fed induction motor	Directly connect the active filter to the AC side so that the CM transformer is cancelled	Integral multiples of switching frequency	Compared with passive filter, the 5th, 11th, and other higher harmonic currents are compensated
	Zhu et al. [91]	Inverter fed induction motor	Integration of the active filter on the inverter drive circuit, thus eliminating CM transformer	10 kHz to 100 MHz	Average 15 dBμA reduction on CM current in 10 kHz to 0.5 MHz and average 5 dBμA reduction in 0.5 MHz to 10 MHz
	Zhang et al. [92]	Inverter fed AC motor	Proposes a voltage-sensing current-compensating active filter which is based on a CM impedance network	150 kHz to 30 MHz	Average 10 dBμA reduction on DC side CM EMI below 3 MHz and average 5 dBμA reduction up to 30 MHz

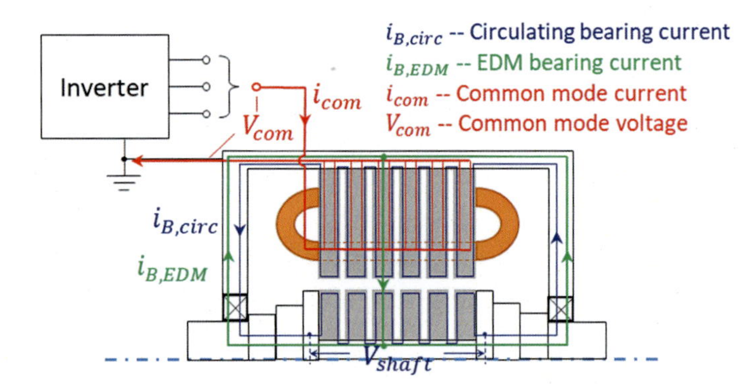

Fig. 7 Illustration of high frequency common mode current and bearing current paths in an electrical machine (lamination thickness has been greatly exaggerated to aid clarity)

and found that the maximum ratio between those two currents is approximately 0.35 [93]. Muetze and Binder also proposed an engineering flowchart to determine the type of bearing current based on the grounding configuration and the machine size, before estimating the magnitude of bearing current and introducing the correlating mitigation techniques [94]. They performed scalability study of inverter-induced bearing currents in electrical machines with power ratings from 0.5 to 750 kW, and found that there are higher bearing current densities in very small machines (<10 kW) and very large machines (>100 kW) than those in the middle range machines. The dominant bearing current types of the very small machines and very large machines are EDM and high-frequency circuit bearing currents respectively [95]. Then, they also proposed the corresponding mitigation techniques. Ahola, Muetze et al. calculated the rotor-to-frame and the stator winding-to-rotor capacitances based on electrostatic FEA, which can improve the accuracy in the prediction of the bearing voltage ratio and EDM bearing currents [96].

Regarding the bearing fault detection, Muetze and Strangas reviewed the bearing failure mechanisms, fault diagnosis procedures and thus the prediction of the bearing remaining useful life based on bearing currents and voltages [97, 98]. They also proposed a possible solution to improve the accuracy in bearing remaining useful life estimation, by a combination of data-based model and physical-based model. Romanenko, Muetze and Ahola investigated the influence of electrical discharge on the bearing grease dielectric strength and chemical composition and found that the electrical discharge activity can deteriorate the bearing health by formatting a conducting channel between bearing rolling elements and bearing races [99]. They also measured the machine vibration behaviours due to electric discharge machining bearing currents over a 940 h operation, and found that the vibration at a single frequency observation is not reliable for bearing fault detection. Therefore, they proposed a simultaneous monitoring of multiple frequency observation of the machine vibration signals to detect the bearing fault due to EDM bearing currents [100].

To reduce the bearing currents, Golkhandan and Torkaman proposed a lumped parameter model of a 200 MVA turbo generator to predict the rotor shaft voltage and thus the circulating bearing current, considering the skin effect and stray capacitances. Then, they also tried both passive RC-filter and SEPIC converter to reduce the shaft voltage, and both of them can work while the latter is more effective [101]. Kalaiselvi and Srinivas proposed hybrid PWM switching schemes for an open-end winding induction machine to reduce the common mode voltage. By exploiting the extra degree of freedoms in the switching schemes of the dual inverters for the open-end winding machine, the common mode voltage and EDM bearing currents have been completely eliminated [102]. Vostrov and Pyrhonen investigated the influence of the geometry design of stator winding and slot on the non-circulating bearing current magnitude and found that the stator winding position modification is effective in the reduction of the stray capacitance. Thus, the bearing voltage ratio can be reduced by half, with only a minor change in the stator design [103].

5 Discussion

A. Comparison of different suppressing methods

To better compare the attenuation band and amplitude provided by different methods, the spectral distribution of the maximum attenuation caused by different methods are presented in Fig. 8. Among each method, references providing the highest attenuation frequency and amplitude are included. The superscripted references in Fig. 8 indicate that they have not been applied in motor drives.

As can be seen from Fig. 8, research on the conductive EMI of motor and power converter is not limited to the range of 150 kHz–30 MHz. The main reason lies in:

(a) The degree of electrical integration of equipment is increasing in many applications, such as EV/PHEV, robotics, etc. As a result of pursuing higher efficiency and power density, the layout of electrical system becomes more compact. Hence, the EMI at higher frequency is more likely to interfere other electronic components in the system. In addition, radiated EMI is mainly induced by the high-frequency circulating current flowing along the power cable. Therefore, EMI standards set limits on the EMI of electrical components over a wider range of frequencies. For example, CISPR25 0 stipulates the limits and test configuration for conductive and radiated EMI in EV/PHEV from 150 kHz to 2500 MHz.

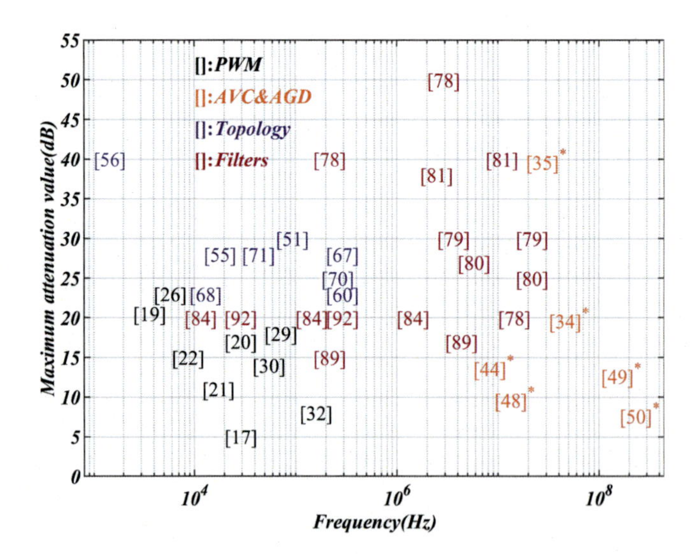

Fig. 8 Spectral distribution of maximum attenuations of different methods

(b) WBG device-based electric drive has become a hotspot in academia and industry 000. Although WBG devices have the potential to significantly improve the efficiency and power density of electric drives, the corner frequencies of EMI spectrum are increased, incurring more severe EMI at high frequencies. A comparative study on conductive CM EMI of motor drives based on Si, SiC and GaN has been carried out in 0.

From the perspective of attenuation band, the attenuation bands of VSFPWM and complex topologies are below 30 MHz, while the attenuation bands of AVC&AGD reach 300 MHz. The reason for this phenomenon is that VSFPWM strategies suppress CM EMI by adjusting the duration d of voltage pulse, while AVC&AGD achieve it by improve the rise/fall transients t_r of voltage pulse. The duration of voltage pulse affects the first corner frequency in the EMI spectrum, while switching transients determine the higher corner frequency. Among VSFPWM strategies, the highest attenuation frequency occurs in [29], which employs model-predictive PWM. The highest attenuation frequency of complex topologies is provided by a dual-inverter fed dual-winding PMSM drive in [67]. In [50], a AGD adjusting R_g from 0.12Ω to 64 Ω achieves the highest attenuation frequency among the reviewed works.

From the perspective of attenuation amplitude, in the low-frequency range, complex topologies achieve higher attenuation than VSFPWM by eliminating ZCMV. The same effect can be achieved through ZCMV PWM in three-phase VSI 00. Generally, EMI attenuation at high frequency is smaller than that at low frequency, but 40 dB attenuation at 20 MHz is achieved in [35] by shaping the transient voltage into an infinitely differentiable Gaussian transient.

In contrast to other methods achieving maximum attenuation in a narrow frequency band, passive filters can reflect EMI in a wider band through cascaded resonant circuits. It can be seen from Fig. 8 that cascaded filters are capable of achieving maximum attenuation at multiple frequencies.

B. Merits and demerits of the reviewed methods

VSFPWM does not require additional hardware, but it also has the lowest attenuation frequency and amplitude. Sole reliance on VSFPWM is likely to be sufficient in order to meet the EMI limits set in standards [32].

Compared with EMI filters, the attenuation bands realized with complex inverter topologies are narrower, while the cost of additional power devices and other associated circuitry is substantially higher. Therefore, suppressing EMI tends not to be the prime purpose of employing complex converter topologies. The adoption of complex topology is primarily concerned with features such as higher rated voltage, higher power, high reliability and other factors, such as the NPC three-level inverters adopted in medium-voltage high-power motor drives, with EMI dividends usually being something of a by-product.

Feedback-based close-loop AVC snf AGD are well suited to high switching frequency motor drives, although WBG devices such as GaN FETs with faster switching transient and lower threshold voltage V_{th}, has yet to be widely adopted in motor drives. Although closed-loop AVC and AGD show good EMI attenuation capability at high frequency, they have, to date, not been applied to motor drives. Nevertheless, an open-loop AGD-based induction motor drive is reported in 0.

At the expense of increasing cost, volume, and weight, passive and active filters provide the highest attenuation and the widest attenuation band. They have been widely applied in different applications, such EV/PHEV, aerospace sector, etc. However, the values of the passive components in filters are always somewhat over-designed to allow for EMI measurement accuracy and device aging concerns.

C. Discussion on future development of EMI suppressing method

Although various EMI suppression methods have been proposed, EMI filters remain an essential component in motor drives. In the design and implementation of passive/active filters, EMI measurement and sensing play critical roles.

In the design of passive filters, the conductive EMI is measured experimentally using equipment such as LISN, radiated frequency probes, spectrum analysers, etc. The structure and values of passive components are determined by the measurement results. However, the inductances and capacitances in filters are prone to ageing under the influence of temperature, thus reducing the performance and efficiency of the filters over time. Furthermore, the aging of the motor and its drive also result in some shifts in the EMI spectrum, which exacerbates the deterioration of EMI filters.

In active EMI filters, conductive EMI is sensed by resonant circuits, and various sensing circuits are summarized in [85]. The compensating effect of active filters is limited by the sensing precision of the resonant frequencies. Furthermore, the effectiveness of traditional active EMI filter for volume reduction is limited due to the use of passive components in sensing circuit 0.

In order to further reduce the cost, volume, and weight of EMI filters, and improve their attenuating ability and efficiency, real-time EMI monitoring with low hardware cost could play a key role in the future development. EMI real-time monitoring aims to sense and predict conductive EMI based on accurate predictive model and system feedbacks. Several advancements could be achieved based on EMI real-time monitoring.

(a) Life cycle dynamic passive filter. Based on EMI real-time monitoring, the parameters and even configuration of passive filters can be dynamically adjusted through low-cost switching circuits to deal with EMI deterioration caused by component ageing.
(b) The active filter without EMI sensing circuit. EMI real-time monitoring can be used to replace the EMI sensing circuit composed of passive components, which can further reduce the volume and weight of active EMI filter.

6 Conclusion

In this paper, representative methods of suppressing conductive EMI in motor drives have been reviewed from several perspectives. The methods are divided into two categories, viz. mitigation of EMI from the source and suppression EMI on its propagation path. Furthermore, the characteristic of each method is discussed and the mechanisms are illustrated. The application cases, the effective frequency range and the suppression level are summarized. Finally, the attenuation band and attenuation level of each method are compared, and the future development opportunities for EMI suppressing methods are discussed.

References

1. Electromagnetic Interference Control Requirements for Composite Airplanes, EMI D6–16050–5 (2004)
2. Muetze A, Binder A (2007) Calculation of motor capacitances for prediction of the voltage across the bearings in machines of inverter-based drive systems. IEEE Trans Ind Appl 43(3):665–672
3. Naik R, Nondahl TA, Melfi MJ, Schiferl R, Wang J (2003) Circuit model for shaft voltage prediction in induction motors fed by PWM-based AC drives. IEEE Trans Ind Appl 39(5):1294–1299
4. Busse D, Erdman J, Kerkman RJ, Schlegel D, Skibinski G (1997) System electrical parameters and their effects on bearing currents. IEEE Trans Ind Appl 33(2):577–584
5. Moreira AF, Lipo TA, Venkataramanan G, Bernet S (2002) High-frequency modeling for cable and induction motor overvoltage studies in long cable drives. IEEE Trans Ind Appl 38(5):1297–1306
6. Marlier C, Videt A, Idir N (2015) NIF-based frequency-domain modeling method of three-wire shielded energy cables for EMC simulation. IEEE Trans Electromagn Compat 57(1):145–155
7. Cataliotti A, Cara DD, Marsala G, Pecoraro A, Ragusa A, Tinè G (2016) High-frequency experimental characterization and modeling of six pack IGBTs power modules. IEEE Trans Industr Electron 63(11):6664–6673
8. Nejadpak A, Mohammed OA (2013) Physics-based modeling of power converters from finite element electromagnetic field computations. IEEE Trans Magn 49(1):567–576
9. Zhang H, Wu A (2019) Common-mode noise reduction by parasitic capacitance cancellation in the three-phase inverter. IEEE Trans Electromagn Compat 61(1):295–300
10. Brovont AD, Lemmon AN (2019) Utilization of power module baseplate capacitance for common-mode EMI filter reduction. In: IEEE electric ship technologies symposium (ESTS). Washington, DC, USA, pp 403−408
11. Gong X, Ferreira JA (2014) Investigation of conducted EMI in SiC JFET inverters using separated heat sinks. IEEE Trans Industr Electron 61(1):115–125
12. Fang Z, Jiang D, Shen Z, Qu R (2017) Impact of application of SiC devices in motor drive on EMI. In: IEEE applied power electronics conference and exposition (APEC). Tampa, FL, pp 652−658
13. Yang L, Wang S, Feng J (2017) Advances in electromagnetic interference modeling and noise reduction for adjustable speed motor drive systems. In: 2017 IEEE international symposium on electromagnetic compatibility & signal/power integrity (EMCSI). Washington, DC, pp 249−254
14. Zhang B, Wang S (2020) A survey of EMI research in power electronics systems with wide-bandgap semiconductor devices. IEEE J Emerg Sel Top Power Electron 8(1):626–643

15. Amin A, Choi S (2019) A review on recent characterization effort of CM EMI in power electronics system with emerging wide band gap switch. In: 2019 IEEE electric ship technologies symposium (ESTS). Washington, DC, USA, pp 241−248
16. Natarajan S, Sudhakar Babu T, Balasubramanian K, Subramaniam U, Almakhles DJ (2020) A state-of-the-art review on conducted electromagnetic interference in non-isolated DC to DC converters. IEEE Access 8:2564–2577
17. Pu T, Bu F, Huang W, Zhu L (2017) Implementation of random SVPWM strategy for three-phase voltage source inverter based on FPGA. In: 2017 20th international conference on electrical machines and systems (ICEMS). Sydney, NSW, pp 1–4
18. Sivarani TS, Jawhar SJ, Kumar CA (2016) Intensive random carrier pulse width modulation for induction motor drives based on hopping between discrete carrier frequencies. IET Power Electron 9(3):417–426
19. Zhang Y, Yin Z, Liu J, Zhang R, Sun X (2020) IPMSM sensorless control using high-frequency voltage injection method with random switching frequency for audible noise improvement. IEEE Trans Industr Electron 67(7):6019–6030
20. Lee K, Shen G, Yao W, Lu Z (2017) Performance characterization of random pulse width modulation algorithms in industrial and commercial adjustable-speed drives. IEEE Trans Ind Appl 53(2):1078–1087
21. Lezynski P (2018) Random modulation in inverters with respect to electromagnetic compatibility and power quality. IEEE J Emerg Sel Top Power Electron 6(2):782–790
22. Kumar ACB, Narayanan G (2016) Variable-switching frequency PWM technique for induction motor drive to spread acoustic noise spectrum with reduced current ripple. IEEE Trans Ind Appl 52(5):3927–3938
23. Santolaria A, Balcells J, Gonzalez D (2002) Theoretical and experimental results of power converter frequency modulation. In: IEEE 2002 28th annual conference of the industrial electronics society, IECON 02, vol 1. Sevilla, pp 193–197
24. Santolaria A, Balcells J, Díez, David, Gago J (2003) Evaluation of switching frequency modulation in EMI emissions reduction applied to power converters, vol 3. pp 2306−2311. https://doi.org/10.1109/IECON.2003.1280604
25. Balcells J, Santolaria A, Orlandi A, Gonzalez D, Gago J (2005) EMI reduction in switched power converters using frequency Modulation techniques. IEEE Trans Electromagn Compat 47(3):569–576
26. Yongxiang X, Qingbing Y, Jibin Z, Hao W (2014) Influence of periodic carrier frequency modulation on inverter loss of permanent magnet synchronous motor drive system. In: 2014 17th international conference on electrical machines and systems (ICEMS). Hangzhou, pp 2101–2106
27. Gonzalez D et al (2007) Conducted EMI reduction in power converters by means of periodic switching frequency modulation. IEEE Trans Power Electron 22(6):2271–2281
28. Yongxiang X, Qingbing Y, Jibin Z, Hao W (2014) Influence of periodic carrier frequency modulation on stator steel core loss and rotor eddy current loss of permanent magnet synchronous motor. In: 2014 17th international conference on electrical machines and systems (ICEMS). Hangzhou, pp 2094–2100
29. Yongxiang X, Qingbing Y, Jibin Z, Baochao W, Junlong L (2015) Periodic carrier frequency modulation in reducing low-frequency electromagnetic interference of permanent magnet synchronous motor drive system. IEEE Trans Magn 51(11):1–4
30. Jiang D, Wang F (2013) Variable switching frequency PWM for three-phase converters based on current ripple prediction. IEEE Trans Power Electron 28(11):4951–4961
31. Jiang D, Wang F (2014) Current-ripple prediction for three-phase PWM converters. IEEE Trans Ind Appl 50(1):531–538
32. Jiang D, Li Q, Shen Z (2017) Model predictive PWM for AC motor drives. IET Electr Power Appl 11(5):815–822
33. Oswald N, Stark BH, Holliday D, Hargis C, Drury B (2011) Analysis of shaped pulse transitions in power electronic switching waveforms for reduced EMI generation. IEEE Trans Ind Appl 47(5):2154–2165

34. Oswald N, Anthony P, McNeill N, Stark BH (2014) An experimental investigation of the tradeoff between switching losses and EMI generation with hard-switched All-Si, Si-SiC, and All-SiC device combinations. IEEE Trans Power Electron 29(5):2393–2407
35. Yang X, Yuan Y, Zhang X, Palmer PR (2015) Shaping high-power IGBT switching transitions by active voltage control for reduced EMI generation. IEEE Trans Ind Appl 51(2):1669–1677
36. Yang X, Long Z, Wen Y, Huang H, Palmer PR (2016) Investigation of the trade-off between switching losses and EMI generation in Gaussian S-shaping for high-power IGBT switching transients by active voltage control. IET Power Electronics 9(9):1979–1984
37. Yang X, Otsuki M, Palmer PR (2015) Physics-based insulated-gate bipolar transistor model with input capacitance correction. IET Power Electronics 8(3):417–427
38. Idir N, Bausiere R, Franchaud JJ (2006) Active gate voltage control of turn-on di/dt and turn-off dv/dt in insulated gate transistors. IEEE Trans Power Electron 21(4):849–855
39. Cheng YS, Mannen T, Wada K, Miyazaki K, Takamiya M, Sakurai T (2019) Optimization platform to find a switching pattern of digital active gate drive for reducing both switching loss and surge voltage. IEEE Trans Ind Appl 55(5):5023–5031
40. Wang R et al (2020) Self-adaptive active gate driver for IGBT switching performance optimization based on status monitoring. IEEE Trans Power Electron 35(6):6362–6372
41. Zhang Z, Wang Z, Wang F, Tolbert LM, Blalock BJ (2015) Reliability-oriented design of gate driver for SiC devices in voltage source converter. In: 2015 IEEE international workshop on integrated power packaging (IWIPP). Chicago, IL, pp 20−23
42. Choudhury A (2018) Present status of SiC based power converters and gate drivers–a review. In: 2018 international power electronics conference (IPEC-Niigata 2018-ECCE Asia). Niigata, pp 3401−3405
43. Liu Y, Yang Y (2019) Review of SiC MOSFET drive circuit. In: 2019 IEEE international conference on electron devices and solid-state circuits (EDSSC). Xi'an, China, pp 1–3
44. Paredes A, Sala V, Ghorbani H, Romeral L (2016) A novel active gate driver for silicon carbide MOSFET. In: IECON 2016-42nd annual conference of the IEEE industrial electronics society. Florence, pp 3172–3177
45. Camacho AP, Sala V, Ghorbani H, Martinez JLR (2017) A novel active gate driver for improving SiC MOSFET switching trajectory. IEEE Trans Industr Electron 64(11):9032–9042
46. P Nayak K Hatua (2016) Active gate driving technique for a 1200 V SiC MOSFET to minimize detrimental effects of parasitic inductance in the converter layout. In: 2016 IEEE energy conversion congress and exposition (ECCE). Milwaukee, WI, pp 1−8
47. Nayak P, Hatua K (2018) Active gate driving technique for a 1200 V SiC MOSFET to minimize detrimental effects of parasitic inductance in the converter layout. IEEE Trans Ind Appl 54(2): 1622–1633
48. Yang Y, Wen Y, Gao Y (2019) A novel active gate driver for improving switching performance of high-power SiC MOSFET modules. IEEE Trans Power Electron 34(8):7775–7787
49. Bau P, Cousineau M, Cougo B, Richardeau F, Rouger N (2020) CMOS active gate driver for closed-loop dv/dt control of GaN transistors. IEEE Trans Power Electron 35(12):13322–13332
50. Dymond HCP et al (2018) A 6.7-GHz active gate driver for GaN FETs to combat overshoot, ringing, and EMI. IEEE Trans Power Electron 33(1):581–594
51. Julian AL, Oriti G, Lipo TA (1999) Elimination of common-mode voltage in three-phase sinusoidal power converters. IEEE Trans Power Electron 14(5):982–989
52. Zhang R, Boroyevich D, Prasad VH, Mao H, Lee FC, Dubovsky S (1997) A three-phase inverter with a neutral leg with space vector modulation. In: Proceedings of APEC 97-applied power electronics conference, vol 2. Atlanta, GA, USA, pp 857–863
53. Zhang R, Prasad VH, Boroyevich D, Lee FC (2002) Three-dimensional space vector modulation for four-leg voltage-source converters. IEEE Trans Power Electron 17(3):314–326
54. Garg P, Essakiappan S, Krishnamoorthy HS, Enjeti PN (2015) A fault-tolerant three-phase adjustable speed drive topology with active common-mode voltage suppression. IEEE Trans Power Electron 30(5):2828–2839

55. Chen R et al (2019) Investigation of fourth-leg for common-mode noise reduction in three-level neutral point clamped inverter fed motor drive. In: IEEE applied power electronics conference and exposition (APEC). Anaheim, CA, USA, pp 2582−2588
56. Zhang H, Von Jouanne A, Dai S, Wallace AK, Wang F (2000) Multilevel inverter modulation schemes to eliminate common-mode voltages. IEEE Trans Ind Appl 36(6):1645–1653
57. Duang-upra M, Kamsuwan Y (2019) Three-segment switching sequences for a space-vector modulated three-level inverter to eliminate common-mode voltages. In: IECON 2019-45th annual conference of the IEEE industrial electronics society. Lisbon, Portugal, pp 2082–2087
58. Nguyen TT, Nguyen N, Prasad NR (2017) Novel eliminated common-mode voltage PWM sequences and an online algorithm to reduce current ripple for a three-level inverter. IEEE Trans Power Electron 32(10):7482–7493
59. Kai L, Zhao J, Wu W, Li M, Ma L, Zhang G (2016) Performance analysis of zero common-mode voltage pulse-width modulation techniques for three-level neutral point clamped inverters. IET Power Electronics 9(14):2654–2664
60. Chen J, Jiang D, Sun W, Shen Z, Zhang Y (2019) An improved variable switching frequency modulation strategy for three-level converters with reduced conducted EMI. In: 2019 IEEE energy conversion congress and exposition (ECCE). Baltimore, MD, USA, pp 6937−6942
61. Xing X, Chen A, Zhang Z, Chen J, Zhang C (2016) Model predictive control method to reduce common-mode voltage and balance the neutral-point voltage in three-level T-type inverter. In: 2016 IEEE applied power electronics conference and exposition (APEC) long beach. CA, pp 3453−3458
62. Zhang D, Wang F, Burgos R, Lai R, Boroyevich D (2010) Impact of interleaving on AC passive components of paralleled three-phase voltage-source converters. IEEE Trans Ind Appl 46(3):1042–1054
63. Zhang D, Wang F, Burgos R, Boroyevich D (2012) Total flux minimization control for integrated inter-phase inductors in paralleled, interleaved three-phase two-level voltage-source converters with discontinuous space-vector modulation. IEEE Trans Power Electron 27(4):1679–1688
64. Zhang D, Fred Wang F, Burgos R, Boroyevich D (2011) Common-mode circulating current control of paralleled interleaved three-phase two-level voltage-source converters with discontinuous space-vector modulation. IEEE Trans Power Electron 26(12):3925–3935
65. Jiang D, Shen Z (2016) Paralleled inverters with zero common-mode voltage. In: 2016 IEEE energy conversion congress and exposition (ECCE). Milwaukee, WI, pp 1−8
66. Jiang D, Shen Z, Wang F (2018) Common-mode voltage reduction for paralleled inverters. IEEE Trans Power Electron 33(5):3961–3974
67. Shen Z, Jiang D, Zou T, Qu R (2019) Dual-segment three-phase PMSM with dual inverters for leakage current and common-mode EMI reduction. IEEE Trans Power Electron 34(6):5606–5619
68. von Jauanne A, Zhang H (1999) A dual-bridge inverter approach to eliminating common-mode voltages and bearing and leakage currents. IEEE Trans Power Electron 14(1):43–48
69. Zhang H, von jouanne A, Dai S (2001) A reduced-switch dual-bridge inverter topology for the mitigation of bearing currents, EMI, and DC-link voltage variations. IEEE Trans Ind Appl 37(5):1365–1372
70. Shen Z, Jiang D, Liu Z, Ye D, Li J (2020) Common-mode voltage elimination for dual two-level inverter-fed asymmetrical six-phase PMSM. IEEE Trans Power Electron 35(4):3828–3840
71. Han X, Jiang D, Zou T, Qu R, Yang K (2019) Two-segment three-phase PMSM drive with carrier phase-shift PWM for torque ripple and vibration reduction. IEEE Trans Power Electron 34(1):588–599
72. Ye D et al (2020) Variable switching sequence PWM strategy of dual three-phase machine drive for high-frequency current harmonic suppression. IEEE Trans Power Electron 35(5):4984–4995
73. Jettanasen C (2012) Analysis of conducted electromagnetic interference generated by PWM inverter Fed-AC motor drives. In: 2012 15th international conference on electrical machines and systems (ICEMS). Sapporo, pp 1–6

74. Akagi H, Doumoto T (2005) A passive EMI filter for preventing high-frequency leakage current from flowing through the grounded inverter heat sink of an adjustable-speed motor drive system. IEEE Trans Ind Appl 41(5):1215–1223
75. Akagi H, Oe T (2008) A specific filter for eliminating high-frequency leakage current from the grounded heat sink in a motor drive with an active front end. IEEE Trans Power Electron 23(2):763–770
76. Akagi H, Tamura S (2006) A passive EMI filter for eliminating both bearing current and ground leakage current from an inverter-driven motor. IEEE Trans Power Electron 21(5):1459–1469
77. Akagi H, Shimizu T (2008) Attenuation of conducted EMI emissions from an inverter-driven motor. IEEE Trans Power Electron 23(1):282–290
78. Xue J, Wang F, Chen W (2013) A study of motor-end EMI filter on output common-mode noise suppression in DC-fed motor drive system. In: 2013 twenty-eighth annual IEEE applied power electronics conference and exposition (APEC) long beach. CA, pp 1556−1561
79. Xue J, Wang F, Zhang X, Boroyevich D, Mattavelli P (2012) Design of output passive EMI filter in DC-fed motor drive. In: 2012 twenty-seventh annual IEEE applied power electronics conference and exposition (APEC). Orlando, FL, pp 634−640
80. Xue J, Wang F, Guo B (2014) EMI noise mode transformation due to propagation path unbalance in three-phase motor drive system and its implication to EMI filter design. In: 2014 IEEE applied power electronics conference and exposition-APEC 2014 fort worth, TX 2014, pp 806–811
81. Chen H, Yan Y, Zhao H (2016) Extraction of common-mode impedance of an inverter-fed induction motor. IEEE Trans Electromagn Compat 58(2):599–606
82. Chen H, Ye S (2018) Modeling of common-mode impedance of an inverter-fed induction motor from online measurement. IEEE Trans Electromagn Compat 60(5):1581–1589
83. Ogasawara S, Ayano H, Akagi H (1998) An active circuit for cancellation of common-mode voltage generated by a PWM inverter. IEEE Trans Power Electron 13(5):835–841
84. Ogasawara S, Ayano H, Akagi H (1997) An active circuit for cancellation of common-mode voltage generated by a PWM inverter, PESC97. In: Record 28th annual IEEE power electronics specialists conference. Formerly power conditioning specialists conference 1970–71. Power processing and electronic specialists conference 1972, vol 2. Saint Louis, MO, USA, pp 1547–1553
85. Narayanasamy B, Luo F (2019) A survey of active EMI filters for conducted EMI noise reduction in power electronic converters. IEEE Trans Electromagn Compat 61(6):2040–2049
86. Asiminoael L, Blaabjerg F, Hansen S (2007) Detection is key-harmonic detection methods for active power filter applications. IEEE Ind Appl Mag 13(4):22–33
87. Sergej K, Asiminoaei L, Hansen S (2009) Harmonic detection methods of active filters for adjustable speed drive applications. In: 2009 13th European conference on power electronics and applications. Barcelona, pp 1–10
88. Di Piazza MC, Tine G, Vitale G (2008) An improved active common-mode voltage compensation device for induction motor drives. IEEE Trans Industr Electron 55(4):1823–1834
89. Akagi H, Hatada T (2009) Voltage balancing control for a three-level diode-clamped converter in a medium-voltage transformerless hybrid active filter. IEEE Trans Power Electron 24(3):571–579
90. Akagi H, Kondo R (2010) A transformerless hybrid active filter using a three-level pulsewidth modulation (PWM) converter for a medium-voltage motor drive. IEEE Trans Power Electron 25(6):1365–1374
91. Zhu C, Hubing TH (2014) An active cancellation circuit for reducing electrical noise from three-phase AC motor drivers. IEEE Trans Electromagn Compat 56(1):60–66
92. Zhang Y, Li Q, Jiang D (2020) A motor CM impedance based transformerless active EMI filter for DC-side common-mode EMI suppression in motor drive system. IEEE Trans Power Electron 35(10):10238–10248
93. Muetze A, Binder A (2007) Calculation of circulating bearing currents in machines of inverter-based drive systems. IEEE Trans Industr Electron 54(2):932–938

94. Muetze A, Binder A (2007) Practical rules for assessment of inverter-induced bearing currents in inverter-fed AC motors up to 500 kW. IEEE Trans Industr Electron 54(3):1614–1622
95. Binder A, Muetze A (2008) Scaling effects of inverter-induced bearing currents in AC machines. IEEE Trans Ind Appl 44(3):769–776
96. Ahola J, Muetze A, Niemelä M, Romanenko A (2019) Normalization-based approach to electric motor BVR related capacitances computation. IEEE Trans Ind Appl 55(3):2770–2780
97. Muetze A, Strangas EG (2014) On inverter induced bearing currents, bearing maintenance scheduling, and prognosis. In: 2014 international conference on electrical machines (ICEM), 2–5 Sept. 2014. pp 1915–1921
98. Muetze A, Strangas EG (2016) The useful life of inverter-based drive bearings: methods and research directions from localized maintenace to prognosis. IEEE Ind Appl Mag 22(4):63–73
99. Romanenko A, Muetze A, Ahola J (2016) Effects of electrostatic discharges on bearing grease dielectric strength and composition. IEEE Trans Ind Appl 52(6):4835–4842
100. Romanenko A, Muetze A, Ahola J (2017) Incipient bearing damage monitoring of 940-h variable speed drive system operation. IEEE Trans Energy Convers 32(1):99–110
101. Golkhandan RK, Torkaman H (2020) Reduction of induced shaft voltage and bearing current in turbo generators: modeling, compensation, and practical test. IEEE Trans Ind Electron 1–1
102. Kalaiselvi J, Srinivas S (2015) Bearing currents and shaft voltage reduction in dual-inverter-fed open-end winding induction motor with reduced CMV PWM methods. IEEE Trans Industr Electron 62(1):144–152
103. Vostrov K, Pyrhonen J, Niemelä M, Ahola J, Lindh P (2020) Mitigating noncirculating bearing currents by a correct stator magnetic circuit and winding design. IEEE Trans Ind Electron 1–1
104. Han X, Jiang D, Zou T, Qu R, Yang K (2018) Two-segment three-phase PMSM drive with carrier phase-shift PWM. In: 2018 IEEE applied power electronics conference and exposition (APEC). San Antonio, TX, pp 848−854
105. Vehicles boats and internal combustion engines-radio disturbance characteristics-limits and methods of measurement for the protection of on-board receivers CISPR 25 Ed. 4.0 (2016)
106. Morya AK et al (2019) Wide bandgap devices in AC electric drives: opportunities and challenges. IEEE Trans Transp Electrification 5(1):3–20
107. Yang M et al (2020) Resonance suppression and EMI reduction of GaN-based motor drive with sine wave filter. IEEE Trans Ind Appl 56(3):2741–2751
108. Morris CT, Han D, Sarlioglu B (2016) Comparison and evaluation of common mode EMI filter topologies for GaN-based motor drive systems. In: 2016 IEEE applied power electronics conference and exposition (APEC) long beach. CA, pp 2950−2956
109. Han D, Li S, Wu Y, Choi W, Sarlioglu B (2017) Comparative analysis on conducted CM EMI emission of motor drives: WBG versus Si devices. IEEE Trans Industr Electron 64(10):8353–8363
110. Hava AM, Ün E A High-performance PWM algorithm for common-mode voltage reduction in three-phase voltage source inverters. IEEE Trans Power Electron 26(7):1998–2008
111. Huang Y et al Analytical characterization of CM and DM performance of three-phase voltage-source inverters under various PWM patterns. IEEE Trans Power Electron. https://doi.org/10.1109/TPEL.2020.3024836
112. Sukhatme Y, Titus J, Nayak P, Hatua K (2017) Digitally controlled active gate driver for SiC MOSFET based induction motor drive switching at 100 kHz. In: 2017 IEEE transportation electrification conference (ITEC-India). Pune, pp 1−5. https://doi.org/10.1109/ITEC-India.2017
113. Kotny J, Duquesne T, Idir N (2017) Influence of temperature on the EMI filter efficiency for embedded SiC power converters. In: 2017 IEEE vehicle power and propulsion conference (VPPC). Belfort, pp 1−6
114. Hami F, Boulzazen H, Kadi M (2017) High-frequency characterization and modeling of EMI filters under temperature variations. IEEE Trans Electromagn Compat 59(6):1906–1915
115. Narayanasamy B, Peng H, Yuan Z, Emon AI, Luo F (2020) Modeling and analysis of a differential mode active EMI filter with an analog twin circuit. IEEE Trans Electromagn Compat 62(4):1591–1600

Fault-Tolerant Control for Permanent Magnet Motors

Qian Chen and Wenxiang Zhao

Abstract Five-phase permanent magnet synchronous motor (PMSM) has been widely studied and applied in many industrial fields because of its high efficiency, high power density, and wide speed range. With the improvement of safety and reliability requirements in these fields, it is of great significance to study the multi-phase motor drive system and its fault-tolerant control. Firstly, this chapter takes five-phase PMSM as the research object and makes a brief introduction to the common fault types and control methods. Then, two different fault-tolerant ideas based on the principle of minimum torque ripple and invariable magnetomotive force respectively are analyzed. Focusing on the latter, two methods of fault-tolerant and reduced-order matrix are derived. Furthermore, model predictive control (MPC), vector control (VC), and direct torque control (DTC) are used to realize fault-tolerant control according to the two different matrices. In order to achieve the optimal running state of the motor, the maximum torque per ampere (MTPA) control method based on the fault-tolerant algorithm is also indispensable. The MTPA control methods based on signal injection have high tracking accuracy and good dynamic performance, but the injected high-frequency signal will bring extra torque fluctuation. Therefore, virtual signal injection (VSI) control is more suitable for fault-tolerant algorithms. Finally, VSI-MTPA is adopted to improve the torque output capacity and operation efficiency of the motor under fault-tolerant operation.

Q. Chen · W. Zhao (✉)
Jiangsu University, Zhenjiang, China
e-mail: zwx@ujs.edu.cn

Q. Chen
e-mail: chenqian0501@ujs.edu.cn

C. H. T. Lee (ed.), *Emerging Technologies for Electric and Hybrid Vehicles*,
Green Energy and Technology, https://doi.org/10.1007/978-981-99-3060-9_5

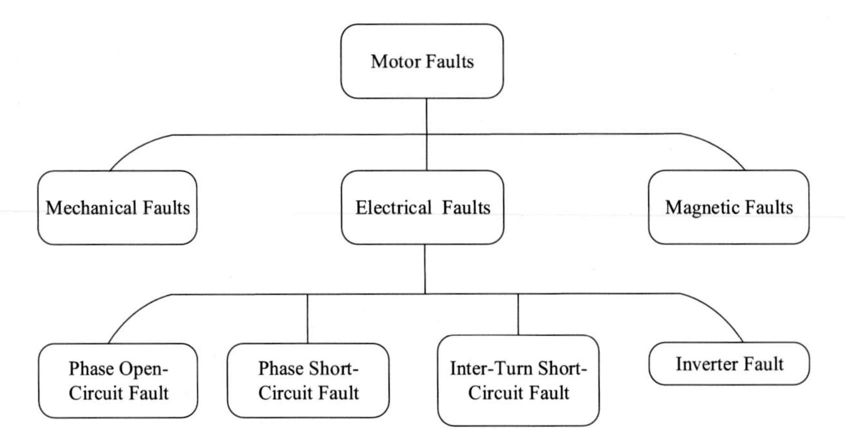

Fig. 1 Classification of motor fault types

1 Background

In recent years, permanent magnet synchronous motor has attracted more and more attention because of its advantages of high power density and high efficiency, which has become the key and core of advanced manufacturing, electrified transportation, aerospace, and other strategic emerging industries.

In applications with high requirements for on-device security and reliability, faults must be discovered and solved in time, and even the system is required to operate with fault tolerance. For three-phase motor, the premise of fault tolerance control is that the inverter has fault tolerance capacity [1]. Compared with the traditional three-phase motor, the multi-phase motor has more control degrees of freedom, so it does not need to change the driver structure, and the fault-tolerant control algorithm can realize the trouble-free operation after failure. Therefore, the research on fault-tolerant control of motor is basically aimed at multi-phase motor [2, 3].

As is shown in Fig. 1, motor faults can include mechanical faults, magnetic faults, and electrical faults, depending on the location of fault occurrence. Among them, electrical faults can be divided into phase open-circuit fault, phase short-circuit fault, inter-turn short-circuit fault, and inverter fault.

In comparison, the research on electrical faults of multi-phase motor is relatively less, which has become a hot topic for scholars. Therefore, this chapter will introduce fault-tolerant methods for five-phase PMSM.

2 Control Methods

The existing control strategies can be roughly divided into vector control (VC), direct torque control (DTC), and model predictive control (MPC).

At the end of 1960, K. Hasse of TU Darmstadt firstly proposed a new idea of vector control. The basic principle of vector control is very simple, which is to control the excitation current and torque current of the motor respectively according to the principle of magnetic field orientation. The core idea of vector control is to realize the decoupling control of torque and magnetic field through the Clark matrix and Park rotation matrix. Meanwhile, the alternating current (AC) control of motor stator current is transformed into dq axis direct current (DC) control, which is conducive to the realization of the control algorithm. Concretely, the stator magnetic field, rotor magnetic field or air gap magnetic field of the motor can be controlled by controlling the d-axis current (excitation component of the stator current), and then the torque can be controlled by controlling the q-axis current [4, 5]. For an m-phase motor drive system, m decoupled currents can be obtained after the phase current is transformed by generalized Clark and Park transformation [6]. The transformation principle is shown in Fig. 2. Furthermore, harmonic plane component should be considered in vector control of multi-phase motor. The common control strategy is to carry out closed-loop control for harmonic current. And the current control block diagram of five-phase motor is depicted in Fig. 3.

Vector control indirectly controls the torque by controlling the stator current. Compared with VC, DTC adopts hysteresis controller to control stator flux constant, and uses hysteresis controller to control torque, so there is no need for complex coordinate transformation and rotor position information θ_e. In addition, the control structure is simple, which further optimizes the torque dynamic response performance [7, 8]. In recent years, the traditional DTC strategy using switching state table has attracted more attention in the field of multiphase motor drive. And the block diagram of direct torque control system of five-phase motor is shown in Fig. 4.

Model predictive control was first proposed by Holtz in 1980. The core idea of this method lies in how to carry out prediction. The whole control block diagram mainly includes predictive models, feedback correction, and rolling optimization [9, 10], which is simple and easy to be realized online. Figures 5 and 6 show the control flow chart and control block diagram respectively.

The specific steps are described as follows: (a) Detect the motor speed, stator current, and dc bus voltage through sensors; (b) The calculated n effective voltage vectors are substituted into the discrete prediction model to obtain the predicted

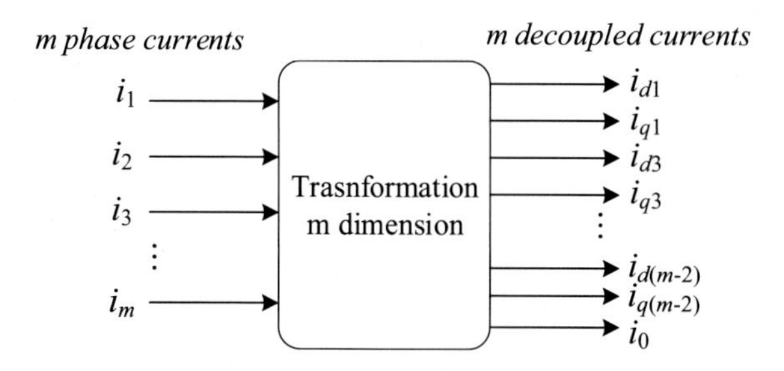

Fig. 2 M phase motor current decoupling transformation

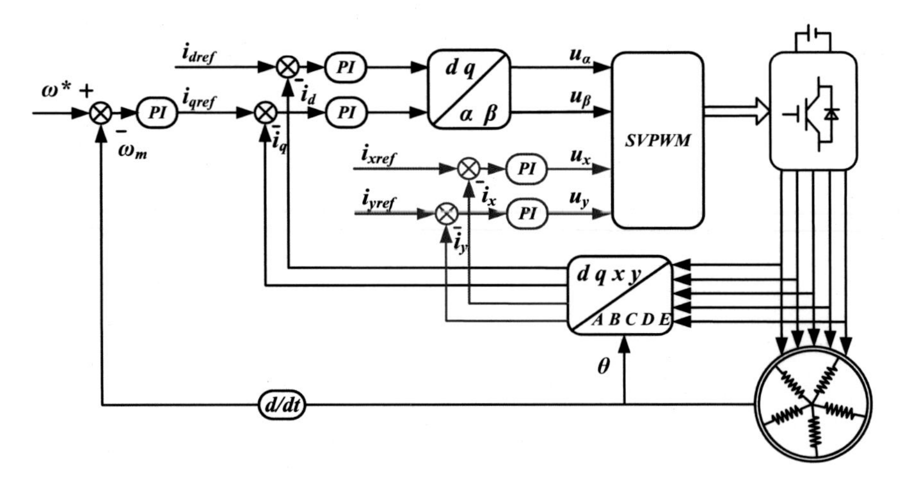

Fig. 3 Four-dimensional current control block diagram of five-phase motor

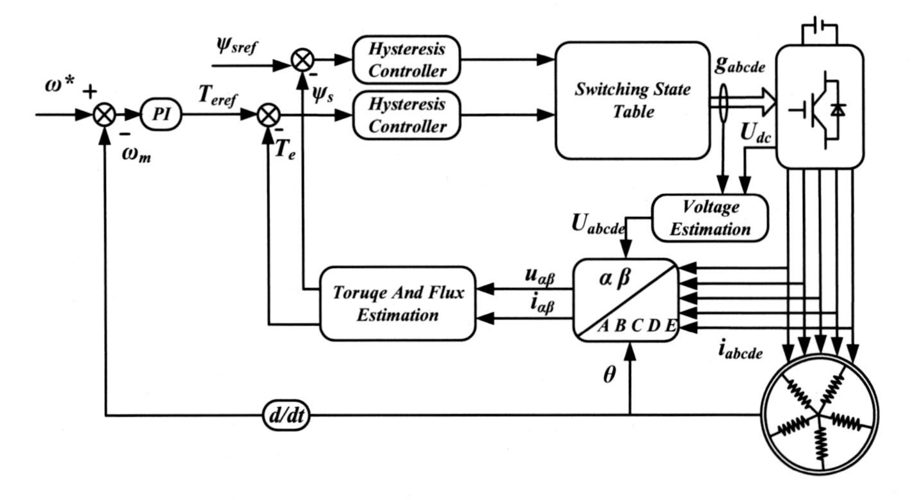

Fig. 4 Direct torque control block diagram of five-phase motor

current value at the next moment; (c) The value of the predicted current is inserted into the value function and compared with the given reference value to obtain the corresponding value of the value function G; (d) The n G values calculated are compared, and the switching state corresponding to the minimum value is selected as the optimal control output of the inverter.

PWM driving technology can be roughly divided into hysteresis control, SVPWM control, Carried Pulse Width Modulation (CPWM) control. Current hysteresis control [11–13] was first proposed by H. A. Toliyat and T. A. Lipo et al. in 1990. Its basic idea is to obtain the *dq* axis current according to the mathematical model and the reference value of each phase current through coordinate inverse transformation. Then,

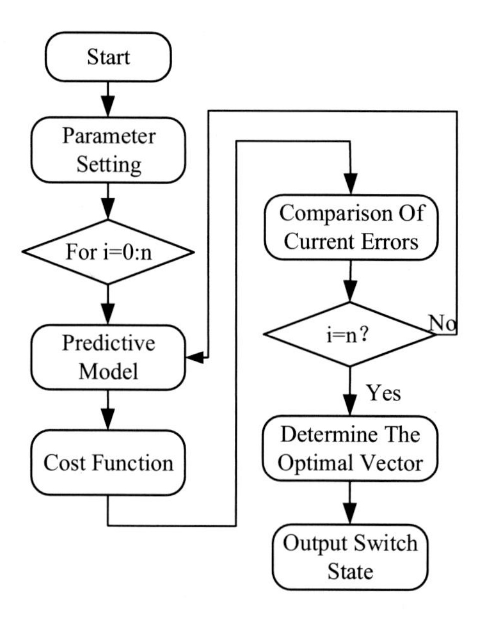

Fig. 5 Control flow chart of model predictive control

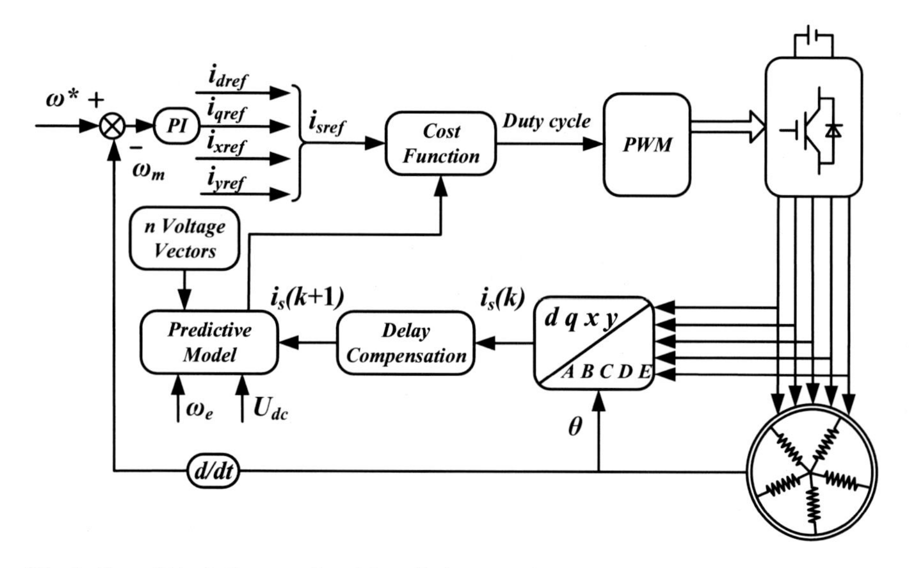

Fig. 6 Control block diagram of model predictive control

compare the reference phase current with the feedback of each phase current to obtain the PWM signal for output to the driver control motor. The advantage of this method is that it can be applied to multiphase motors with arbitrary number of phases, but the problem is that the switching frequency is not fixed under current hysteresis control and the noise is large. SVPWM drive control is actually a process of selecting the appropriate reference voltage vector to synthesize the desired stator voltage vector. According to the selected appropriate reference voltage vectors, it can be subdivided into adjacent maximum two vectors, adjacent four vectors, and minimum switching loss driving mode [14–16]. CPWM controller is mainly composed of modulation wave and carrier generator. In a carrier cycle, the inverter outputs the equivalent square wave voltage according to the principle of voltage-vs-balance. Also, CPWM algorithm needs to determine the appropriate harmonic components to improve bus voltage utilization.

Generally speaking, the existing drive control technology under normal operation conditions is becoming mature, and how to build a new model under motor failure will be the focus of this chapter.

3 Open-Circuit Fault-Tolerant Theory

In the 1990s, fault-tolerant control has attracted wide attention and many scholars published a large number of research papers on fault-tolerant control of multi-phase motor. Taking five-phase motor as an example, open-circuit fault is the most common fault type. Generally speaking, the research on open-circuit faults can be roughly divided into two kinds of fault-tolerant control methods: using constant decoupling matrix and reducing decoupling order. Among them, the former realizes undisturbed operation by calculating the given fault-tolerant current, so the performance of current controller directly affects the fault-tolerant operation effect. The latter tends to require additional compensation strategies in order to eliminate the imbalance caused by faults.

3.1 Fault-Tolerant Control with Minimum Torque Ripple Principle

The control strategy with the minimum torque ripple achieves fault-tolerant operation by directly adjusting the currents given in the natural coordinate system, and hysteresis controller is generally used to realize current control. Therefore, this section will introduce different fault-tolerant current reference generation methods.

(1) Lagrange operator method

(1.1) Optimal torque control

In 2003, J. Wang proposed an optimal torque fault-tolerant control strategy [17]. The objective function is constructed with the constraint of minimum torque ripple, and the fault-tolerant control in constant torque region and constant power region is realized with the given fault-tolerant current. Equation (1) is the objective function. The first item on the right side of the formula represents the purpose of minimum copper loss and wide speed range; the second item represents the constraint condition of minimum torque ripple. However, Ref. [17] only verifies the feasibility of the algorithm through simulation.

$$F_a = \sum_{j=1}^{m} \left(Li_j + \omega\psi_j\right)^2 + \lambda\left(T_d - \sum_{j=1}^{m} a_j(\theta)i_j\right) \tag{1}$$

$$\omega = \omega_0 = \frac{\Omega - \Omega_b}{\Omega_b} \cdot \frac{T_d}{T_{er}} \tag{2}$$

where L is the motor phase inductance; i_j and ψ_j are the current and flux linkage of phase j respectively; λ is the Lagrange multiplier; m is the number of motor phases; $a_j(\theta)i_j$ is the instantaneous torque of phase j at the rotor position θ; ω is the field weakening weight coefficient; Ω is the rotor speed; Ω_b is the motor base speed; T_d is the given torque; T_{er} is the rated torque.

In 2010, Ref. [18] improved the fault-tolerant strategy by adding the estimated instantaneous torque error into the objective function to improve the magnetic flux weakening degree, and experimentally verified the improved control strategy. However, this scheme was only suitable for the motor driven by H-bridge. In 2015, Ref. [19] realized the open-circuit and short-circuit fault-tolerant control of five-phase Y-connected motor in flux weakening region by adding the constraint condition that the sum of phase currents is zero into the objective function. In order to improve the current tracking effect, a current controller in a static coordinate system is proposed [19] to track the non-sinusoidal current generated by the optimal torque control.

(1.2) Optimal fault-tolerant control

In 2008, L. Parsa proposed a fault-tolerant strategy based on instantaneous power theory for the five-phase permanent magnet motor whose back EMF contains the third harmonic [20]. In order to achieve the minimum copper loss after fault, the constraints of minimum torque ripple and no zero-sequence current are established. The objective function constructed in Ref. [20] can be expressed as:

$$f(\boldsymbol{i}, p_1, p_2) = \frac{1}{2}\boldsymbol{i}^{\mathrm{T}}\boldsymbol{i} + p_1(\boldsymbol{T}^* - \boldsymbol{K}^{\mathrm{T}}\boldsymbol{Q}\boldsymbol{i}) + p_2\boldsymbol{F}^{\mathrm{T}}\boldsymbol{i} \tag{3}$$

where p_1 and p_2 are Lagrange operators; $\boldsymbol{i}$ is the stator current; $\boldsymbol{T}^*$ is the given torque; $\boldsymbol{Q}$ is diagonal matrix; $\boldsymbol{K}$ is the velocity normalized back EMF.

In Eq. (3), the first item on the right side indicates that the optimal control needs to satisfy the minimum copper loss; the second item represents the constraint condition of the minimum torque ripple; the third item represents the constraint condition that the sum of phase currents is zero. In 2014, L. Parsa's research group applied the fault-tolerant strategy of optimal control to five-phase motors with different winding connections [21]. In 2015, the fault-tolerant control was further improved [22] to achieve fault-tolerant control under motor open circuit and short circuit faults.

(2) Instantaneous torque method

In 2006, N. Bianchi proposed a fault-tolerant current solution method for open-circuit and short-circuit faults of five-phase permanent magnet motor based on an analytical model [23]. When a five-phase motor fails, the 2nd and 4th ripple components appear in the torque. By adjusting the fundamental current, the 2nd order ripple component in the torque can be eliminated and the torque pulsation can be reduced; the 3rd order harmonic current can be used to eliminate the 4th order pulsation component in the torque. The derivation of this algorithm is based on the fact that the mutual inductance of the motor is zero. Therefore, the motor torque can be regarded as the superposition of the torque of each phase, and the fault-tolerant current reference can be obtained through the instantaneous torque balance formula. However, the algorithm does not use an optimization algorithm, and fails to consider the minimum copper loss.

(3) Average power method

In view of the shortcomings that the current solved by the optimal control strategy in literature [24] contains high-order harmonics and unequal current amplitudes, L. Parsa's research group proposed a fault-tolerant control strategy based on mirror symmetry for the five-phase motor with trapezoidal wave back EMF in 2011 [22]. The torque ripple is reduced by injecting third harmonic current. This method divides the torque after the fault into the average torque component and the disturbance torque component, which realizes the disturbance-free operation of the motor after fault by adjusting the amplitude and phase of the given current. In 2013, L. Parsa applied the method to five-phase permanent magnet motors with different winding connections [25]. It should be pointed out that since the pentagonal winding connection does not require the constraint of zero sequence current, it has an additional degree of freedom, which can be used to reduce torque ripple, increase average torque, or reduce copper loss.

3.2 Fault-Tolerant Control with Constant Magnetomotive Forces Principle

Since the multi-phase motor has more degrees of freedom, adjusting the normal phase current after the motor fault can still ensure that the magneto electromotive force remains unchanged before and after fault.

(1) Normal operation

Firstly, the mathematical model of the motor under normal working conditions is introduced. The mathematical model of five-phase motor in natural coordinate system is complicated. In order to simplify the analysis, it is usually transformed to a rotating coordinate system, at which time the corresponding transformation matrix is needed. The principle of matrix transformation is to ensure the invariable magnetic force in different coordinate systems. Therefore, before solving the transformation matrix, the magnetomotive force of the five-phase permanent magnet synchronous motor is analyzed.

When symmetric sinusoidal current is applied to the motor, taking axis A as the reference coordinate, the rotating magnetic force of each phase can be expressed as:

$$\begin{cases} \mathrm{MMF}_A = 0.5N_5 i_A \cos\gamma = 0.5N_5 I_m \cos\theta_e \cos\gamma \\ \mathrm{MMF}_B = 0.5N_5 i_B \cos(\gamma - 0.4\pi) = 0.5N_5 I_m \cos(\theta_e - \alpha)\cos(\gamma - 0.4\pi) \\ \mathrm{MMF}_C = 0.5N_5 i_C \cos(\gamma - 0.8\pi) = 0.5N_5 I_m \cos(\theta_e - 2\alpha)\cos(\gamma - 0.8\pi) \\ \mathrm{MMF}_D = 0.5N_5 i_D \cos(\gamma - 1.2\pi) = 0.5N_5 I_m \cos(\theta_e - 3\alpha)\cos(\gamma - 1.2\pi) \\ \mathrm{MMF}_E = 0.5N_5 i_E \cos(\gamma - 1.6\pi) = 0.5N_5 I_m \cos(\theta_e - 4\alpha)\cos(\gamma - 1.6\pi) \end{cases} \tag{4}$$

where I_m is the current amplitude; γ is the electric angle of the stator at one point centered on the axis of phase A; θ_e is the electric angle of the rotor; $\alpha = 0.4\pi$; i_A, i_B, i_C, i_D, i_E

By expanding and adding Eq. (4), the synthetic rotating magnetomotive force of five-phase motor under normal operation conditions can be obtained:

$$\begin{aligned} \mathrm{MMF}_s &= \mathrm{MMF}_A + \mathrm{MMF}_B + \mathrm{MMF}_C + \mathrm{MMF}_D + \mathrm{MMF}_E \\ &= 1.25N_5 I_m(\cos\theta_e \cos\gamma + \sin\theta_e \sin\gamma) \end{aligned} \tag{5}$$

The space vector diagram of the magnetomotive force of five-phase motor is shown in Fig. 7.

Where α_1–β_1 and α_3–β_3 space are fundamental and cubic static coordinate systems under five-phase motor respectively.

The synthesis of the magnetic motive force in the natural coordinate system and the static coordinate system is equal, so it can be concluded as:

$$N_2 i_{\alpha 1} = N_5 i_A + N_5 i_B \cos\alpha + N_5 i_C \cos 2\alpha + N_5 i_D \cos 3\alpha + N_5 i_E \cos 4\alpha$$

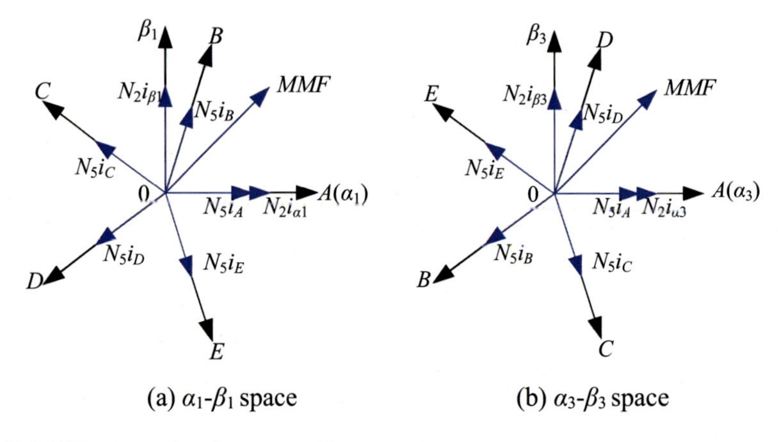

Fig. 7 MMFs of natural and static coordinates

$$= N_5(i_A + i_B \cos\alpha + i_C \cos 2\alpha + i_D \cos 3\alpha + i_E \cos 4\alpha) \tag{6}$$

$$\begin{aligned} N_2 i_{\beta 1} &= N_5 i_A + N_5 i_B \sin\alpha + N_5 i_C \sin 2\alpha + N_5 i_D \sin 3\alpha + N_5 i_E \sin 4\alpha \\ &= N_5(i_B \sin\alpha + i_C \sin 2\alpha + i_D \sin 3\alpha + i_E \sin 4\alpha) \end{aligned} \tag{7}$$

$$\begin{aligned} N_2 i_{\alpha 3} &= N_5 i_A + N_5 i_B \cos 3\alpha + N_5 i_C \cos 6\alpha + N_5 i_D \cos 9\alpha + N_5 i_E \cos 12\alpha \\ &= N_5(i_A + i_B \cos 3\alpha + i_C \cos 6\alpha + i_D \cos 9\alpha + i_E \cos 12\alpha) \end{aligned} \tag{8}$$

$$\begin{aligned} N_2 i_{\beta 3} &= N_5 i_A + N_5 i_B sin3\alpha + N_5 i_C sin6\alpha + N_5 i_D sin9\alpha + N_5 i_E sin12\alpha \\ &= N_5(i_B sin3\alpha + i_C sin6\alpha + i_D sin9\alpha + i_E sin12\alpha) \end{aligned} \tag{9}$$

Because the stator windings of the five-phase motor adopt Y-connected mode, it should be satisfied that $i_A + i_B + i_C + i_D + i_E = 0$. Combining Eqs. (6)–(9), Clarke transformation matrix in the five-phase natural coordinate system can be expressed as:

$$T_{\text{clarke}} = \frac{N_5}{N_2}\begin{bmatrix} 1 & \cos\alpha & \cos 2\alpha & \cos 3\alpha & \cos 4\alpha \\ 0 & \sin\alpha & \sin 2\alpha & \sin 3\alpha & \sin 4\alpha \\ 1 & \cos 3\alpha & \cos 6\alpha & \cos 9\alpha & \cos 12\alpha \\ 0 & \sin 3\alpha & \sin 6\alpha & \sin 9\alpha & \sin 12\alpha \\ 1 & 1 & 1 & 1 & 1 \end{bmatrix} \tag{10}$$

where $N_5/N_2 = 2/5$ according to the principle of constant amplitude; $N_5/N_2 = \sqrt{2/5}$ according to the principle of constant power. And this chapter adopts the principle of constant amplitude. The first and second rows represent the d_1–q_1 subspace

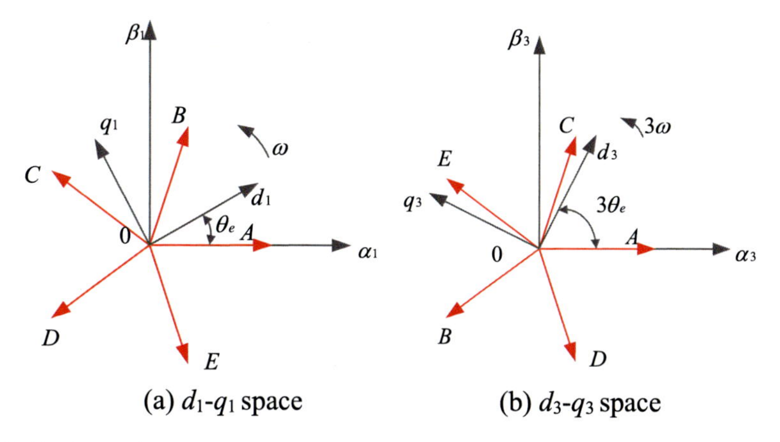

Fig. 8 Fundamental and cubic space rotation coordinate systems

participating in the energy conversion of the motor. The third and fourth rows correspond to the d_3–q_3 subspace, which does not participate in electromechanical energy transformation. And finally, the last row represents zero-sequence space.

As shown in Fig. 8, d_1–q_1 and d_3–q_3 are fundamental and cubic space rotation coordinate systems respectively, where the rotational speeds are ω and 3ω respectively.

The corresponding Park rotation transformation matrix can be expressed as:

$$T_{\text{park}} = \begin{bmatrix} \cos\theta_e & \sin\theta_e & 0 & 0 & 0 \\ -\sin\theta_e & \cos\theta_e & 0 & 0 & 0 \\ 0 & 0 & 1 & 0 & 0 \\ 0 & 0 & 0 & 1 & 0 \\ 0 & 0 & 0 & 0 & 1 \end{bmatrix} \tag{11}$$

The extended transformation matrix from the natural coordinate system to the rotating coordinate system can be concluded as:

$$T(\theta_e) = \frac{2}{5}\begin{bmatrix} \cos\theta_e & \cos(\theta_e+\alpha) & \cos(\theta_e+2\alpha) & \cos(\theta_e+3\alpha) & \cos(\theta_e+4\alpha) \\ -\sin\theta_e & \sin(\theta_e+\alpha) & \sin(\theta_e+2\alpha) & \sin(\theta_e+3\alpha) & \sin(\theta_e+4\alpha) \\ 1 & \cos 3\alpha & \cos 6\alpha & \cos 9\alpha & \cos 12\alpha \\ 0 & \sin 3\alpha & \sin 6\alpha & \sin 9\alpha & \sin 12\alpha \\ 1 & 1 & 1 & 1 & 1 \end{bmatrix} \tag{12}$$

Under normal working conditions, the stator voltage equation of five-phase motor can be expressed as:

$$U_s = R_s I_s + \frac{d\psi_s}{dt} \tag{13}$$

The voltage equation is composed of stator phase resistance voltage drop and flux change rate, where U_s, I_s, R_s, ψ_s are stator phase voltage, current, resistance, and flux matrix respectively. $U_s = [U_a\ U_b\ U_c\ U_d\ U_e]^{\mathrm{T}}, I_s = [i_a\ i_b\ i_c\ i_d\ i_e]^{\mathrm{T}}, R_s = \mathrm{diag}[R_s\ R_s\ R_s\ R_s\ R_s]$.

The stator flux equation consists of armature winding flux and permanent magnet flux:

$$\psi_s = L_s I_s + \psi_m \tag{14}$$

where ψ_m is permanent magnet flux linkage; L_s is phase inductance. ψ_m can be expressed as:

$$\psi_m = \psi_f [\cos\theta_e\ \cos(\theta_e - \alpha)\ \cos(\theta_e - 2\alpha)\ \cos(\theta_e - 3\alpha)\ \cos(\theta_e - 4\alpha)]^T \tag{15}$$

where ψ_f is the amplitude of permanent magnet flux linkage.

Where L_s can be expressed as:

$$L_s = \begin{bmatrix} L_{AA}\ L_{AB}\ L_{AC}\ L_{AD}\ L_{AE} \\ L_{BA}\ L_{BB}\ L_{BC}\ L_{BD}\ L_{BE} \\ L_{CA}\ L_{CB}\ L_{CC}\ L_{CD}\ L_{CE} \\ L_{DA}\ L_{DB}\ L_{DC}\ L_{DD}\ L_{DE} \\ L_{EA}\ L_{EB}\ L_{EC}\ L_{ED}\ L_{EE} \end{bmatrix} \tag{16}$$

According to the principle of electromechanical energy conversion, magnetic common energy is expressed as:

$$W = \frac{1}{2} I_s^T \mathbf{L}_s I_s + I_s^T \boldsymbol{\psi}_m \tag{17}$$

The torque equation of the motor is obtained by taking the partial derivative of the mechanical angular displacement by the magnetic common energy:

$$T_e = \frac{\partial W}{\partial \theta_m} = p \frac{\partial W}{\partial \theta_e} = \frac{1}{2} p I_s^T \frac{\partial}{\partial \theta_e} (\mathbf{L}_s I_s) + p I_s^T \frac{\partial \boldsymbol{\psi}_m}{\partial \theta_e} \tag{18}$$

The expression of the voltage equation in the rotating coordinate system can be obtained by using the transformation matrix:

$$U_{d1q1d3q3} = T(\theta_e) U_s = R I_{d1q1d3q3} + T(\theta_e) \frac{d(L_s I_e)}{dt} + T(\theta_e) \frac{d\psi_m}{dt} \tag{19}$$

According to Eq. (16), no-load EMF can be expressed as:

$$E_{d1q1d3q3} = T(\theta_e)\frac{d\psi_m}{dt} = \begin{bmatrix} e_{d1} \\ e_{q1} \\ e_{d3} \\ e_{q3} \end{bmatrix} = \omega_e \psi_f \begin{bmatrix} 0 \\ 1 \\ 0 \\ 0 \end{bmatrix} \tag{20}$$

The induced voltage component can be expressed as:

$$\begin{aligned} E_{L\_d1q1d3q3} &= T(\theta_e)\frac{d(L_s I_s)}{dt} = \frac{d[T(\theta_e)L_s I_s]}{dt} - \frac{dT(\theta_e)}{dt}L_s I_s \\ &= \frac{d\left(L_{d1q1d3q3}\right)}{dt} I_{d1q1d3q3} + L_{d1q1d3q3}\frac{d\left(I_{d1q1d3q3}\right)}{dt} \\ &\quad - \Omega L_{d1q1d3q3} I_{d1q1d3q3} \end{aligned} \tag{21}$$

where $I_{d1q1d3q3}$, $L_{d1q1d3q3}$

$$I_{d1q1d3q3} = T(\theta_e)I_s = \left[i_{d1}\ i_{q1}\ i_{d3}\ i_{q3}\right]^T \tag{22}$$

$$L_{d1q1d3q3} = T(\theta_e)L_s T^{-1}(\theta_e) = \begin{bmatrix} L_d & 0 & 0 & 0 & 0 \\ 0 & L_d & 0 & 0 & 0 \\ 0 & 0 & L_{ls} & 0 & 0 \\ 0 & 0 & 0 & L_{ls} & 0 \\ 0 & 0 & 0 & 0 & L_{ls} \end{bmatrix} \tag{23}$$

where $L_d = L_{ls} + 2.5(L_m - L_\theta)$, $L_q = L_{ls} + 2.5(L_m + L_\theta)$; L_{ls} is the leakage inductance; L_m is the self-induced DC component; L_θ is the amplitude of the second harmonic component of the phase inductance

The velocity matrix is expressed as:

$$\Omega = \frac{dT(\theta_e)}{dt}T^{-1}(\theta_e) = \omega_e \begin{bmatrix} 0 & 1 & 0 & 0 & 0 \\ -1 & 0 & 0 & 0 & 0 \\ 0 & 0 & 0 & 0 & 0 \\ 0 & 0 & 0 & 0 & 0 \\ 0 & 0 & 0 & 0 & 0 \end{bmatrix} \tag{24}$$

Finally, the voltage equation in the rotating coordinate system can be obtained as:

$$\begin{cases} u_{d1} = i_{d1}R_s + L_d\frac{di_{d1}}{dt} - \omega_e L_q i_{q1} \\ u_{q1} = i_{q1}R_s + L_q\frac{di_{q1}}{dt} + \omega_e L_d i_{d1} + \psi_f \\ u_{d3} = i_{d3}R_s + L_{ls}\frac{di_{d3}}{dt} \\ u_{q3} = i_{q3}R_s + L_{ls}\frac{di_{q3}}{dt} \end{cases} \tag{25}$$

The torque equation is obtained as:

$$T_e = \frac{5}{2}p\left[\psi_f i_{q1} + \left(L_d - L_q\right)i_{d1}i_{q1}\right] \tag{26}$$

When the open-circuit fault occurs in five-phase motor, the balance of the motor is broken due to the loss of the fault phase and the phase distribution of the motor is no longer spatially symmetrical. In order to realize magnetic directional control under faults, the reduced-order matrix of five-phase motor under open faults needs to be reconstructed. The construction principle of the reduced-order matrix is to ensure that the fundamental magnetic force after the motor failure is consistent with the normal operation. Many kinds of reduced-order matrices have been constructed by scholars. This chapter mainly chooses two kinds of reduced-order matrices to make a brief introduction. In order to distinguish the two different matrices, subscripts of 1 and 2 are added to the corresponding variables in this part to distinguish them.

(2) The first kind of matrix

(2.1) Single-phase open-circuit fault

Single-phase open circuit fault is firstly analyzed. Assuming that the fault phase is A, the fault-tolerant currents need to satisfy the following equation:

$$\begin{aligned} &\frac{5}{2}NI_m e^{j(\theta_e+\pi/2)} = N\left(i_A + \lambda i_B + \lambda^2 i_C + \lambda^3 i_D + \lambda^4 i_E\right) \\ &i_A \equiv 0 \end{aligned} \tag{27}$$

where λ is the rotation factor, $\lambda = e^{j\alpha}$.

In a Y-connected five-phase motor system, the sum of the neutral current still needs to be constrained to zero. Therefore, the current component in the α–β coordinate system after open-circuit fault can be expressed as:

$$\frac{2}{5}\begin{bmatrix} \cos\alpha & \cos 2\alpha & \cos 3\alpha & \cos 4\alpha \\ \sin\alpha & \sin 2\alpha & \sin 3\alpha & \sin 4\alpha \\ \cos 3\alpha & \cos 6\alpha & \cos 9\alpha & \cos 12\alpha \\ \sin 3\alpha & \sin 6\alpha & \sin 9\alpha & \sin 12\alpha \\ 1 & 1 & 1 & 1 \end{bmatrix}\begin{bmatrix} i_B \\ i_C \\ i_D \\ i_E \end{bmatrix} = \begin{bmatrix} i_{\alpha 1} \\ i_{\beta 1} \\ i_{\alpha 3} \\ i_{\beta 3} \\ 0 \end{bmatrix} \tag{28}$$

where i_α, i_β and $i_{\alpha 3}$, $i_{\beta 3}$ are the current components in fundamental and cubic space α–β coordinate systems respectively.

Since the third-row vector in Eq. (28) is not orthogonal to other rows, the third-row vector is deleted. To make the first-row vector orthogonal to the last-row zero-current vector, Eq. (28) can be modified as follows:

$$\frac{2}{5}\begin{bmatrix} \cos\alpha + x & \cos 2\alpha + x & \cos 3\alpha + x & \cos 4\alpha + x \\ \sin\alpha & \sin 2\alpha & \sin 3\alpha & \sin 4\alpha \\ \sin 3\alpha & \sin 6\alpha & \sin 9\alpha & \sin 12\alpha \\ 1 & 1 & 1 & 1 \end{bmatrix}\begin{bmatrix} i_B \\ i_C \\ i_D \\ i_E \end{bmatrix} = \begin{bmatrix} i_{\alpha 1} \\ i_{\beta 1} \\ i_{31} \\ 0 \end{bmatrix} \tag{29}$$

where i_{31} is the current component of fault-tolerant current in cubic space; x is the correction factor.

References [26, 27] established a transformation matrix with correction factor x of $-1/4$ under the constraint of matrix orthogonal, and proposed relevant fault-tolerant strategies. However, the modified Clarke transformation matrix needs to be combined with additional compensation matrix to achieve decoupling of the motor in the two-phase rotating coordinate system. In Refs. [28, 29], the correction factor x is determined to be -1 based on the principle of equal back EMF before and after faults. Under the Clarke transformation matrix, the motor can achieve complete decoupling in the rotating coordinate system. Therefore, the correction factor in this chapter is all determined to be -1.

The reduced Clarke transform under phase A open-circuit fault can be expressed as:

$$T_{\text{Clarke1}}^{A} = \frac{2}{5}\begin{bmatrix} \cos\alpha-1 & \cos 2\alpha-1 & \cos 3\alpha-1 & \cos 4\alpha-1 \\ \sin\alpha & \sin 2\alpha & \sin 3\alpha & \sin 4\alpha \\ \sin 3\alpha & \sin 6\alpha & \sin 9\alpha & \sin 12\alpha \\ 1 & 1 & 1 & 1 \end{bmatrix} \tag{30}$$

Using the reduced Clarke transform matrix shown in Eq. (30), the no-load back EMF in the fundamental space α–β coordinate system can be expressed as:

$$\begin{aligned} E_{\alpha 1} &= \cos\alpha E_B + \cos 2\alpha E_C + \cos 3\alpha E_D + \cos 4\alpha E_E + x(E_B + E_C + E_D + E_E) \\ &= -xE_A + \cos\alpha E_B + \cos 2\alpha E_C + \cos 3\alpha E_D + \cos 4\alpha E_E \end{aligned} \tag{31}$$

$$E_{\beta 1} = 0E_A + \sin\alpha E_B + \sin 2\alpha E_C + \sin 3\alpha E_D + \sin 4\alpha E_E \tag{32}$$

where E_A, E_B, E_C, E_D, and E_E are no-load back EMF of phase A, B, C, D, and E.

The reduced-order Park transformation matrix in the case of single-phase open circuit fault is shown as:

$$T_{\text{Park}}^{A} = \begin{bmatrix} \cos\theta_e & \sin\theta_e & 0 & 0 \\ -\sin\theta_e & \cos\theta_e & 0 & 0 \\ 0 & 0 & 1 & 0 \\ 0 & 0 & 0 & 1 \end{bmatrix} \tag{33}$$

Substituting Eqs. (30) and (33) into Eq. (13), the stator voltage equation of the motor with phase A open-circuit fault can be expressed as:

$$\begin{aligned} U_{dq1}^{A} &= T_{\text{Park}}^{A} T_{\text{Clarke1}}^{A} U_{s1}^{A} = T_{\text{Park}}^{A} T_{\text{Clarke1}}^{A} R_{s1}^{A} I_{s1}^{A} + T_{\text{Park}}^{A} T_{\text{Clarke1}}^{A} \frac{d\psi_s^A}{dt} \\ &= T_{\text{Park}}^{A} T_{\text{Clarke1}}^{A} R_{s1}^{A} \left(T_{\text{Park}}^{A} T_{\text{Clarke1}}^{A}\right)^{-1} T_{\text{Park}}^{A} T_{\text{Clarke1}}^{A} I_{s1}^{A} \end{aligned}$$

$$
\begin{aligned}
&+ T_{\text{Park}}^{A} T_{\text{Clarke1}}^{A} \frac{d\left(L_{s1}^{A} I_{s1}^{A}\right)}{dt} + T_{\text{Park}}^{A} T_{\text{Clarke1}}^{A} \frac{d\psi_{m}^{A}}{dt} \\
&= R_{dq1}^{A} I_{dq1}^{A} + T_{\text{Park}}^{A} T_{\text{Clarke1}}^{A} \frac{d\left(L_{s1}^{A} I_{s1}^{A}\right)}{dt} + T_{\text{Park}}^{A} T_{\text{Clarke1}}^{A} \frac{d\psi_{m}^{A}}{dt}
\end{aligned} \tag{34}
$$

where U_{dq1}^{A} is stator voltage matrix in the rotating coordinate system under phase A open-circuit fault; U_{s1}^{A} is stator voltage matrix in the natural coordinate system under phase A open-circuit fault; R_{s1}^{A}, L_{s1}^{A}, I_{s1}^{A} are resistance matrix, inductance matrix and current matrix in the natural coordinate system under phase A fault respectively; R_{dq1}^{A}, L_{dq1}^{A}, I_{dq1}^{A} are resistance matrix, inductance matrix and current matrix in the rotating coordinate system under phase A fault respectively; ψ_{m}^{A} is permanent magnet flux linkage matrix in the natural coordinate system under phase A fault.

The no-load back EMF in the rotating coordinate system under phase A fault can be expressed as:

$$
E_{dq1}^{A} = T_{\text{Park}}^{A} T_{\text{Clarke1}}^{A} \frac{d\psi_{m}^{A}}{dt} = \begin{bmatrix} E_{d1}^{A} \\ E_{q1}^{A} \\ E_{31}^{A} \\ E_{01}^{A} \end{bmatrix} = \omega_{e} \psi_{f} \begin{bmatrix} 0 \\ 1 \\ 0 \\ 0.4 \sin \theta_{e} \end{bmatrix} \tag{35}
$$

The induced voltage generated by the stator windings can be expressed as:

$$
\begin{aligned}
E_{L-dq1}^{A} &= T_{\text{Park}}^{A} T_{\text{Clarke1}}^{A} \frac{d\left(L_{s1}^{A} I_{s1}^{A}\right)}{dt} \\
&= \frac{d\left(T_{\text{Park}}^{A} T_{\text{Clarke1}}^{A} L_{s1}^{A} I_{s1}^{A}\right)}{dt} - \frac{d\left(T_{\text{Park}}^{A} T_{\text{Clarke1}}^{A}\right)}{dt} L_{s1}^{A} I_{s1}^{A} \\
&= \frac{d\left(L_{dq1}^{A}\right)}{dt} I_{dq1}^{A} + L_{dq1} \frac{d\left(I_{dq1}^{A}\right)}{dt} - \Omega_{1}^{A} L_{dq1}^{A} I_{dq1}^{A}
\end{aligned} \tag{36}
$$

where,

$$
\begin{aligned}
I_{dq1}^{A} &= T_{\text{Park}}^{A} T_{\text{Clarke1}}^{A} I_{s1}^{A} \\
&= \left[i_{d1}^{A} \; i_{q1}^{A} \; i_{31}^{A} \; i_{0} \right]^{T}
\end{aligned} \tag{37}
$$

$$
L_{dq1}^{A} = T_{\text{Park}}^{A} T_{\text{Clarke1}}^{A} L_{s1}^{A} \left(T_{\text{Clarke1}}^{A}\right)^{-1} \left(T_{\text{park}}^{A}\right)^{-1} \tag{38}
$$

$$
\Omega_{1}^{A} = \frac{d\left(T_{\text{park}}^{A}\right)}{dt} \left(T_{\text{park}}^{A}\right)^{-1} = \omega_{e} \begin{bmatrix} 0 & 1 & 0 & 0 \\ -1 & 0 & 0 & 0 \\ 0 & 0 & 0 & 0 \\ 0 & 0 & 0 & 0 \end{bmatrix} \tag{39}
$$

$$\begin{cases} L_d = L_{ls} + 2.5(L_m - L_\theta) \\ L_q = L_{ls} + 2.5(L_m + L_\theta) \end{cases} \tag{40}$$

Substituting Eqs. (37), (38) and (39) into Eq. (34), the stator voltage equation in the rotating coordinate system can be expressed as:

$$\begin{bmatrix} u_{d1}^A \\ u_{q1}^A \\ u_{31}^A \end{bmatrix} = R \begin{bmatrix} i_{d1}^A \\ i_{q1}^A \\ i_{31}^A \end{bmatrix} + \begin{bmatrix} L_d & 0 & 0 \\ 0 & L_q & 0 \\ 0 & 0 & L_{ls} \end{bmatrix} \frac{d}{dt} \begin{bmatrix} i_{d1}^A \\ i_{q1}^A \\ i_{31}^A \end{bmatrix} - \omega_e \begin{bmatrix} 0 & L_q & 0 \\ -L_d & 0 & 0 \\ 0 & 0 & 0 \end{bmatrix} \begin{bmatrix} i_{d1}^A \\ i_{q1}^A \\ i_{31}^A \end{bmatrix} + \omega_e \begin{bmatrix} 0 \\ \psi_f \\ 0 \end{bmatrix} \tag{41}$$

Substituting Eqs. (35) and (37) into Eq. (18), the output torque under phase A open-circuit fault can be obtained according to the magnetic energy method:

$$T_e^A = \frac{5}{2} P \left[i_{q1}^A \psi_m^A + i_{d1}^A i_{q1}^A (L_d - L_q) \right] \tag{42}$$

According to the above Eq. (42), when the fault-tolerant current remains unchanged on the d–q axis which means $i_{d1}^A = i_{d1}$ $i_{q1}^A = i_{q1}$, the output torque of the motor before and after the failure can be guaranteed to remain unchanged [30]. The third-order space current i_{31}^A does not participate in electromagnetic energy conversion, but it can improve the fault-tolerant control performance, so as to realize the lowest joule losses (LJL) and equal joule losses (EJL) in fault-tolerant operation. EJL control strategy, namely equal amplitude of residual fault-tolerant current, can improve the output torque capacity of fault-tolerant operation under the current limit of inverter.

(a) LJL

In fundamental space, the copper consumption of the motor is $R_s\left(i_{d1}^{A2} + i_{q1}^{A2} + i_{31}^{A2}\right)$. When the current component in the d–q–3 coordinate system satisfies Eq. (43), the copper consumption is the lowest.

$$i_{d1}^A = i_{d1}\,,\; i_{q1}^A = i_{q1}\,,\; i_{31}^A = 0 \tag{43}$$

According to the inverse transformation of the reduced-order matrix and Eq. (43), the remaining phase fault-tolerant current can be obtained as follows:

$$\begin{cases} i_{B1}^A = -1.468 I_m \sin(\theta_e - 0.224\pi + \beta) \\ i_{C1}^A = -1.263 I_m \sin(\theta_e - 0.846\pi + \beta) \\ i_{D1}^A = -1.263 I_m \sin(\theta_e + 0.846\pi + \beta) \\ i_{E1}^A = -1.468 I_m \sin(\theta_e + 0.224\pi + \beta) \end{cases} \tag{44}$$

where, β is the current angle, $\beta = \arctan(i_d/i_q)$.

(b) EJL

When EJL control strategy is adopted, the fault-tolerant current amplitude of the remaining normal phase is equal, and the remaining phase current needs to meet additional constraints: $i_B = -i_D$ and $i_C = -i_E$. At this point, the current component of the β axis and the third harmonic space should satisfy:

$$\begin{aligned} i_{\beta 1}^{A} &= \sin\alpha i_B + \sin 2\alpha i_C + \sin 3\alpha i_D + \sin 4\alpha i_E \\ &= (i_B + i_C)(\sin\alpha + \sin 2\alpha) \end{aligned} \tag{45}$$

$$\begin{aligned} i_{31}^{A} &= \sin 3\alpha i_B + \sin 6\alpha i_C + \sin 9\alpha i_D + \sin 12\alpha i_E \\ &= \frac{\sin\alpha - \sin 2\alpha}{\sin\alpha + \sin 2\alpha} i_{\beta 1}^{A} = 0.236\left(i_{d1}^{A}\sin\theta_e + i_{q1}^{A}\cos\theta_e\right) \end{aligned} \tag{46}$$

In order to EJL control strategy, the current component in the *d*–*q*–3 coordinate system needs to meet the following requirements:

$$i_{d1}^{A} = i_{d1},\ i_{q1}^{A} = i_{q1},\ i_{31}^{A} = 0.236\left(i_{d1}^{A}\sin\theta_e +_{q1}^{A}\cos\theta_e\right) \tag{47}$$

Similarly, according to the inverse transformation of the reduced-order matrix and Eq. (47), the remaining phase fault-tolerant current can be obtained as follows:

$$\begin{cases} i_{B1}^{A} = -1.382 I_m \sin(\theta_e - 0.2\pi + \beta) \\ i_{C1}^{A} = -1.382 I_m \sin(\theta_e - 0.8\pi + \beta) \\ i_{D1}^{A} = -1.382 I_m \sin(\theta_e + 0.8\pi + \beta) \\ i_{E1}^{A} = -1.382 I_m \sin(\theta_e + 0.2\pi + \beta) \end{cases} \tag{48}$$

(2.2) Two-phase open-circuit fault

When the adjacent-phase open-circuit fault occurs in the five-phase motor, assuming that the open-circuit fault occurs in phase A and phase B, the fault-tolerant currents after fault shall meet the following requirements:

$$\begin{aligned} &\frac{5}{2} N I_m e^{j\left(\theta_e + \pi/2\right)} = N\left(i_A + \lambda i_B + \lambda^2 i_C + \lambda^3 i_D + \lambda^4 i_E\right) \\ &i_A \equiv 0,\ i_B \equiv 0 \end{aligned} \tag{49}$$

When open-circuit fault occurs in phase A and phase B, the current component in the α–β coordinate system after open-circuit fault can be expressed as:

$$\frac{2}{5}\begin{bmatrix} \cos 2\alpha & \cos 3\alpha & \cos 4\alpha \\ \sin 2\alpha & \sin 3\alpha & \sin 4\alpha \\ 1 & 1 & 1 \end{bmatrix}\begin{bmatrix} i_C \\ i_D \\ i_E \end{bmatrix} = \begin{bmatrix} i_{\alpha 1} \\ i_{\beta 1} \\ 0 \end{bmatrix} \tag{50}$$

After open-circuit fault, the permanent magnet flux component in the fundamental space α–β coordinate system can be expressed as

$$\begin{bmatrix} \psi_\alpha^{AB} \\ \psi_\beta^{AB} \\ \psi_0^{AB} \end{bmatrix} = \frac{2}{5}\begin{bmatrix} \cos 2\alpha & \cos 3\alpha & \cos 4\alpha \\ \sin 2\alpha & \sin 3\alpha & \sin 4\alpha \\ 1 & 1 & 1 \end{bmatrix}\begin{bmatrix} \psi_C \\ \psi_D \\ \psi_E \end{bmatrix} = \begin{bmatrix} \psi_\alpha - 0.4(\psi_A + \cos\alpha\psi_B) \\ \psi_\beta - 0.4\sin\alpha\psi_B \\ 0.6472\psi_f \cos(\theta_e - \pi - 0.5\alpha) \end{bmatrix} \quad (51)$$

where $\psi AB\alpha$, $\psi AB\beta$, $\psi AB0$ are components of permanent magnet flux linkage in α–β–0 coordinate system under open-circuit fault of phase A and phase B respectively; ψ_A, ψ_B, ψ_C, ψ_D, ψ_E are permanent magnet flux linkage of phase A, B, C, D, E; ψ_α, ψ_β are the component of permanent magnet flux in the α–β axis under normal operation, which can be expressed as:

$$\psi_\alpha = 0.4(\psi_A + \cos\alpha\psi_B + \cos 2\alpha\psi_C + \cos 3\alpha\psi_D + \cos 4\alpha\psi_E) = \psi_f \cos\theta_e \quad (52)$$

$$\psi_\beta = 0.4(\sin\alpha\psi_B + \sin 2\alpha\psi_C + \sin 3\alpha\psi_D + \sin 4\alpha\psi_E) = \psi_f \sin\theta_e \quad (53)$$

ψ_α, ψ_β can form a circular flux track in normal operation. When open circuit faults occur in phase A and B, the trajectory formed by ψ_α^{AB} and ψ_β^{AB} is shown in Fig. 9a, which is an ellipse. Figure 9b shows the components of zero-sequence planar permanent magnet in α–β axis ($\psi_{0\_\alpha}^{AB}$ and $\psi_{0\_\beta}^{AB}$). In order to achieve fault-tolerant operation, it is necessary to ensure that the flux component of α–β axis can still form a circle after fault. The flux component in Fig. 9a is modified to obtain the flux component of α–β axis as shown in Fig. 9c.

The components of zero-sequence planar permanent magnet in α–β axis ($\psi_{0\_\alpha}^{AB}$ and $\psi_{0\_\beta}^{AB}$) as shown in Fig. 9b is expressed as:

$$\begin{bmatrix} \psi_{0\_\alpha}^{AB} \\ \psi_{0\_\beta}^{AB} \end{bmatrix} = \psi_0^{AB}\begin{bmatrix} \cos(\pi + 0.5\alpha) \\ \sin(\pi + 0.5\alpha) \end{bmatrix} = -0.6472\cos(\theta_e - \pi - 0.5\alpha)\begin{bmatrix} \cos 0.5\alpha \\ \sin 0.5\alpha \end{bmatrix} \quad (54)$$

After modification, the permanent magnet flux of α–β axis shown in Fig. 9c can be expressed as:

$$\begin{bmatrix} \psi_\alpha^{AB} \\ \psi_\beta^{AB} \\ \psi_0^{AB} \end{bmatrix} = \begin{bmatrix} \psi_\alpha - 0.4(\psi_A + \cos\alpha\psi_B) \\ \psi_\beta - 0.4\sin\alpha\psi_B \\ 0.6472\psi_f \cos(\theta_e - \pi - 0.5\alpha) \end{bmatrix} + \begin{bmatrix} x\psi_{0\_\alpha}^{AB} \\ y\psi_{0\_\beta}^{AB} \\ 0 \end{bmatrix}$$

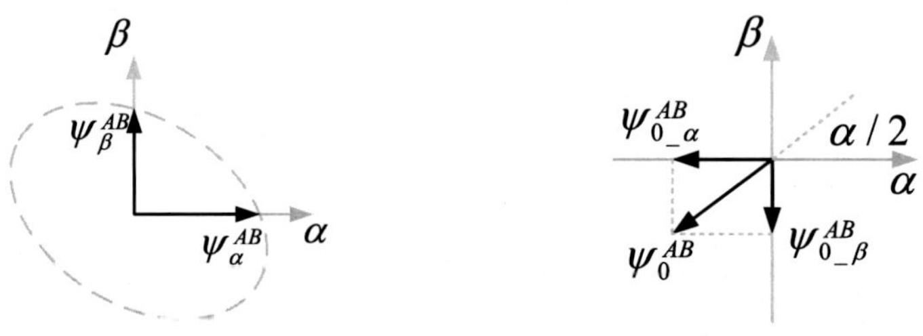

(a) Permanent magnet flux after fault (b) Zero sequence plane permanent magnet flux

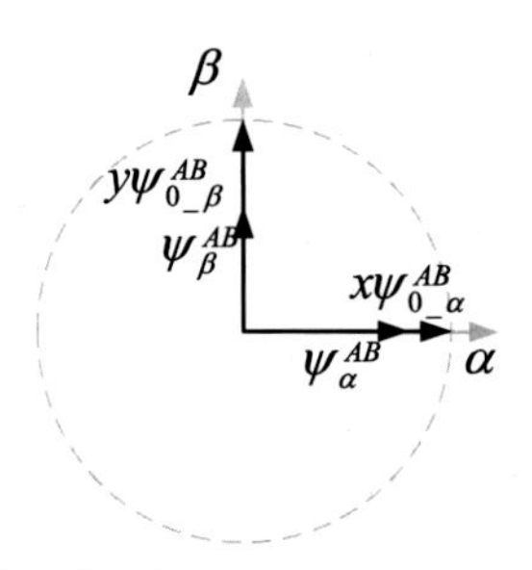

(c) Modified permanent magnet flux

Fig. 9 The trajectory of permanent magnet flux linkage

$$
= \begin{bmatrix} \psi_\alpha \\ \psi_\beta \\ 0 \end{bmatrix} + \begin{bmatrix} -0.4(\psi_A + \cos\alpha\psi_B) + x\psi_{0\_\alpha}^{AB} \\ -0.4\sin\alpha\psi_B + y\psi_{0\_\beta}^{AB} \\ 0.6472\psi_f\cos(\theta_e - \pi - 0.5\alpha) \end{bmatrix} \tag{55}
$$

According to Eqs. (52), (53), and (54), forming a circular flux trajectory needs to be satisfied:

$$
\begin{bmatrix} -0.4(\psi_A + \cos\alpha\psi_B) + x\psi_{0\_\alpha}^{AB} \\ -0.4\sin\alpha\psi_B + y\psi_{0\_\beta}^{AB} \end{bmatrix} \equiv k\begin{bmatrix} \cos\theta_e \\ \sin\theta_e \end{bmatrix} \tag{56}
$$

x, y, and k can be solved by Eq. (56):

$$
x = y = \frac{\cos\alpha}{\cos 0.5\alpha},\ k = 0.4(\cos\alpha - 1) \tag{57}
$$

Substitute Eq. (57) into Eq. (55), Eq. (51) can be re-expressed as:

$$
\begin{bmatrix} \psi_\alpha^{AB} \\ \psi_\beta^{AB} \\ \psi_0^{AB} \end{bmatrix} = \frac{2}{5}\begin{bmatrix} \cos 2\alpha - \cos\alpha & \cos 3\alpha - \cos\alpha & \cos 4\alpha - \cos\alpha \\ \sin 2\alpha - \tan\frac{\delta}{2}\cos\alpha & \sin 3\alpha - \tan\frac{\alpha}{2}\cos\alpha & \sin 4\alpha - \tan\frac{\alpha}{2}\cos\alpha \\ 1 & 1 & 1 \end{bmatrix}\begin{bmatrix} \psi_C \\ \psi_D \\ \psi_E \end{bmatrix}
$$

$$= \psi_f \begin{bmatrix} (0.6 + 0.4\cos\alpha)\cos\theta_e \\ (0.6 + 0.4\cos\alpha)\sin\theta_e \\ -0.6472\psi_f\cos(\theta_e - 0.5\alpha) \end{bmatrix} \tag{58}$$

Therefore, the reduced-order Clarke matrix under open-circuit faults of phase A and B can be defined as:

$$T_{\text{Clarke1}}^{AB} = \frac{2}{5}\begin{bmatrix} \cos 2\alpha - \cos\alpha & \cos 3\alpha - \cos\alpha & \cos 4\alpha - \cos\alpha \\ \sin 2\alpha - \tan\frac{\alpha}{2}\cos\alpha & \sin 3\alpha - \tan\frac{\alpha}{2}\cos\alpha & \sin 4\alpha - \tan\frac{\alpha}{2}\cos\alpha \\ 1 & 1 & 1 \end{bmatrix} \tag{59}$$

The reduced-order Park transformation matrix in the case of two-phase open circuit fault can be defined as:

$$T_{\text{Park}}^{2} = \begin{bmatrix} \cos\theta_e & \sin\theta_e & 0 \\ -\sin\theta_e & \cos\theta_e & 0 \\ 0 & 0 & 1 \end{bmatrix} \tag{60}$$

Similarly, when nonadjacent-phase open circuit faults occur in five-phase motor, taking open circuit faults in A and C phases as an example, the reduced Clarke matrix can be expressed as:

$$T_{\text{Clarke1}}^{AC} = \frac{2}{5}\begin{bmatrix} \cos\alpha - \cos 2\alpha & \cos 3\alpha - \cos 2\alpha & \cos 4\alpha - \cos 2\alpha \\ \sin\alpha - \tan\alpha\cos 2\alpha & \sin\alpha - \tan\alpha\cos 2\alpha & \sin 4\alpha - \tan\alpha\cos 2\alpha \\ 1 & 1 & 1 \end{bmatrix} \tag{61}$$

(3) The second kind of matrix

(3.1) Single-phase open-circuit fault

Assuming that open-circuit fault occurs in phase A, the current of phase A is zero and the currents of the remaining phases can be expressed as:

$$\begin{cases} i_{B2}^{A} = x_b I_m \cos(\theta_e + \alpha_b) \\ i_{C2}^{A} = x_c I_m \cos(\theta_e + \alpha_c) \\ i_{D2}^{A} = x_d I_m \cos(\theta_e + \alpha_d) \\ i_{E2}^{A} = x_e I_m \cos(\theta_e + \alpha_e) \end{cases} \tag{62}$$

At this point, the magnetomotive force generated by the remaining phases can be expressed as:

$$\begin{cases} \mathrm{MMF}_B = 0.5N_5 i_{B2}^A \cos(\gamma - 0.4\pi) \\ \mathrm{MMF}_C = 0.5N_5 i_{C2}^A \cos(\gamma - 0.8\pi) \\ \mathrm{MMF}_D = 0.5N_5 i_{D2}^A \cos(\gamma - 1.2\pi) \\ \mathrm{MMF}_E = 0.5N_5 i_{E2}^A \cos(\gamma - 1.6\pi) \end{cases} \tag{63}$$

By expanding and adding Eq. (63), the synthetic magneto-motive force of five-phase motor under single-phase open-circuit fault can be obtained:

$$\begin{aligned} \mathrm{MMF}_s &= \mathrm{MMF}_B + \mathrm{MMF}_C + \mathrm{MMF}_D + \mathrm{MMF}_E \\ &= 0.5N_5\left(0.309 i_{B2}^A - 0.809 i_{C2}^A - 0.809 i_{D2}^A + 0.309 i_{E2}^A\right)\cos\gamma \\ &\quad + 0.5N_5\left(-0.951 i_{B2}^A - 0.588 i_{C2}^A + 0.588 i_{D2}^A + 0.951 i_{E2}^A\right)\sin\gamma \end{aligned} \tag{64}$$

Due to the Y-connection of the motor, the sum of the phase currents should be zero. In order to maximize the torque output during fault operation under the current limit of inverter, the amplitude of the current of each phase is equal. Therefore, the current shown in Eq. (62) should also satisfy the following limiting conditions [31]:

$$\begin{cases} i_{B2}^A + i_{C2}^A + i_{D2}^A + i_{E2}^A = 0 \\ x_1 = x_2 = x_3 = x_4 \end{cases} \tag{65}$$

According to the above constraints and the invariable principle of the magnetic motive force, the fault-tolerant state of each phase current can be obtained as:

$$\begin{cases} i_{B2}^A = 1.382 I_m \cos(\theta_e - 0.5\alpha) \\ i_{C2}^A = 1.382 I_m \cos(\theta_e - 2\alpha) \\ i_{D2}^A = 1.382 I_m \cos(\theta_e - 3\alpha) \\ i_{E2}^A = 1.382 I_m \cos(\theta_e - 4.5\alpha) \end{cases} \tag{66}$$

Figure 10 shows the current vector contrast under normal working conditions and single-phase open-circuit fault-tolerant working conditions.

The current expression can be expressed in the form of d–q axis current as Eq. (67):

$$\begin{bmatrix} i_{B2}^A \\ i_{C2}^A \\ i_{D2}^A \\ i_{E2}^A \end{bmatrix} = 1.382 \begin{bmatrix} \cos 0.5\alpha & \sin 0.5\alpha \\ \cos 2\alpha & \sin 2\alpha \\ \cos 3\alpha & \sin 3\alpha \\ \cos 4.5\alpha & \sin 4.5\alpha \end{bmatrix} \begin{bmatrix} \cos\theta_e & -\sin\theta_e \\ \sin\theta_e & \cos\theta_e \end{bmatrix} \begin{bmatrix} i_{d2}^A \\ i_{q2}^A \end{bmatrix} \tag{67}$$

For the Y-connected five-phase motor system, when open-circuit fault occurs in phase A winding, the controllable degree of freedom of the motor is 3. Two degrees of freedom can be distributed in α–β subspace. In order to ensure the precision of the control, it is necessary to control another degree of freedom. At the same time, the subspace of the degree of freedom must be orthogonal to the α–β subspace to ensure the decoupling of the system.

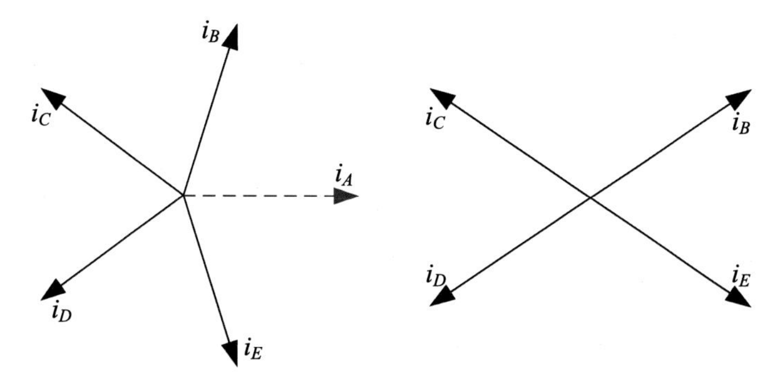

Fig. 10 The current vector contrast under normal working conditions and single-phase open-circuit fault-tolerant working conditions

Firstly, two vector bases are defined according to Eq. (67):

$$\begin{cases} T_1 = \left[\frac{\cos 0.5\alpha}{3.618} \; \frac{\cos 2\alpha}{3.618} \; \frac{\cos 3\alpha}{3.618} \; \frac{\cos 4.5\alpha}{3.618} \right] \\ T_2 = \left[\frac{\sin 0.5\alpha}{1.91} \; \frac{\sin 2\alpha}{1.91} \; \frac{\sin 3\alpha}{1.91} \; \frac{\sin 4.5\alpha}{1.91} \right] \end{cases} \tag{68}$$

The subspace where the third degree of freedom resides is defined as the z subspace, which satisfies the following constraints:

$$\begin{cases} T_1 Z_1^T = T_2 Z_1^T = 0 \\ Z_1 I_s = 0 \end{cases} \tag{69}$$

In order to ensure the reversibility of the matrix, the orthogonal reduced-order transformation matrix can be obtained as:

$$T_{\text{Clarke2}}^A = \begin{bmatrix} \cos 0.5\alpha/3.618 & \cos 2\alpha/3.618 & \cos 3\alpha/3.618 & \cos 4.5\alpha/3.618 \\ \sin 0.5\alpha/1.91 & \sin 2\alpha/1.91 & \sin 3\alpha/1.91 & \sin 4.5\alpha/1.91 \\ \sin\alpha/5 & \sin 4\alpha/5 & \sin 6\alpha/5 & \sin 9\alpha/5 \\ 1 & 1 & 1 & 1 \end{bmatrix} \tag{70}$$

The Park transformation matrix is consistent with Eq. (33). ψ_m^A can be expressed as:

$$\psi_m^A = \begin{bmatrix} \psi_B \\ \psi_C \\ \psi_D \\ \psi_E \end{bmatrix} = \psi_f \begin{bmatrix} \cos(\theta_e - \alpha) \\ \cos(\theta_e - 2\alpha) \\ \cos(\theta_e - 3\alpha) \\ \cos(\theta_e - 4\alpha) \end{bmatrix} \tag{71}$$

The no-load back EMF of each phase can be expressed as:

$$E_{s2}^{A} = \begin{bmatrix} e_B \\ e_C \\ e_D \\ e_E \end{bmatrix} = -\omega_e \psi_f \begin{bmatrix} \sin(\theta_e - \alpha) \\ \sin(\theta_e - 2\alpha) \\ \sin(\theta_e - 3\alpha) \\ \sin(\theta_e - 4\alpha) \end{bmatrix} \tag{72}$$

The reconstructed Clark and Park transformation matrices are used to calculate the stator voltage equation in the rotating space after phase A open-circuit fault, which can be expressed as:

$$\begin{aligned} U_{dq2}^{A} &= T_{\text{Park}}^{A} T_{\text{Clarke2}}^{A} U_{s2}^{A} = T_{\text{Park}}^{A} T_{\text{Clarke2}}^{A} R_{s2}^{A} (T_{\text{Park}}^{A} T_{\text{Clarke2}}^{A})^{-1} T_{\text{Park}}^{A} T_{\text{Clarke2}}^{A} I_{s2}^{A} \\ &\quad + T_{\text{Park}}^{A} T_{\text{Clarke2}}^{A} \frac{d(L_{s2}^{A} I_{s2}^{A})}{dt} + T_{\text{Park}}^{A} T_{\text{Clarke2}}^{A} \frac{d\psi_m^{A}}{dt} \\ &= R_{dq2}^{A} I_{dq2}^{A} + T_{\text{Park}}^{A} T_{\text{Clarke2}}^{A} \frac{d(L_{s2}^{A} I_{s2}^{A})}{dt} + T_{\text{Park}}^{A} T_{\text{Clarke2}}^{A} \frac{d\psi_m^{A}}{dt} \end{aligned} \tag{73}$$

According to Eq. (73), the no-load back EMF in the rotating space can be expressed as:

$$E_{dq2}^{A} = \begin{bmatrix} E_{d2}^{A} \\ E_{q2}^{A} \\ E_{z2}^{A} \\ E_{02}^{A} \end{bmatrix} = T_{\text{Park}}^{A} T_{\text{Clarke2}}^{A} \cdot \omega_e \psi_f \begin{bmatrix} -\sin(\theta_e - \alpha) \\ -\sin(\theta_e - 2\alpha) \\ -\sin(\theta_e - 3\alpha) \\ -\sin(\theta_e - 4\alpha) \end{bmatrix} \tag{74}$$

Considering that the matrix constructed cannot guarantee that the back EMF is consistent with the normal working condition after transformation, the voltage equation of five-phase motor in the rotating coordinate system can be obtained as:

$$\begin{cases} u_{d2}^{A} - E_{d2}^{A} = i_{d2}^{A} R_s + L_d \frac{di_{d2}^{A}}{dt} - \omega_e L_q i_{q2}^{A} \\ u_{q2}^{A} - E_{q2}^{A} = i_{q2}^{A} R_s + L_q \frac{di_{q2}^{A}}{dt} + \omega_e L_d i_{d2}^{A} \\ u_{z2}^{A} - E_{z2}^{A} = i_{z2}^{A} R_s + L_{ls} \frac{di_{z2}^{A}}{dt} \\ u_{02}^{A} - E_{02}^{A} = 0 \end{cases} \tag{75}$$

At this point, the torque equation can be expressed as:

$$T_e = \frac{5}{2} p \left[\psi_f i_{q2}^{A} + (L_d - L_q) i_{d2}^{A} i_{q2}^{A} \right] + 0.955 p \psi_f i_{z2}^{A} \cos\theta_{\text{e}} \tag{76}$$

It can be seen from Eq. (76) that the torque expression consists of two parts, the first half is consistent with that under the normal working condition. Therefore, in control, the average torque of output can be guaranteed as long as the given current of *d–q* axis before and after the failure remains unchanged. The latter part can be regarded as an AC pulsation component. In order to avoid the torque pulsation caused by the AC term, the influence of pulsation can be eliminated by controlling the z-axis current to zero.

(3.2) Two-phase open-circuit fault

When adjacent-phase open circuit occurs in phase C and D, the currents of faulty phases are zero and the remaining phase current can be expressed as:

$$\begin{cases} i_{A2}^{CD} = x_a I_m \cos(\theta_e + \alpha_a) \\ i_{B2}^{CD} = x_b I_m \cos(\theta_e + \alpha_b) \\ i_{E2}^{CD} = x_e I_m \cos(\theta_e + \alpha_e) \end{cases} \tag{77}$$

The magneto-motive force generated by the remaining phases can be expressed as:

$$\begin{cases} \mathrm{MMF}_A = 0.5N_5 i_{A2}^{CD} \cos\gamma \\ \mathrm{MMF}_B = 0.5N_5 i_{B2}^{CD} \cos(\gamma - 0.4\pi) \\ \mathrm{MMF}_E = 0.5N_5 i_{E2}^{CD} \cos(\gamma - 1.6\pi) \end{cases} \tag{78}$$

By expanding and adding Eq. (78), the synthetic magneto-motive force of five-phase motor under two adjacent open-circuit fault can be obtained:

$$\begin{aligned} \mathrm{MMF}_s &= \mathrm{MMF}_A + \mathrm{MMF}_B + \mathrm{MMF}_E \\ &= 0.5N_5\left(i_{A2}^{CD} + 0.309 i_{B2}^{CD} + 0.309 i_{E2}^{CD}\right)\cos\gamma \\ &\quad + 0.5N_5\left(-0.951 i_{B2}^{CD} + 0.951 i_{E2}^{CD}\right)\sin\gamma \end{aligned} \tag{79}$$

It is also necessary to satisfy the principle that the sum of current at neutral points is zero as Eq. (80).

$$i_{A2}^{CD} + i_{B2}^{CD} + i_{E2}^{CD} = 0 \tag{80}$$

According to the above constraints and the invariable principle of magneto-motive force, each phase current in the fault-tolerant state can be obtained as:

$$\begin{cases} i_{A2}^{CD} = 3.618 I_m \cos\theta_e \\ i_{B2}^{CD} = 2.236 I_m \cos(\theta_e - 0.8\pi) \\ i_{E2}^{CD} = 2.236 I_m \cos(\theta_e + 0.8\pi) \end{cases} \tag{81}$$

Figure 11 shows the current vector contrast under normal working conditions and adjacent-phase open-circuit fault-tolerant working conditions.

The current expression can be expressed in the form of d–q axis current as Eq. (82):

$$\begin{bmatrix} i_{A2}^{CD} \\ i_{B2}^{CD} \\ i_{E2}^{CD} \end{bmatrix} = \begin{bmatrix} 3.618 & 0 \\ 2.236\cos 0.8\pi & 2.236\sin 0.8\pi \\ 2.236\cos 0.8\pi & -2.236\sin 0.8\pi \end{bmatrix} \begin{bmatrix} \cos\theta_e & -\sin\theta_e \\ \sin\theta_e & \cos\theta_e \end{bmatrix} \begin{bmatrix} i_{d2}^{CD} \\ i_{q2}^{CD} \end{bmatrix} \tag{82}$$

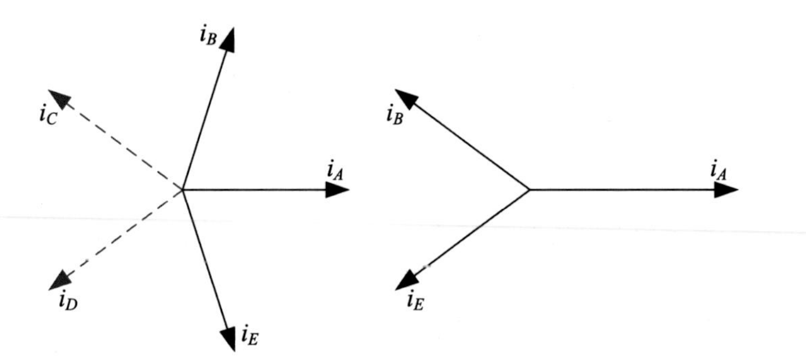

Fig. 11 The current vector contrast under normal working conditions and adjacent-phase open-circuit fault-tolerant working conditions

In order to ensure the orthogonality of the transformation matrix, the Clark transformation matrix can be modified as:

$$T_{\text{Clarke2}}^{CD} = \frac{2}{5}\begin{bmatrix} 1 - 0.539 & \cos\alpha - 0.539 & \cos 4\alpha - 0.539 \\ 0 & \sin\alpha & \sin 4\alpha \\ 1 & 1 & 1 \end{bmatrix} \tag{83}$$

The Park transformation matrix is consistent with Eq. (60).

Similarly, when nonadjacent-phase open-circuit fault occurs in phase B and phase E, the synthetic magneto-motive force of five-phase motor can be obtained as:

$$\begin{aligned} \text{MMF}_s &= \text{MMF}_A + \text{MMF}_C + \text{MMF}_D \\ &= 0.5N_5\left(i_{A2}^{BE} - 0.809 i_{C2}^{BE} - 0.809 i_{D2}^{BE}\right)\cos\gamma \\ &\quad + 0.5N_5\left(-0.588 i_{C2}^{BE} + 0.588 i_{D2}^{BE}\right)\sin\gamma \end{aligned} \tag{84}$$

Each phase current in the fault-tolerant state can be obtained as follows:

$$\begin{cases} i_{A2}^{CD} = 3.618 I_m \cos\theta_e \\ i_{B2}^{CD} = 2.236 I_m \cos(\theta_e - 0.8\pi) \\ i_{E2}^{CD} = 2.236 I_m \cos(\theta_e + 0.8\pi) \end{cases} \tag{85}$$

Figure 12 shows the current vector contrast under normal working conditions and adjacent-phase open-circuit fault-tolerant working conditions.

The current expression can be expressed in the form of d–q axis current as Eq. (82):

$$\begin{bmatrix} i_{A2}^{BE} \\ i_{C2}^{BE} \\ i_{D2}^{BE} \end{bmatrix} = \begin{bmatrix} 1.382 & 0 \\ 2.236\cos 0.6\pi & 2.236\sin 0.6\pi \\ 2.236\cos 0.6\pi & -2.236\sin 0.6\pi \end{bmatrix} \begin{bmatrix} \cos\theta_e & -\sin\theta_e \\ \sin\theta_e & \cos\theta_e \end{bmatrix} \begin{bmatrix} i_{d2}^{BE} \\ i_{q2}^{BE} \end{bmatrix} \tag{86}$$

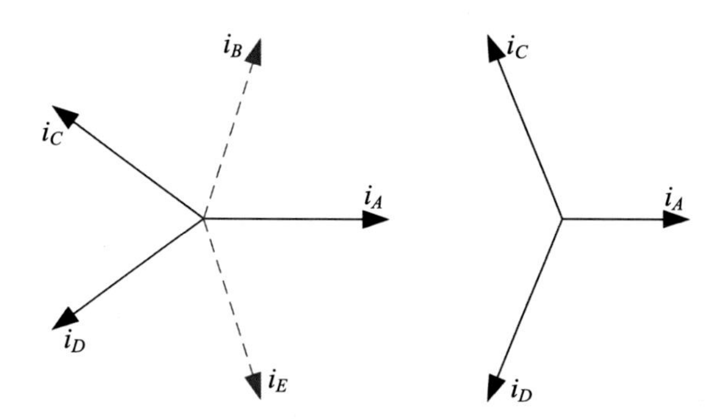

Fig. 12 The current vector contrast under normal working conditions and nonadjacent-phase open-circuit fault-tolerant working conditions

In order to ensure the orthogonality of the transformation matrix, the Clark transformation matrix can be modified as:

$$T_{\text{Clarke2}}^{BE} = \frac{2}{5}\begin{bmatrix} 1 - 0.206 & \cos\alpha - 0.206 & \cos 4\alpha - 0.206 \\ 0 & \sin\alpha & \sin 4\alpha \\ 1 & 1 & 1 \end{bmatrix} \tag{87}$$

The Park transformation matrix is consistent with Eq. (60). Permanent magnet flux after fault ψ_m^{CD} can be expressed as:

$$\psi_m^{CD} = \begin{bmatrix} \psi_A \\ \psi_B \\ \psi_E \end{bmatrix} = \psi_f \begin{bmatrix} \cos\theta_e \\ \cos(\theta_e - \alpha) \\ \cos(\theta_e - 4\alpha) \end{bmatrix} \tag{88}$$

The reconstructed Clark and Park transformation matrices are used to calculate the stator voltage equation in the rotating space after two-phase open-circuit fault, which can be expressed as:

$$\begin{aligned} U_{dq2}^{CD} &= T_{\text{Park}}^2 T_{\text{Clarke2}}^{CD} U_{s2}^{CD} = T_{\text{Park}}^2 T_{\text{Clarke2}}^{CD} R_{s2}^{CD} \left(T_{\text{Park}}^2 T_{\text{Clarke2}}^{CD}\right)^{-1} T_{\text{Park}}^2 T_{\text{Clarke2}}^{CD} I_{s2}^{CD} \\ &+ T_{\text{Park}}^2 T_{\text{Clarke2}}^{CD} \frac{d\left(L_{s2}^{CD} I_{s2}^{CD}\right)}{dt} + T_{\text{Park}}^2 T_{\text{Clarke2}}^{CD} \frac{d\psi_m^{CD}}{dt} \\ &= R_{dq2}^{CD} I_{dq2}^{CD} + T_{\text{Park}}^2 T_{\text{Clarke2}}^{CD} \frac{d\left(L_{s2}^{CD} I_{s2}^{CD}\right)}{dt} + T_{\text{Park}}^2 T_{\text{Clarke2}}^{CD} \frac{d\psi_m^{CD}}{dt} \end{aligned} \tag{89}$$

The no-load back EMF in the rotating space can be expressed as:

$$E_{dq2}^{CD} = \begin{bmatrix} E_{d2}^{CD} \\ E_{q2}^{CD} \\ E_{02}^{CD} \end{bmatrix} = T_{\text{Park}}^2 T_{\text{Clarke2}}^{CD} \cdot \omega_e \psi_f \begin{bmatrix} -\sin\theta_e \\ -\sin(\theta_e - \alpha) \\ -\sin(\theta_e - 4\alpha) \end{bmatrix} \tag{90}$$

According to the analysis of two-phase open circuit fault, the voltage of each phase of the five-phase motor has been reconstructed. The voltage equation in the rotating coordinate system is finally obtained as follows:

$$\begin{cases} u_{d2}^{CD} - E_{d2}^{CD} = i_{d2}^{CD} R_s + L_d \dfrac{di_{d2}^{CD}}{dt} - \omega_e L_q i_{q2}^{CD} \\ u_{q2}^{CD} - E_{q2}^{CD} = i_{q2}^{CD} R_s + L_q \dfrac{di_{q2}^{CD}}{dt} + \omega_e L_d i_{d2}^{CD} \end{cases} \tag{91}$$

The torque equation can be obtained as:

$$T_e = \frac{5}{2} p \left[\psi_f i_{q2}^{CD} + (L_d - L_q) i_{d2}^{CD} i_{q2}^{CD} \right] \tag{92}$$

Similarly, the voltage equation in the rotating coordinate system when nonadjacent-phase fault occurs in phase B and phase E and torque equation are obtained as:

$$\begin{cases} u_{d2}^{CD} - E_{d2}^{CD} = i_{d2}^{CD} R_s + L_d \dfrac{di_{d2}^{CD}}{dt} - \omega_e L_q i_{q2}^{CD} \\ u_{q2}^{CD} - E_{q2}^{CD} = i_{q2}^{CD} R_s + L_q \dfrac{di_{q2}^{CD}}{dt} + \omega_e L_d i_{d2}^{CD} \end{cases} \tag{93}$$

$$T_e = \frac{5}{2} p \left[\psi_f i_{q2}^{BE} + (L_d - L_q) i_{d2}^{BE} i_{q2}^{BE} \right] \tag{94}$$

According to Eqs. (92) and (94), as long as the given cross-axis current is consistent with the normal working condition, stable output torque can be obtained and the average torque remains unchanged.

4 Open-Circuit Fault-Tolerant Methods

In this section, the fault-tolerant control based on different control methods will be implemented according to the fault-tolerant theory introduced in Sect. 6.3.

4.1 Open-Circuit Fault-Tolerant Method Based on Model Predictive Control

The detailed model of predictive control under open-circuit faults will be introduced. In this part, the second kind of fault-tolerant matrix is adopted.

(1) Normal operation analysis

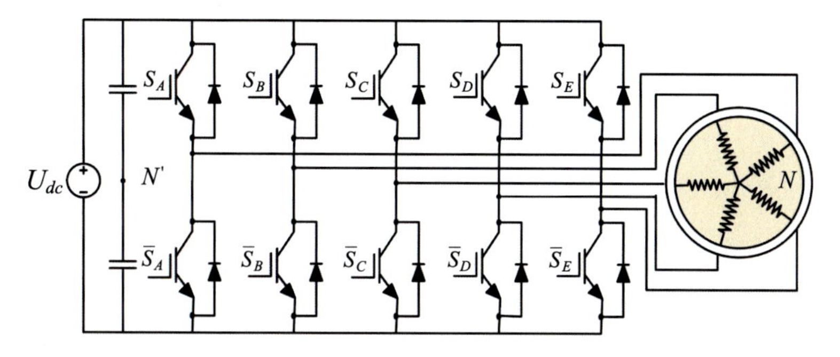

Fig. 13 Five-phase motor and voltage source inverter

The five-phase motor control system is driven by voltage source inverter and its simplified circuit. As shown in Fig. 13, A, B, C, D, and E are inverter drive signal output terminals. N is the neutral point of the motor; N' is the imaginary neutral point of DC bus. The expression of stator voltage of each phase of the motor can be expressed by Eq. (95). Since the motor winding is Y-connected, the stator phase voltage also meets Eq. (96).

$$\begin{cases} U_A = U_{AN'} + U_{N'N} = \frac{U_{dc}}{2}(2S_A - 1) + U_{N'N} \\ U_B = U_{BN'} + U_{N'N} = \frac{U_{dc}}{2}(2S_B - 1) + U_{N'N} \\ U_C = U_{CN'} + U_{N'N} = \frac{U_{dc}}{2}(2S_C - 1) + U_{N'N} \\ U_D = U_{DN'} + U_{N'N} = \frac{U_{dc}}{2}(2S_D - 1) + U_{N'N} \\ U_E = U_{EN'} + U_{N'N} = \frac{U_{dc}}{2}(2S_E - 1) + U_{N'N} \end{cases} \tag{95}$$

where U_{dc} is the DC bus voltage; U_x (x = A, B, C, D, E) is the stator phase voltage of each phase; $U_{xN'}$ is the voltage between the output terminal of the inverter and the imaginary neutral point; $U_{N'N}$ is the voltage between the imaginary neutral point of the inverter and the neutral point of the motor; S_x is the switching state of each bridge arm, '1' indicates that the upper bridge arm is open, '0' indicates that the lower bridge arm is open.

Through Eq. (95), the phase voltage vector of the inverter in the static coordinate system under normal working conditions can be worked out. By using the transformation matrix (12) mentioned above, it can be transformed into two two-dimensional orthogonality α_1–β_1 and α_3–β_3 subspaces, which can be obtained as:

$$\begin{cases} U_{\alpha 1\beta 1} = \frac{2}{5}(U_A + U_B\, e^{j\alpha} + U_C\, e^{j2\alpha} + U_D\, e^{j3\alpha} + U_E\, e^{j4\alpha}) \\ U_{\alpha 3\beta 3} = \frac{2}{5}(U_A + U_C\, e^{j\alpha} + U_E\, e^{j2\alpha} + U_B\, e^{j3\alpha} + U_D\, e^{j4\alpha}) \end{cases} \tag{96}$$

When the switching state of the inverter is brought into Eq. (96), 32 space voltage vectors can be obtained including two zero vectors and 30 effective vectors. The effective vector can also be large vector, medium vector, and small vector according

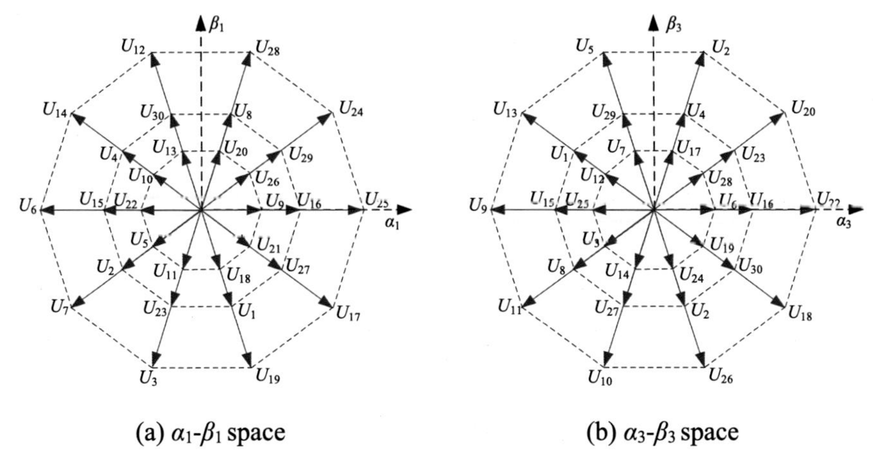

(a) α_1-β_1 space (b) α_3-β_3 space

Fig. 14 Space voltage vector distribution under normal working condition

to the amplitude. They divide two two-dimensional orthogonal subspaces into ten regions equally. The distribution law of amplitude and phase angle is shown in Fig. 14.

(2) Space voltage vector construction

(2.1) Single-phase open-circuit fault

When single-phase open-circuit fault occurs in the winding of phase A, the bridge arm of the inverter corresponding to the A-phase winding no longer provides the driving voltage for the motor. At this time, only four bridge arms of the inverter operate normally. In order to maintain the normal power supply state after fault, the power supply mode of the inverter needs to be adjusted. It is necessary to reconstruct the space voltage vector provided by the inverter, including its quantity, amplitude, and phase. In this section, the second kind of matrix is adopted.

As shown in Fig. 12, when the phase a winding of the motor is open circuit, the stator phase voltage of the remaining phases can be expressed as:

$$\begin{cases} U_B = U_{BN'} + U_{N'N} = \frac{U_{dc}}{2}(2S_B - 1) + U_{N'N} \\ U_C = U_{CN'} + U_{N'N} = \frac{U_{dc}}{2}(2S_C - 1) + U_{N'N} \\ U_D = U_{DN'} + U_{N'N} = \frac{U_{dc}}{2}(2S_D - 1) + U_{N'N} \\ U_E = U_{EN'} + U_{N'N} = \frac{U_{dc}}{2}(2S_E - 1) + U_{N'N} \end{cases} \tag{97}$$

Substituting Eq. (95) into Eq. (97), the potential difference of the two neutral points at this time can be obtained as:

$$U_{NN'} + \frac{U_{dc}}{2} = \frac{U_{dc}}{4}(S_B + S_C + S_D + S_E) + \frac{1}{4}U_A \tag{98}$$

Table 1 Voltage vectors in α–β subspace under single-phase fault

Voltage vector	Amplitude	S_B, S_C, S_D, S_E
U_{15}, U_0, U_{10}, U_5	0	1111, 0000, 1010, 0101
U_8, U_{13}, U_4, U_{14} U_2, U_7, U_1, U_{11}	$0.3804U_{dc}$	1000, 1101, 0100, 1110 0010, 0111, 0001, 1011
U_6, U_9	$0.4472U_{dc}$	0110, 1001
U_3, U_{12}	$0.6145U_{dc}$	0011, 1100

Table 2 Voltage vectors in z subspace under single-phase fault

Voltage vector	Amplitude	S_B, S_C, S_D, S_E
U_0, U_3, U_6 U_9, U_{12}, U_{15}	0	0000, 0011, 0110 1001, 1100, 1111
U_1, U_4, U_7, U_{13} U_2, U_8, U_{11}, U_{14}	$0.1902U_{dc}$	0001, 0100, 0111, 1101 0010, 1000, 1011, 1110
U_5, U_{10}	$0.3804U_{dc}$	0101, 1010

Using the Clarke transformation matrix (70), α–β calculation equation of voltage vector in α–β subspace and z subspace can be obtained as [32]:

$$\begin{bmatrix} u_{\alpha2}^{A} \\ u_{\beta2}^{A} \\ u_{z2}^{A} \end{bmatrix} = T_{\text{Clarke2}}^{A} \begin{bmatrix} U_B \\ U_C \\ U_D \\ U_E \end{bmatrix} = \begin{bmatrix} 0.2236U_{dc}(S_B - S_C - S_D + S_E) \\ 0.3077U_{dc}(S_B + S_C - S_D - S_E) \\ 0.1902U_{dc}(S_B - S_C + S_D - S_E) \end{bmatrix} \tag{99}$$

Substituting the switching state of the inverter into Eq. (99), the voltage vectors in α–β subspace and z subspace are shown in Tables 1 and 2.

The amplitude and phase of each vector in Tables 1 and 2 are given in the form of the coordinate system to obtain the corresponding spatial voltage vector distribution diagram, as shown in Fig. 15. It should be noted that although the amplitudes of switching vectors U_5 and U_{10} are zero, according to the switching state of the two voltage vectors, they are not real zero vectors. Therefore, they are abandoned when selecting effective switching vectors. On the other hand, considering that the switching vectors U_8 and U_{13}, U_4 and U_{14}, U_2 and U_7, U_1 and U_{11} have the same effect respectively. In order to reduce the switching frequency and calculation amount of the inverter, only U_8, U_4, U_2, U_1, U_9, U_{12}, U_6, and U_3 are selected as effective vectors.

(2.2) Two-phase open-circuit fault

Assuming that open-circuit fault occurs in phase C and phase D, the stator phase voltages of the remaining phases are expressed as:

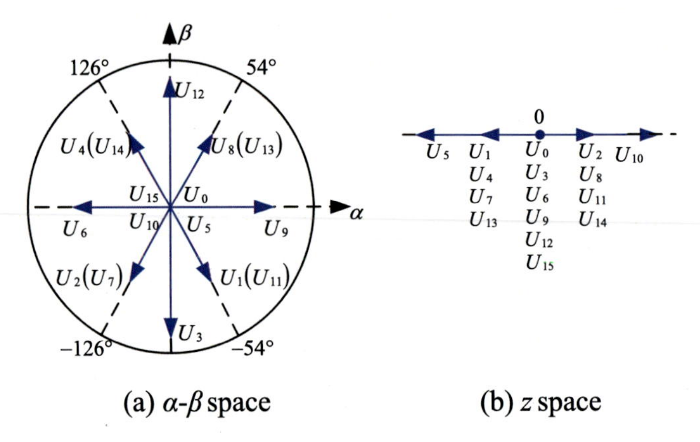

(a) α-β space (b) z space

Fig. 15 Voltage vectors distribution under single-phase fault

$$\begin{cases} U_A = U_{AN'} + U_{N'N} = \frac{U_{dc}}{2}(2S_A - 1) + U_{N'N} \\ U_B = U_{BN'} + U_{N'N} = \frac{U_{dc}}{2}(2S_B - 1) + U_{N'N} \\ U_E = U_{EN'} + U_{N'N} = \frac{U_{dc}}{2}(2S_E - 1) + U_{N'N} \end{cases} \tag{100}$$

Substituting Eq. (95) into Eq. (100), the potential difference between the two neutral points at this time can be obtained as:

$$U_{NN'} + \frac{U_{dc}}{2} = \frac{U_{dc}}{3}(S_A + S_B + S_E) + \frac{1}{3}(U_C + U_D) \tag{101}$$

Using the Clarke transformation matrix (83), α–β calculation equation of voltage vector in α–β subspace can be obtained as:

$$\begin{bmatrix} U^{CD}_{\alpha 2} \\ U^{CD}_{\beta 2} \end{bmatrix} = T^{CD}_{\text{Clarke2}} \begin{bmatrix} U_A \\ U_B \\ U_E \end{bmatrix} = \begin{bmatrix} 0.0920U_{dc}(2S_A - S_B - S_E) \\ 0.3805U_{dc}(S_B - S_E) \end{bmatrix} \tag{102}$$

Substituting the switching state of the inverter into Eq. (102), the voltage vectors in α–β subspace will be obtained, which are shown in Table 3. Figure 16 shows the corresponding space voltage vector distribution.

Assuming that open-circuit fault occurs in phase B and phase E, the stator phase voltages of the remaining phases are expressed as:

Table 3 Voltage vectors under adjacent-phase fault

Voltage vector	Amplitude	S_A, S_B, S_E
U_7, U_0	0	111, 000
U_1, U_2, U_5, U_6	$0.3915U_{dc}$	001, 010, 101, 110
U_3, U_4	$0.1804U_{dc}$	011, 100

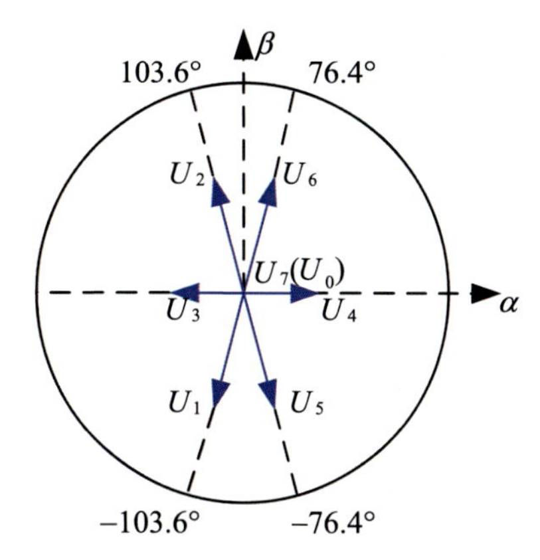

Fig. 16 Voltage vectors distribution under adjacent-phase fault

$$\begin{cases} U_A = U_{AN'} + U_{N'N} = \frac{U_{dc}}{2}(2S_A - 1) + U_{N'N} \\ U_C = U_{CN'} + U_{N'N} = \frac{U_{dc}}{2}(2S_C - 1) + U_{N'N} \\ U_D = U_{DN'} + U_{N'N} = \frac{U_{dc}}{2}(2S_D - 1) + U_{N'N} \end{cases} \tag{103}$$

Substituting Eq. (95) into Eq. (103), the potential difference between the two neutral points at this time can be obtained as:

$$U_{NN'} + \frac{U_{dc}}{2} = \frac{U_{dc}}{3}(S_A + S_C + S_D) + \frac{1}{3}(U_B + U_E) \tag{104}$$

Using the Clarke transformation matrix (87), α–β calculation equation of voltage vector in α–β subspace can be obtained as:

$$\begin{bmatrix} U_{\alpha 2}^{BE} \\ U_{\beta 2}^{BE} \end{bmatrix} = T_{\text{Clarke2}}^{BE} \begin{bmatrix} U_A \\ U_C \\ U_D \end{bmatrix} = \begin{bmatrix} 0.2411U_{dc}(2S_A - S_C - S_D) \\ 0.2352U_{dc}(S_C - S_D) \end{bmatrix} \tag{105}$$

Substituting the switching state of the inverter into Eq. (105), the voltage vectors in α–β subspace will be obtained, which are shown in Table 4. Figure 17 shows the corresponding space voltage vector distribution.

(3) Predictive model

(3.1) Single-phase open-circuit fault

Table 4 Voltage vectors under nonadjacent-phase fault

Voltage vector	Amplitude	S_A, S_C, S_D
U_7, U_0	0	111, 000
U_1, U_2, U_5, U_6	$0.3368U_{dc}$	001, 010, 101, 110
U_3, U_4	$0.4822U_{dc}$	011, 100

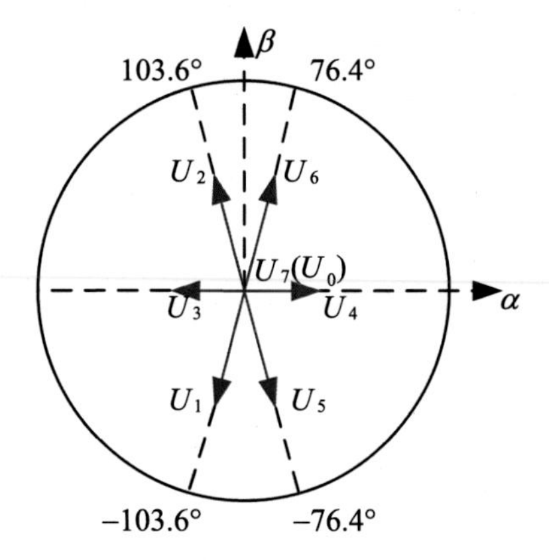

Fig. 17 Voltage vectors distribution under nonadjacent-phase fault

There are many methods to realize model discretization, the most commonly used is the forward Euler method. Discretizing Eq. (75) by forward Euler method, the prediction model at this time can be obtained as follows:

$$\begin{cases} i_{d2}^{A}(k+1) = i_{d2}^{A}(k) + \frac{Ts}{Ld}(u_{d2}^{A}(k) - R_s i_{d2}^{A}(k) + \omega_e L_q i_{q2}^{A}(k) - e_{d2}^{A}(k)) \\ i_{q2}^{A}(k+1) = i_{q2}^{A}(k) + \frac{Ts}{Lq}(u_{q2}^{A}(k) - R_s i_{q2}^{A}(k) - \omega_e L_d i_{d}^{A}(k) - e_{q2}^{A}(k)) \\ i_{z2}^{A}(k+1) = i_{z2}^{A}(k) + \frac{Ts}{Lls}(u_{z2}^{A}(k) - R_s i_{z2}^{A}(k) - e_{z2}^{A}(k)) \end{cases} \tag{106}$$

where T_s is the sampling period; $i_{d2}^{A}(k)$, $i_{q2}^{A}(k)$ and $i_{z2}^{A}(k)$ are the currents in d–q–z axes at the current moment respectively; $i_{d2}^{A}(k+1)$, $i_{q2}^{A}(k+1)$ and $i_{z2}^{A}(k+1)$ are the predicted currents in d–q–z axes at $(k+1)$th instant respectively; $u_{d2}^{A}(k)$, $u_{q2}^{A}(k)$, $u_{z2}^{A}(k)$ and $e_{d2}^{A}(k)$, $e_{q2}^{A}(k)$, $e_{z2}^{A}(k)$ are the stator phase voltage and back EMF in d–q–z axes respectively.

It can be seen from Eq. (106) that after the single-phase open-circuit fault of the motor, there are three current prediction terms in the prediction equation. Therefore, the cost function can be set as:

$$g_1 = \left|i_{dref2}^{A} - i_{d2}^{A}(k+1)\right|^2 + \left|i_{qref2}^{A} - i_{q2}^{A}(k+1)\right|^2 + \left|i_{zref2}^{A} - i_{z2}^{A}(k+1)\right|^2 \tag{107}$$

where i_{dref2}^{A}, i_{qref2}^{A}, and i_{zref2}^{A} respectively represent the given currents in d–q–z axes.

(3.2) Two-phase open-circuit fault

Similarly, taking adjacent-phase open-circuit fault as an example, the prediction model discretized by forward Euler method can be obtained as:

$$\begin{cases} i_{d2}^{CD}(k+1) = i_{d2}^{CD}(k) + \dfrac{Ts}{Ld}(u_{d2}^{CD}(k) - R_s i_{d2}^{CD}(k) + \omega_e L_q i_{q2}^{CD}(k) - e_{d2}^{CD}(k)) \\ i_{q2}^{CD}(k+1) = i_{q2}^{CD}(k) + \dfrac{Ts}{Lq}(u_{q2}^{CD}(k) - R_s i_{q2}^{CD}(k) - \omega_e L_d i_{d}^{CD}(k) - e_{q2}^{CD}(k)) \end{cases} \tag{108}$$

where T_s is the sampling period; $i_{d2}^{CD}(k)$ and $i_{q2}^{CD}(k)$ are the currents in d–q axes at the current moment respectively; $i_{d2}^{CD}(k+1)$ and $i_{q2}^{CD}(k+1)$ are the predicted currents in d–q axes at $(k+1)$th instant respectively; $u_{d2}^{CD}(k)$, $u_{q2}^{CD}(k)$ and $e_{d2}^{CD}(k)$, $e_{q2}^{CD}(k)$ are the stator phase voltage and back EMF in d–q–z axes respectively.

It can be seen from Eq. (109) that after adjacent-phase open circuit fault, there are two current prediction terms in the prediction equation. Therefore, the cost function can be set as:

$$g_2 = \left|i_{dref2}^{CD} - i_{d2}^{CD}(k+1)\right|^2 + \left|i_{qref2}^{CD} - i_{q2}^{CD}(k+1)\right|^2 \tag{109}$$

where *iA d*ref2 and *iA q*ref2 respectively represent the given currents in d–q axes.

(4) Delay compensation

In the previous section, the predictive model and cost function in model predictive control are analyzed. It can be seen that in the prediction process, the current, speed, and other variables of the motor at the current time need to be detected in real time as feedback and then the optimal switching action of the inverter is obtained through the value function. However, due to the influence of sampling delay and other factors the switching action of the inverter often lags one beat. Therefore, the optimal control action predicted and judged at time k can only act on the system at time $k+1$. At this time, the motor current has changed from is (k) to is $(k+1)$ and the control accuracy of the system has been seriously affected.

Figure 18 shows the schematic diagram of the delay link. x^* is the reference value; $x(k)$, $x(k+1)$, and $x(k+2)$ are the values obtained by sampling the variable x at the current time. It can be seen that the corresponding sampling data at t_1 time is $x(k)$. Due to the influence of calculation and delay link, the optimal control action is judged at t_2 time, and will not act on the system until t_3 time. In the design time (t_1), the variable k should be controlled for the actual operation. In order to solve the problem caused by delay, take the variable at $k+1$ time as the initial value, and use the prediction equation to predict the variable at $k+2$ time. Finally, substitute the value at $k+2$ time is substituted into the value function for calculation. At this time, the value function can be rewritten as:

$$g_1 = \left|i_{dref2}^{A} - i_{d2}^{A}(k+2)\right|^2 + \left|i_{qref2}^{A} - i_{q2}^{A}(k+2)\right|^2 + \left|i_{zref2}^{A} - i_{z2}^{A}(k+2)\right|^2 \tag{110}$$

$$g_2 = \left|i_{dref2}^{CD} - i_{d2}^{CD}(k+2)\right|^2 + \left|i_{qref2}^{CD} - i_{q2}^{CD}(k+2)\right|^2 \tag{111}$$

(5) Control block diagram

Figures 19 and 20 show the control block diagram of model predictive control under single-phase and adjacent-phase open-circuit fault respectively.

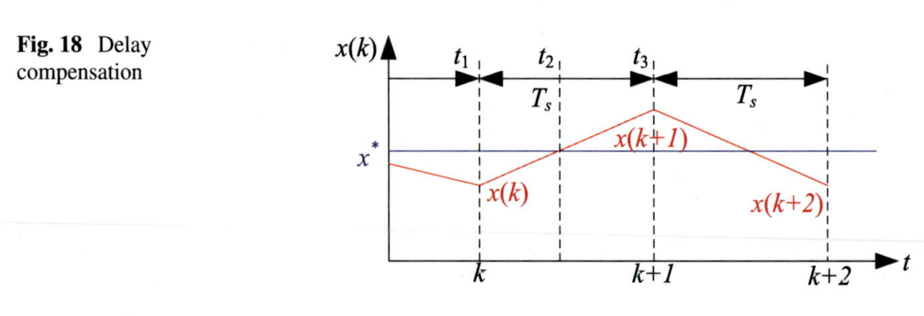

Fig. 18 Delay compensation

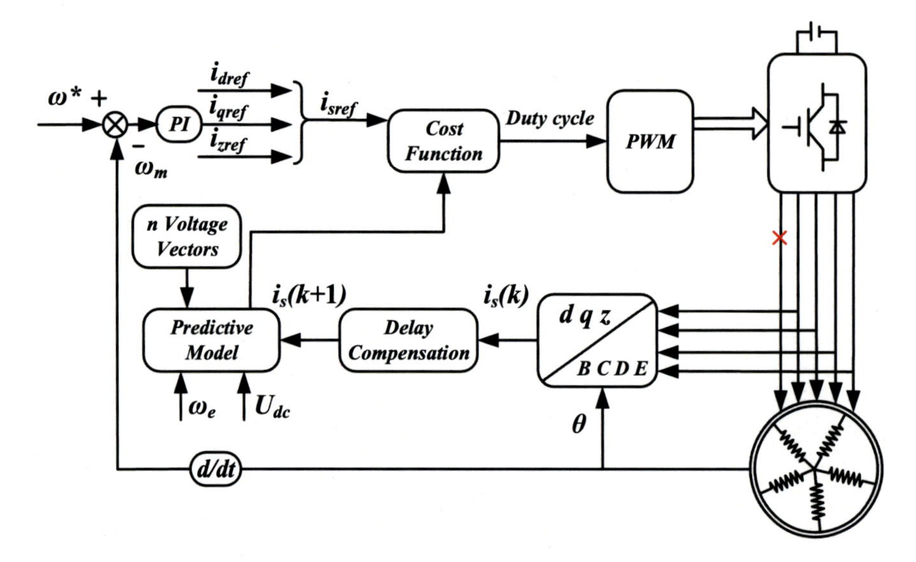

Fig. 19 Control block diagram of MPC under single-phase open-circuit fault

4.2 Open-Circuit Fault-Tolerant Method Based on Vector Control

The detailed SVPWM modulation process under open-circuit faults will be introduced. In this part, the second kind of fault-tolerant matrix is adopted.

(1) Single-phase open-circuit fault

The voltage vectors have been constructed in Tables 1 and 2, and Fig. 15. Then, Fig. 21 illustrates the synthesis of the reference voltage vector in sector I under single-phase open-circuit fault. The synthesized reference voltage vector can be expressed as:

$$\begin{cases} U_{\alpha\text{ref}} = (T_1 U_{8\alpha} + T_2 U_{9\alpha} + T_3 U_{13\alpha}) / T_s \\ U_{\beta\text{ref}} = (T_1 U_{8\beta} + T_2 U_{9\beta} + T_3 U_{13\beta}) / T_s \end{cases} \tag{112}$$

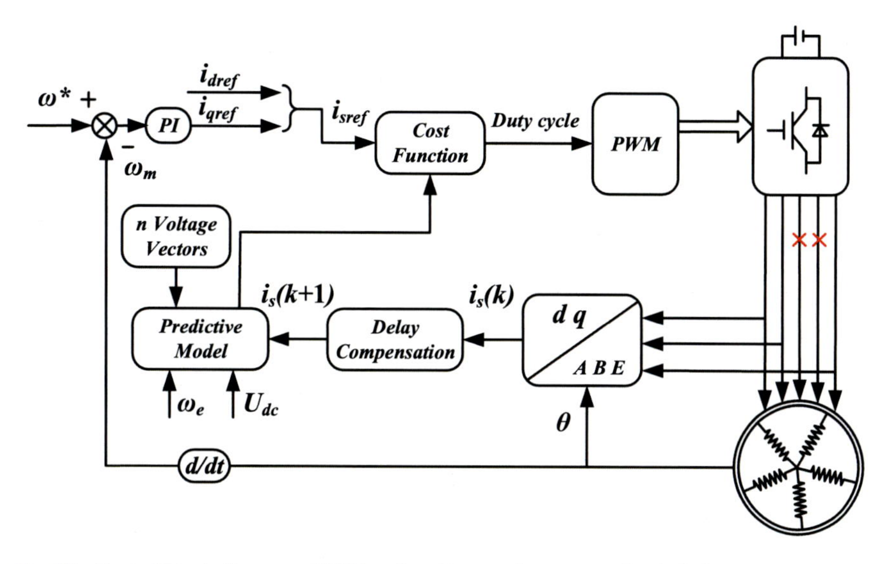

Fig. 20 Control block diagram of MPC under adjacent-phase open-circuit fault

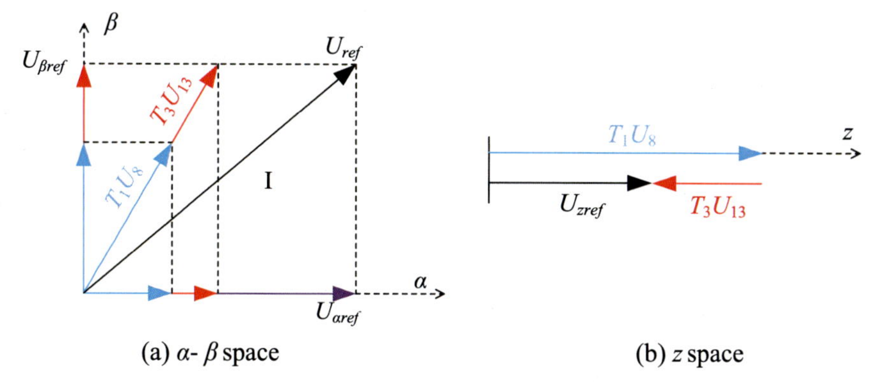

Fig. 21 Realization of the reference voltage vector located in sector I (single-phase open-circuit fault)

where T_1, T_2, and T_3 are the action time of U_8, U_9, and U_{13} respectively; T_s is the sampling time; $U_{\alpha\text{ref}}$ and $U_{\beta\text{ref}}$ are the voltage components of the vector in α–β axes. Then the amplitude relationship corresponding to Fig. 21a can be expressed as:

$$\begin{cases} U_{8\alpha} = 0.2236U_{dc}\,;\ U_{8\beta} = 0.3077U_{dc} \\ U_{9\alpha} = 0.4472U_{dc}\,;\ U_{9\beta} = 0 \\ U_{13\alpha} = 0.2236U_{dc}\,;\ U_{13\beta} = 0.3077U_{dc} \end{cases} \tag{113}$$

Similarly, the z-subspace synthesized reference voltage vector in Fig. 21b can be expressed as:

$$\begin{aligned} U_{zref} &= (T_1 U_{8z} + T_3 U_{13z}) / T_s \\ &= (0.1902 U_{dc} T_1 - 0.1902 U_{dc} T_3) / T_s \end{aligned} \tag{114}$$

According to Eqs. (112), (132), and (114), the synthesized reference voltage vector can be expressed as:

$$\begin{bmatrix} U_{\alpha ref} \\ U_{\beta ref} \\ U_{zref} \end{bmatrix} = \frac{U_{dc}}{T_s} M_1 \begin{bmatrix} T_1 \\ T_2 \\ T_3 \end{bmatrix} \tag{115}$$

$$M_1 = \begin{bmatrix} 0.2236 & 0.4472 & 0.2236 \\ 0.3077 & 0 & 0.3077 \\ 0.1902 & 0 & -0.1902 \end{bmatrix} \tag{116}$$

According to the inverse transformation of Eq. (115), the action time of each voltage vector can be obtained as:

$$\begin{bmatrix} T_1 \\ T_2 \\ T_3 \end{bmatrix} = \frac{T_s}{U_{dc}} M_1^{-1} \begin{bmatrix} U_{\alpha ref} \\ U_{\beta ref} \\ U_{zref} \end{bmatrix} \tag{117}$$

In order to ensure that the sum of vector action time T_1, T_2, and T_3 is less than the total sampling time, it should satisfy the relationship as follows:

$$\begin{bmatrix} T_1 \\ T_2 \\ T_3 \end{bmatrix} = \frac{T_s}{T_1 + T_2 + T_3} \begin{bmatrix} T_1 \\ T_2 \\ T_3 \end{bmatrix} \tag{118}$$

As for the zero voltage, vector action time T_0 can be shown as follows:

$$T_0 = (T_s - T_1 - T_2 - T_3) / 2 \tag{119}$$

The reference voltage vector action time corresponding to each voltage vector under the open circuit of phase A can be given by:

$$\begin{cases} a = \frac{2.2361 U_{\alpha ref}}{U_{dc}} T_s + \frac{-1.6250 U_{\beta ref}}{U_{dc}} T_s, b = \frac{-2.2358 U_{\alpha ref}}{U_{dc}} T_s + \frac{1.6247 U_{\beta ref}}{U_{dc}} T_s \\ c = \frac{2.2358 U_{\alpha ref}}{U_{dc}} T_s + \frac{1.6247 U_{\beta ref}}{U_{dc}} T_s, d = \frac{-2.2361 U_{\alpha ref}}{U_{dc}} T_s + \frac{-1.6250 U_{\beta ref}}{U_{dc}} T_s \\ e = \frac{2.2361 U_{\alpha ref}}{U_{dc}} T_s + \frac{2.6288 U_{zref}}{U_{dc}} T_s, f = \frac{2.2361 U_{\alpha ref}}{U_{dc}} T_s + \frac{-2.6288 U_{zref}}{U_{dc}} T_s \\ g = \frac{1.6250 U_{\beta ref}}{U_{dc}} T_s + \frac{2.6288 U_{zref}}{U_{dc}} T_s, h = \frac{1.6250 U_{\beta ref}}{U_{dc}} T_s + \frac{-2.6288 U_{zref}}{U_{dc}} T_s \end{cases} \tag{120}$$

Action time in each sector under phase A open-circuit fault is shown in Table 5.

(2) Two-phase open-circuit fault

Table 5 Action time in each sector under single-phase open-circuit

Vector	1	2	3	4	5	6	7	8
T_1	g	e	$-e$	h	$-h$	$-f$	f	$-g$
T_2	a	d	c	d	$-a$	$-b$	$-c$	$-d$
T_3	h	f	$-f$	g	g	$-e$	e	$-h$

Taking phase C and phase D open circuit as an example, the voltage vectors have been constructed in Table 3 and Fig. 16. Then, Fig. 22 illustrates the synthesis of the reference voltage vector in sector I under adjacent-phase open-circuit fault.

It can be seen from Fig. 22 that, in order to meet the parallelogram rule, the action time of each vector should meet:

$$\frac{U_{\text{ref}}}{\sin(180^\circ - 76.4^\circ)} = \frac{U_6 T_1 / T_s}{\sin\theta} = \frac{U_4 T_2 / T_s}{\sin(76.4^\circ - \theta)} \tag{121}$$

According to Eq. (121), the action time of each vector can be given by:

$$\begin{cases} T_1 = 2.6279 \frac{T_s}{U_{dc}} U_{\beta\text{ref}} \\ T_2 = 5.4348 \frac{T_s}{U_{dc}} U_{\alpha\text{ref}} - 1.3148 \frac{T_s}{U_{dc}} U_{\beta\text{ref}} \end{cases} \tag{122}$$

Five kinds of vector action time are given as follows:

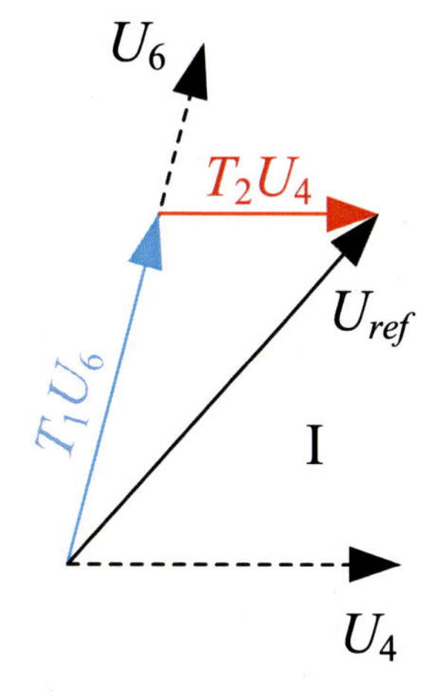

Fig. 22 Realization of the reference voltage vector located in sector I (adjacent-phase open-circuit fault)

Table 6 Action time in each sector under adjacent phase open-circuit

Vector	1	2	3	4	5	6
T_1	a	d	$-c$	$-a$	e	$-b$
T_2	b	e	a	$-b$	$-d$	$-a$

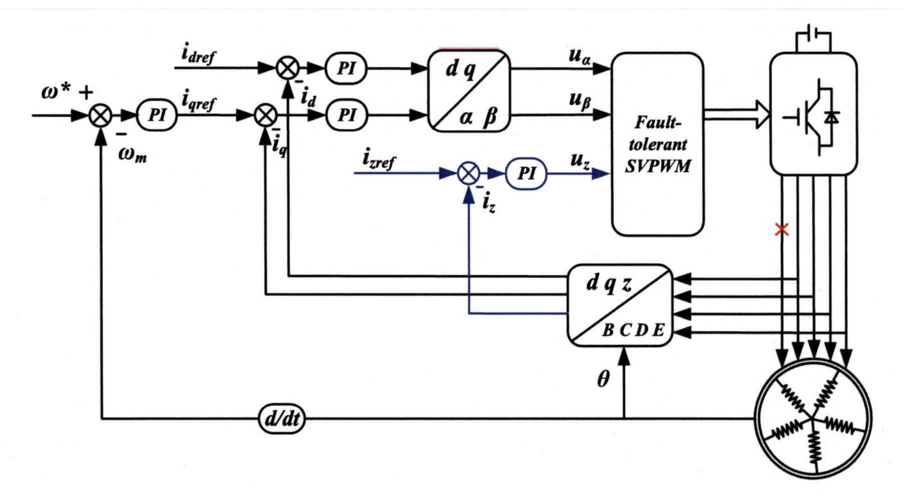

Fig. 23 Control block diagram of VC under single-phase open-circuit fault

$$
\begin{cases}
a = 2.6279\frac{T_s}{U_{dc}}\,U_{\beta\mathrm{ref}} \\
b = 5.4348\frac{T_s}{U_{dc}}\,U_{\alpha\mathrm{ref}} - 1.3148\frac{T_s}{U_{dc}}\,U_{\beta\mathrm{ref}} \\
c = 5.4348\frac{T_s}{U_{dc}}\,U_{\alpha\mathrm{ref}} + 1.3148\frac{T_s}{U_{dc}}\,U_{\beta\mathrm{ref}} \\
d = 5.4314\frac{T_s}{U_{dc}}\,U_{\alpha\mathrm{ref}} + 1.3140\frac{T_s}{U_{dc}}\,U_{\beta\mathrm{ref}} \\
e = 5.4314\frac{T_s}{U_{dc}}\,U_{\alpha\mathrm{ref}} - 1.3140\frac{T_s}{U_{dc}}\,U_{\beta\mathrm{ref}}
\end{cases} \tag{123}
$$

Action time in each sector under adjacent-phase open-circuit fault is shown in Table 6.

(3) Control block diagram

Figure 23 shows the control block diagram of vector control under single-phase open-circuit fault.

4.3 Open-Circuit Fault-Tolerant Method Based on Direct Torque Control

The detailed direct torque control under open-circuit faults will be introduced. In this part, the first kind of fault-tolerant matrix is adopted.

(1) Space voltage vector construction

When open-circuit fault occurs in phase A, the remaining four-phase stator voltage can be obtained as:

$$\begin{bmatrix} U_{B1}^{A} \\ U_{C1}^{A} \\ U_{D1}^{A} \\ U_{E1}^{A} \end{bmatrix} = \frac{U_{dc}}{4} \begin{bmatrix} 3 & -1 & -1 & -1 \\ -1 & 3 & -1 & -1 \\ -1 & -1 & 3 & -1 \\ -1 & -1 & -1 & 3 \end{bmatrix} \begin{bmatrix} S_B \\ S_C \\ S_D \\ S_E \end{bmatrix} \tag{124}$$

By Eq. (124) and transformation matrix (30), the voltage component in the stationary coordinate system can be rewritten as:

$$\begin{bmatrix} u_{\alpha 2}^{A} \\ u_{\beta 2}^{A} \\ u_{z2}^{A} \end{bmatrix} = \begin{bmatrix} 0.2336 & -0.2336 & -0.2336 & 0.2336 \\ 0.3804 & 0.2351 & -0.3804 & -0.2351 \\ -0.2351 & 0.3804 & -0.3804 & 0.2351 \end{bmatrix} \begin{bmatrix} S_B \\ S_C \\ S_D \\ S_E \end{bmatrix} \tag{125}$$

The corresponding vector distribution is shown in Fig. 24.

(2) Fault-tolerant switching table

In order to achieve the best control effect of torque and flux under fault-tolerant condition, the key of fault-tolerant control is to rationally divide sectors and select the best voltage vector. In the case of phase A fault, there are five kinds of voltage

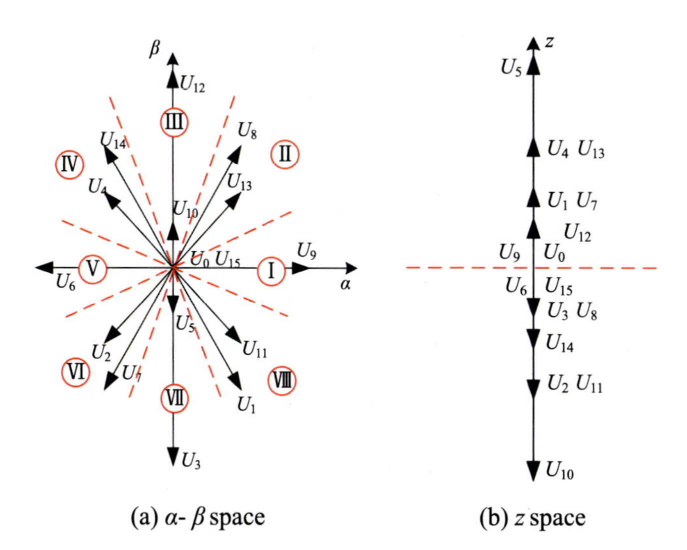

Fig. 24 Voltage vectors under single-phase open-circuit fault

Table 7 Fault-tolerant Switching Table

$\Delta\psi_s$	ΔT_e	Δi_z	Sector							
			I	II	III	IV	V	VI	VII	VIII
+1	+1	≥0	U_8	U_{10}	U_{14}	U_6	U_2	U_3	U_{11}	U_9
		<0	U_{13}	U_{12}	U_4	U_6	U_7	U_5	U_1	U_9
+1	−1	≥0	U_{11}	U_9	U_8	U_{10}	U_{14}	U_6	U_2	U_3
		<0	U_1	U_9	U_{13}	U_{12}	U_4	U_6	U_7	U_5
−1	+1	≥0	U_{14}	U_6	U_2	U_3	U_{11}	U_9	U_8	U_{10}
		<0	U_4	U_6	U_7	U_5	U_1	U_9	U_{13}	U_{12}
−1	−1	≥0	U_2	U_3	U_{11}	U_9	U_8	U_{10}	U_{14}	U_6
		<0	U_7	U_5	U_1	U_9	U_{13}	U_{12}	U_4	U_6

vectors with different sizes in α–β space (zero vector excluded). 8 large voltage vectors can be selected to divide the sectors into 8 sectors, as shown in Fig. 24a. z space is only divided into upper and lower parts [33, 34], as shown in Fig. 24b.

Then, it is necessary to determine the sector where the stator voltage vector is located. And, the appropriate space voltage vector is selected according to the state of ΔT_e and $\Delta\psi_s$ to satisfy the control of torque and flux linkage [35]. Due to the small leakage inductance of PMSM, the stator voltage vector u_{z2}^A is approximately in phase with the third harmonic current vector i_{z2}^A. Therefore, the voltage vector u_{z2}^A selected in the third harmonic space is reversed to the third harmonic current vector i_{z2}^A, so that the third harmonic current error Δi_z can be suppressed to 0. In normal operation, both flux and torque hysteresis comparators are three-stage comparators. Flux and torque have three states: increasing, decreasing, and keeping constant. However, only two-level comparators of increase and decrease states are used for fault tolerance. To simplify the switching table, only a two-stage flux hysteresis comparator and a two-stage torque hysteresis comparator are used, as shown in Table 7.

(3) Control block diagram

Figure 25 shows the control block diagram of direct torque control under single-phase open-circuit fault.

5 Short-Circuit Fault-Tolerant Theory and Methods

For a multi-phase motor drive system, short-circuit faults are relatively common. Especially when short-circuit faults occur at the end of stator windings, it will pose a serious threat to the system, which is manifested in sharp current increasing and severe torque ripples. Taking the probability and harmfulness of fault occurrence into consideration, single-phase short-circuit fault and interphase short-circuit fault are discussed in this paper. As shown in Fig. 26, the impact of short-circuit

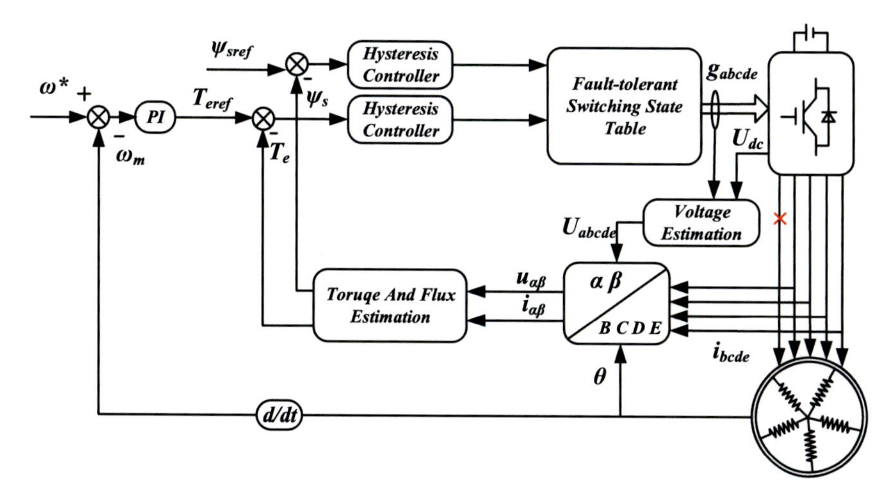

Fig. 25 Control block diagram of DTC under single-phase open-circuit fault

faults on torque can be divided into two parts: torque ripples caused by the lack of normal currents of faulty phases and torque fluctuations resulting from the generation of short-circuit current of faulty phases. Therefore, the short-circuit fault-tolerant currents are the combination of remedial currents derived under open-circuit fault and compensatory currents used to eliminate the influence of short-circuit current. In this section, the second kind of fault-tolerant matrix is adopted.

(1) Mathematical model

Taking single-phase short-circuit fault as an example, the open-circuit fault-tolerant currents are consistent with Eq. (66). On the other hand, to eliminate the influence of single-phase short-circuit current, the compensatory currents should be derived and the corresponding equivalent motor mathematical model is established as Fig. 27.

The compensatory currents added into the remaining non-faulty phases must observe two constraints: First, the sum of the magnetomotive force generated by

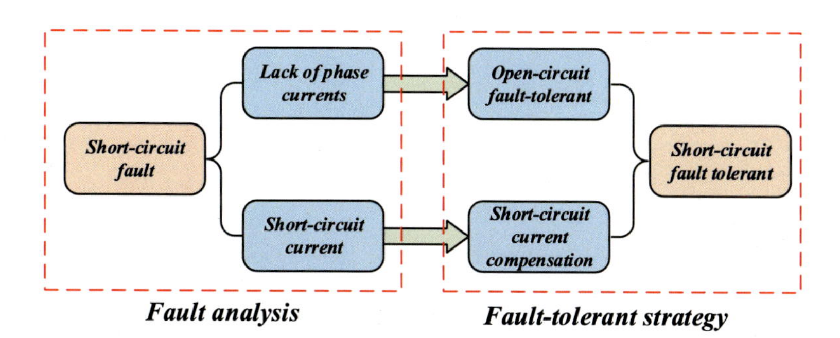

Fig. 26 Short-circuit fault-tolerant method

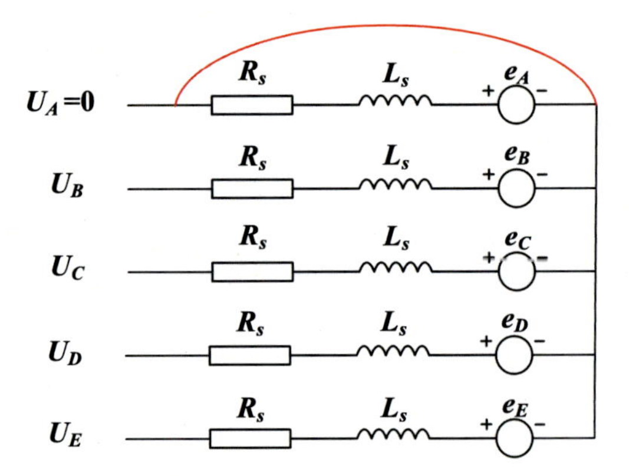

Fig. 27 Equivalent motor mathematical model of single-phase short-circuit fault

the compensatory currents and short-circuit current should be zero [36]. Second, the sum of the extra compensatory currents should be equal to zero due to the star connection topology.

The two constraints under single-phase short-circuit fault can be expressed as:

$$\begin{cases} Ni_{sc}^{A} + \lambda Ni_{B}' + \lambda^2 Ni_{C}' + \lambda^3 Ni_{D}' + \lambda^4 Ni_{E}' = 0 \\ i_{B}' + i_{C}' + i_{D}' + i_{E}' = 0 \end{cases} \tag{126}$$

where I_x' $(x = A, B, C, D, E)$ is the corresponding phase compensatory current of short-circuit fault; I_{sc}^{A} is short-circuit current of phase A.

According to the two constraints, the phase compensatory currents can be calculated as:

$$\begin{cases} i_{B}' = -0.447 I_{sc}^{A} \\ i_{C}' = 0.447 I_{sc}^{A} \\ i_{D}' = 0.447 I_{sc}^{A} \\ i_{E}' = -0.447 I_{sc}^{A} \end{cases} \tag{127}$$

Hence, the total single-phase short circuit fault-tolerant currents can be expressed as the combination of Eqs. (66) and (127), which are indicated as follows:

$$\begin{cases} i_{B}* = 1.382 I_m \cos(\theta_e - 0.5\alpha) - 0.447 I_{sc}^{A} \\ i_{C}* = 1.382 I_m \cos(\theta_e - 2\alpha) + 0.447 I_{sc}^{A} \\ i_{D}* = 1.382 I_m \cos(\theta_e - 3\alpha) + 0.447 I_{sc}^{A} \\ i_{E}* = 1.382 I_m \cos(\theta_e - 4.5\alpha) - 0.447 I_{sc}^{A} \end{cases} \tag{128}$$

To simplify the compensatory process, he calculated phase compensatory currents will directly function on the reference currents. By applying the transformation matrix (70), the compensatory reference currents in α–β space can be obtained as:

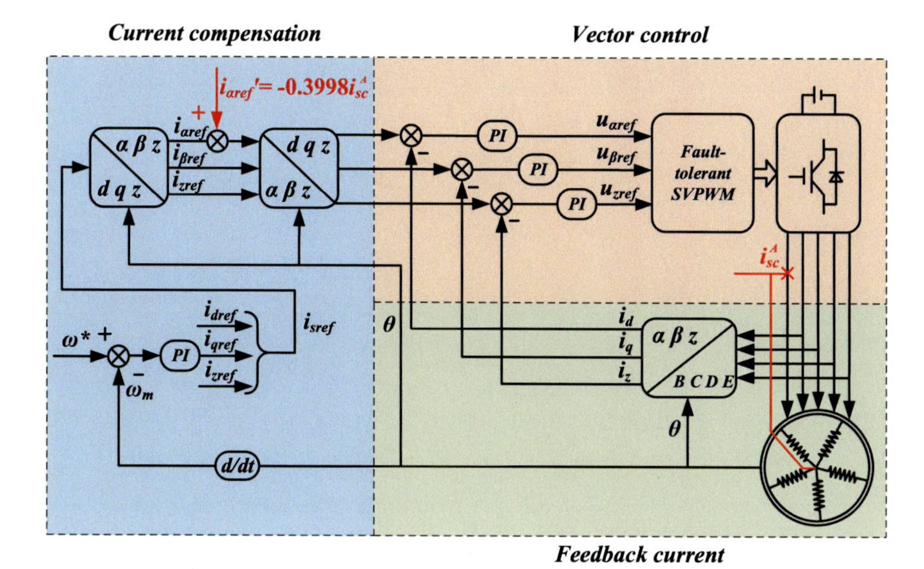

Fig. 28 Control block diagram of VC under single-phase short-circuit fault

$$\begin{cases} i'_{\alpha\text{ref}} = -0.3998 I_{sc}^{A} \\ i'_{\beta\text{ref}} = 0 \end{cases} \tag{129}$$

(2) Control block diagram

When compensatory currents function in the revised reference currents, the impact of short-circuit current will be offset. Combining open-circuit fault-tolerance and short-circuit compensatory currents, short-circuit fault-tolerance can be realized. Taking vector control as an example, Fig. 28 shows the control block diagram under single-phase short-circuit fault.

6 Maximum-Torque-per-Ampere Control

Interior permanent magnet synchronous motor (IPMSM) embedded the permanent magnet inside the motor rotor, which greatly improves the reliability. Moreover, the asymmetric rotator structure makes the *d*–*q* axis inductance of IPMSM unequal. Thus, the output torque of IPMSM contains two components: permanent magnet and reluctance. The permanent magnet torque is related to the permanent magnet and the *q*-axis current, while the reluctance torque is related to the *d*–*q* axis inductance difference and the *d*–*q* axis current. Therefore, when the current amplitude is constant, different output torques will be obtained due to different current vector angles. The largest output point is the target operating condition of Maximum Torque Per Ampere

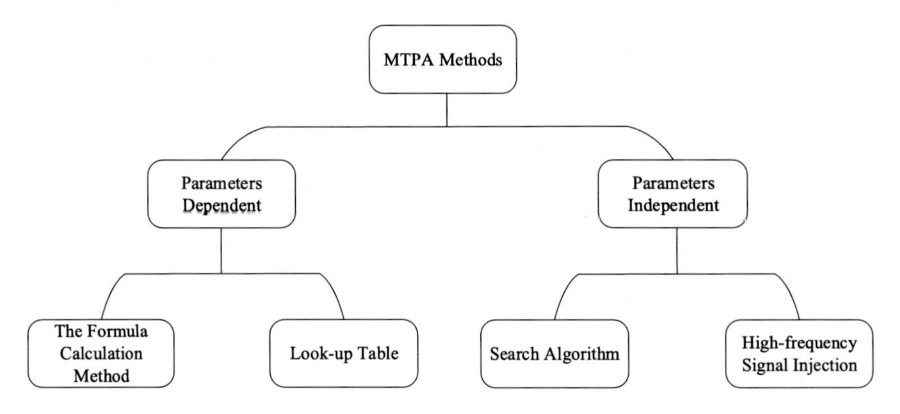

Fig. 29 Classification of the MTPA control methods

(MTPA). The goal of MTPA control is to keep the motor working at this maximum torque output point.

Since IPMSM has no copper consumption of the rotors, MTPA control can minimize the current and therefore the copper consumption under the same torque. The iron and copper consumption of IPMSM is negligible. Therefore, the MTPA point is very close to the maximum efficiency point of the motor, so MTPA control has a great advantage in optimizing the efficiency of the motor.

The MTPA control strategy takes the maximum output torque as the control goal, and realizes the maximum torque output of the same current or the minimum armature current corresponding to the same load by seeking the optimal current vector angle. At present, many scholars have mass research on MTPA control, and the relevant literature is also very rich. Because IPMSM is widely used in industry and MTPA control is excellent, MTPA algorithm also has many mature applications in industry. These methods can be summarized as parameters dependent and parameters independent. The main methods included are shown in Fig. 29.

As a relatively new MTPA control algorithm, high-frequency signal injection [37–46] has been proposed in recent years. The main methods are to inject an additional high-frequency small signal into the current vector to act on the motor, and then analyze the response signal, mainly the response signal in the torque. Compare the injection signal with the response signal to judge the current working condition, so as to adjust the current vector angle to reach the MTPA working condition.

In the actual system, it is difficult to obtain the real-time torque change, and the high-precision sensor is expensive. Therefore, many kinds of literature seek other ways to indirectly represent the change of torque signal. Reference [37] replaces the response signal in torque by measuring the change of speed, but the change of speed is affected by the bandwidth of speed loop, which requires additional consideration for the parameter design of the regulator. In Ref. [38], the change of torque signal is replaced by the change of motor output power. The change of power is affected by both voltage and current, and will be affected by the inherent harmonic in voltage and current. Reference [39] combines high-frequency signal injection with DTC control to realize MTPA control under DTC algorithm. By injecting a signal into the reference flux, the change in torque is replaced by the change in current. This scheme is first implemented in DTC, because DTC adopts bang-bang regulator, which has rich harmonic content, which is not conducive to signal analysis, and the performance of the algorithm is closely related to the accuracy of flux observer. The stability analysis of signal injection algorithm is proved in Refs. [40, 41]. These methods first inject an additional high-frequency small signal, then process and extract the injected signal by using the relationship between torque/flux and current vector angle/current, and finally achieve the operating condition of MTPA by adjusting the control system. It should be noted that when the high-frequency small signal is injected into the motor system as an actual signal, it will bring additional loss and torque ripple. Therefore, Refs. [42, 43] proposed an MTPA control method based on virtual signal injection control (VSI). The signal injection part of the algorithm is a processing module independent of the motor control system. Through separate signal processing, the given value of the motor is finally obtained and acted in the system, which avoids the disadvantages caused by the actual signal injection. Although the control method does not depend on the parameters of the motor, the establishment of mathematical model is very important. Reference [44] introduced the method of space vector signal injection (SVSI), which injects high-frequency signals into the two-phase stationary coordinate system and tracks MTPA points through flux and current amplitude to achieve high-precision tracking of MTPA angles. References [45, 46] introduced SVSI-MTPA method and VSI-MTPA method based on fault-tolerant control, respectively.

The mathematical model of the motor under fault-tolerant control has been introduced. Next, two MTPA control methods based on signal injection are mainly introduced, including space vector signal injection and virtual signal injection. Then, the virtual signal injection method is taken as an example, and the VSI-MTPA control under fault-tolerant will be analyzed.

6.1 Principle of MTPA Algorithm

Figure 30 shows the relationship between current amplitude and current vector angle of five-phase IPMSM. From the constant torque curve, it can be seen that there is a point to minimize the amplitude of stator current. At this time, the ratio of unit torque to current at this point is the largest, that is, MTPA point, the corresponding current vector angle is MTPA angle, and the curve composed of different MTPA points is MTPA trajectory curve. The purpose of MTPA control is to control the current vector angle at the MTPA point under different loads, so that the motor can output the minimum stator current under the same load and obtain the maximum operation efficiency.

The torque expression of five-phase IPMSM in rotating coordinate system is

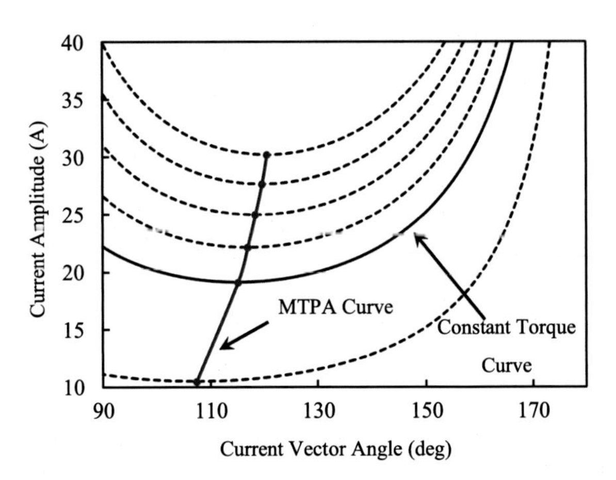

Fig. 30 Constant torque curve of five-phase IPMSM

$$T_e = \frac{5}{2}p\left[\psi_f i_q + (L_d - L_q)i_d i_q\right] \tag{130}$$

where i_d, i_q, and L_d, L_q are the components of stator current and inductance in d–q axes respectively; ψ_f is the permanent magnet flux linkage and p is the pole pairs.

The components of stator current i_d and i_q are expressed in the form of current angle as:

$$\begin{cases} i_d = i_s \cos\beta \\ i_q = i_s \sin\beta \end{cases} \tag{131}$$

where i_s represents the amplitude of stator current; β represents the current vector angle.

Then the torque expression can be written as:

$$T_e = \frac{5}{2}p\left[\psi_f i_s \sin\beta + \frac{1}{2}(L_d - L_q)i_s^2 \sin 2\beta\right] \tag{132}$$

At the same time, it can be seen from Fig. 30 that torque T_e and current vector angle β can fit the following relationships:

$$\begin{cases} \frac{\partial T_e}{\partial\beta} > 0(\beta < \beta_{\mathrm{MTPA}}) \\ \frac{\partial T_e}{\partial\beta} = 0(\beta = \beta_{\mathrm{MTPA}}) \\ \frac{\partial T_e}{\partial\beta} > 0(\beta > \beta_{\mathrm{MTPA}}) \end{cases} \tag{133}$$

When $\partial T_e/\partial\beta = 0$, the MTPA angle can be calculated by Eq. (132)

$$\beta = \cos^{-1}\frac{-\psi_f + \sqrt{\psi_f^2 + 8(L_d - L_q)^2 i_s^2}}{4(L_d - L_q)i_s} \tag{134}$$

It can be seen from Eq. (134) that it includes motor parameters such as permanent magnet flux amplitude and stator inductance. Considering the influence of parameter changes, the tracking accuracy of MTPA point will become worse. Therefore, how to according to torque T_e and current vector angle β. It is very important to obtain the accurate MTPA angle through appropriate methods.

In addition, another MTPA determination method based on flux mainly relies on the electromagnetic torque equation as follows:

$$T_e(\psi, i) = \psi \times i = |\psi| \cdot |i| \sin(\beta - \delta) \tag{135}$$

The output torque is expressed as a vector multiplication of flux and current. Where $|\psi|$ and δ is the amplitude and angle of the stator flux linkage; $|i|$ and β is current vector of the amplitude and angle.

Figure 31 shows the relationship between current vector angle and stator flux amplitude on a constant torque curve [44]. Points A, B, and C represent the three conditions controlled by MTPA. Point A represents $\beta < \beta_{\text{MTPA}}$, which is before the MTPA point, point B represents $\beta = \beta_{\text{MTPA}}$, which is the MTPA point, and point C represents $\beta > \beta_{\text{MTPA}}$, which is after the MTPA point. The current vector angle increases from 90° to 180°, and the stator flux amplitude in the figure decreases from 0.014 to 0.06. Using this property, the current vector angle in the traditional algorithm can be replaced by the stator flux amplitude.

In the three cases controlled by MTPA (point A, B, C in Fig. 31), the partial derivative of current amplitude with respect to flux amplitude can be expressed as:

$$\frac{d|i|}{d|\psi|}\begin{cases} > 0 \;\; (\beta < \beta_{\text{MTPA}}) \\ = 0 \;\; (\beta = \beta_{\text{MTPA}}) \\ < 0 \;\; (\beta > \beta_{\text{MTPA}}) \end{cases} \tag{136}$$

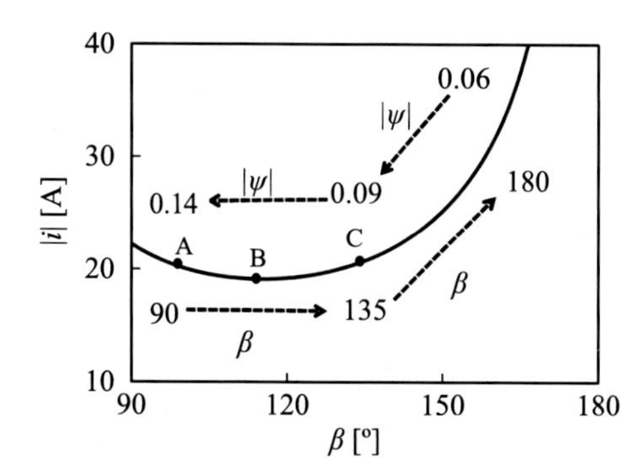

Fig. 31 Relation between stator flux amplitude and current vector angle at constant torque

6.2 Principle of SVSI-MTPA Algorithm

In Ref. [44], IPMSM can be regarded as inductive load. The voltage vector leads the current vector by 90°, ignoring the voltage drop on the resistor, so the angle difference between the current vector and the flux vector is less than 90°. Thus, in steady state, if β changes, the flux amplitude will change oppositely. In other words, as β increases, $\sin(\beta-\delta)$ increases. Under the constraint of constant load, flux amplitude will decrease. Similarly, when β decreases, flux amplitude increases. The same properties as shown in the figure.

When the flux amplitude contains a disturbance signal, assume that the disturbance is represented as follows:

$$\Delta\psi = A\sin(\omega_h t) \tag{137}$$

The amplitude of the disturbed current can be expressed as:

$$|i(\psi + \Delta\psi)| \approx |i| + A\frac{d|i|}{d|\psi|}\sin(\omega_h t) \tag{138}$$

By first-order low-pass filter and first-order high-pass filter with time constant, the flux and current signal of specific frequency ω_h are extracted and multiplied:

$$\begin{aligned} K\frac{d|i|}{d|\psi|}A\sin^2(\omega_h t) &= K\frac{d|i|}{d|\psi|}A\left\{\frac{1}{2}[\cos(0) - \cos(2\omega_h t)]\right\} \\ &= \frac{1}{2}KA\frac{d|i|}{d|\psi|} - \frac{1}{2}KA\frac{d|i|}{d|\psi|}\cos(2\omega_h t) \end{aligned} \tag{139}$$

The MTPA factor ($d|i|/d|\psi|$) appears on the right equation, it is extracted by a low-pass filter with a cut-off frequency below ω_h. Eventually the criterion of signal input is a pure integrator to adjust the current vector angle. The relevant signal processing block diagram is shown in Fig. 32.

Although the SVSI-MTPA has strong robustness and high precision tracking performance, the high-frequency signal injected into the control system will

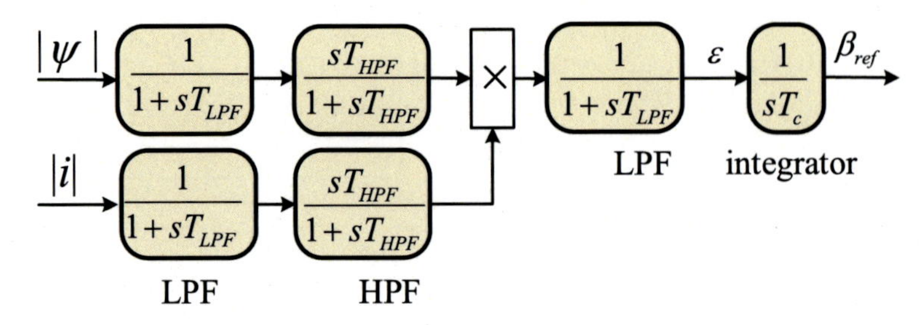

Fig. 32 High-frequency signal processing block diagram

lead to extra torque pulsation and loss, thus affecting the performance of fault-tolerant control. Therefore, the VSI-MTPA control method is more combined with fault-tolerant control.

6.3 *Principle of SVSI-MTPA Algorithm*

In Ref. [42], a high-frequency sinusoidal small signal will be injected into the current vector angle, as shown in Eq. (140):

$$\Delta\beta = A\sin(\omega_h t) \tag{140}$$

where A is the amplitude of the injection signal; ω_h is the frequency of the injected signal.

Then Eq. (130) can be expressed as:

$$T_e^h = \frac{5}{2}p\left[\psi_f i_s\cos(\beta + A\sin(\omega_h)) - \frac{1}{2}(L_d - L_q)i_s^2\sin 2(\beta + A\sin(\omega_h))\right] \tag{141}$$

where T_e^h represents the torque after high-frequency injection.

Using Taylor series expansion, we can get:

$$T_e^h(\beta + A\sin(\omega_h t)) = T_e(\beta) + \frac{\partial T_e}{\partial\beta}A\sin(\omega_h t) + \frac{1}{2}\frac{\partial}{\partial\beta}\left(\frac{\partial T_e}{\partial\beta}\right)A^2\sin^2(\omega_h t) + \cdots \tag{142}$$

where $T_e(\beta)$ represents the torque without high-frequency signal injection. Considering that the high-frequency torque expression of Eq. (141) is affected by motor parameters, it will be expressed by power expression.

When ignoring the iron loss of the motor, the output power of five-phase IPMSM can be expressed as follows:

$$T_e^h(\beta + A\sin(\omega_h t)) = T_e(\beta) + \frac{\partial T_e}{\partial\beta}A\sin(\omega_h t) + \frac{1}{2}\frac{\partial}{\partial\beta}\left(\frac{\partial T_e}{\partial\beta}\right)A^2\sin^2(\omega_h t) + \cdots \tag{143}$$

The output torque of the motor can be expressed as:

$$P_m = \frac{5}{2}[(u_d - R_s i_d)i_d + (u_q - R_s i_q)i_q] \tag{144}$$

$$\begin{cases} \frac{(u_q - R_s i_q)}{\omega_m} = p\psi_f \\ \frac{(u_d - R_s i_d)}{i_q \omega_m} = p(L_d - L_q) \end{cases} \tag{145}$$

The torque expression with the injected high-frequency signal can be expressed as:

$$T_e^h = \frac{5}{2}\left[\frac{(u_q - R_s i_q)}{\omega_m} + \frac{(u_d - R_s i_d)}{i_q \omega_m} i_d^h\right] i_q^h \tag{146}$$

where i_d^h and i_q^h respectively represent the current in d–q axes after injecting the high-frequency signal.

In order to improve the injection accuracy [43], the expression of high-frequency torque after complete signal injection can be obtained:

$$T_e^h = \frac{5}{2}\left[\frac{(u_q - R_s i_q)}{\omega_m} - pL_d i_d + \frac{(u_d - R_s i_d)}{i_q \omega_m} i_d^h + pL_d i_d^h\right] i_q^h \tag{147}$$

Figure 33 shows the signal processing module used to extract $\partial T_e/\partial \beta$. Firstly, a band-pass filter is set to remove the high-frequency torque component. Secondly, the high-frequency torque component is multiplied by the high-frequency small signal, and its expansion is shown in Eq. (148). Among them, MTPA factor can be obtained by a low-pass filter. The fixed relationship between torque and current vector angle has been given in Eq. (133). When the current vector angle output by the system is greater than or less than the MTPA angle, the output result is always a non-zero value, that is, there is a corresponding error. The error can be eliminated through the adjustment of an integrator, so that the current vector angle output by the system is equal to the MTPA angle.

$$k\frac{\partial T_e}{\partial \beta} A \sin^2(\omega_h t) = \frac{1}{2} kA\frac{\partial T_e}{\partial \beta} - \frac{\partial T_e}{\partial \beta} kA\cos(2\omega_h t) \tag{148}$$

Figure 34 shows the overall block diagram of the virtual signal injection module. Because of this method, the signal injection link in the algorithm is a processing module independent of the motor control system. Through separate signal processing, the given value of the motor is finally obtained and acted in the system, avoiding the disadvantages caused by actual signal injection. Therefore, it can be called virtual

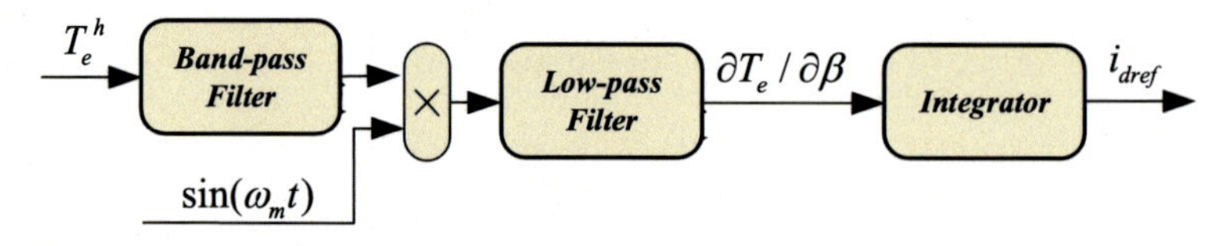

Fig. 33 Schematic of signal processing block to extract $\partial T_e/\partial \beta$

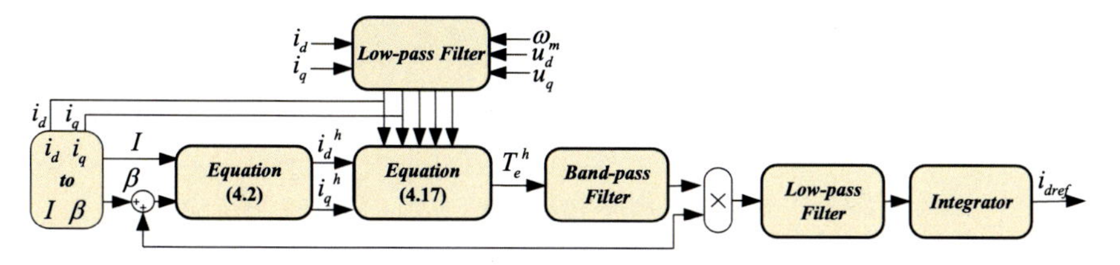

Fig. 34 Overall block diagram of the virtual signal injection module

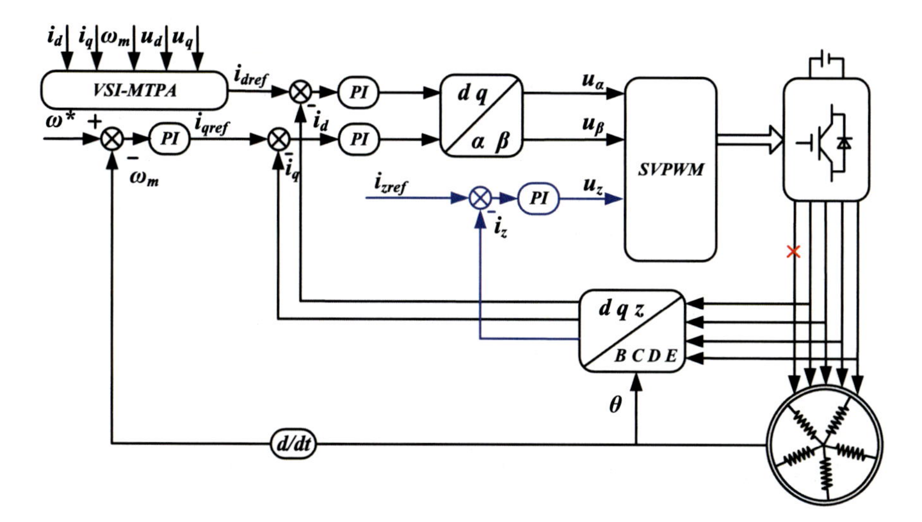

Fig. 35 Fault-tolerant MTPA control scheme under phase-A open-circuit fault

signal injection method. The VSI-MTPA is combined with fault-tolerant control, and the overall control block diagram is shown in Fig. 35.

References

1. Welchko BA, Lipo TA, Jahns TM et al (2004) Fault tolerant three-phase AC motor drive topologies: a comparison of features, cost, and limitations. IEEE Trans Power Electron 19(4):1108–1116
2. Zhao Y, Lipo TA (1996) Modeling and control of a multiphase induction machine with structural unbalance. IEEE Trans Energy Convers 11(3):570–577
3. Fu JR, Lipo TA (1994) Disturbance free operation of a multiphase current regulated motor drive with an opened phase. IEEE Trans Ind Appl 30(5):1267–1274
4. Kumar R, Das S, Syam P et al (2015) Review on model reference adaptive system for sensorless vector control of induction motor drives. IET Electr Power Appl 9(7):496–511
5. Oliveira CMR, Aguiar ML, Monteiro JRBA et al (2016) Vector control of induction motor using an integral sliding mode controller with anti-windup. J Control Autom Electr Syst 27(2):169–178

6. Liu C, Li Y, Zhen Z (2017) Control and drive techniques for multiphase machines: a review. Proc CSEE 32(24):17–29
7. Wang X, Wang Z, Cheng M et al (2017) Remedial strategies of T-NPC three-level asymmetric six-phase PMSM drives based on SVM-DTC. IEEE Trans Ind Electron 64(9):6841–6853
8. Wang Z, Wang X, Cao J et al (2017) Direct torque control of T-NPC inverters fed double-stator-winding PMSM drives With SVM. IEEE Trans Power Electron 33(2):1541–1551
9. Kouro S, Cortes P, Vargas R et al (2009) Model predictive control—a simple and powerful method to control power converters. IEEE Trans Ind Electron 56(6):1826–1838
10. Barrero F, Arahal MR, Gregor R et al (2009) A proof of concept study of predictive current control for VSI-driven asymmetrical dual three-phase AC machines. IEEE Trans Ind Electron 56(6):1937–1954
11. Buso S, Fasolo S, Malesani L, Mattavelli P (2000) A dead-beat adaptive hysteresis current control. IEEE Trans Ind Appl 36(4):1174–1180
12. Pan C, Chang T (1994) An improved hysteresis current controller for reducing switching frequency. IEEE Trans Power Electron 9(1):97–104
13. Dawande MS, Kanetkar VR, Dubey GK (1996) Three-phase switch mode rectifier with hysteresis current control. IEEE Trans Power Electron 11(3):466–471
14. Baranwal R, Basu K, Mohan N (2015) Carrier-Based implementation of SVPWM for dual Two-Level VSI and dual matrix converter with zero common-mode voltage. IEEE Trans Power Electron 30(3):1471–1487
15. Li X, Dusmez S, Akin B, Rajashekara K (2016) A new SVPWM for the phase current reconstruction of three-phase three-level T-type converters. IEEE Trans Power Electron 31(3):2627–2637
16. He K, Li J, Xiao L, Xiong Y, Wu L (2021) Randomized pulse pattern strategy of synchronized SVPWM for low-frequency-ratio applications. IEEE Trans Power Electron 36(6):6404–6414
17. Wang J, Atallah K, Howe D (2003) Optimal torque control of fault-tolerant permanent magnet brushless machines. IEEE Trans Magn 39(5):2962–2964
18. Sun Z, Wang J, Jewell GW et al (2010) Enhanced optimal torque control of fault-tolerant PM machine under flux-weakening operation. IEEE Trans Industr Electron 57(1):344–353
19. Sen B, Wang J (2016) Stationary frame fault-tolerant current control of polyphase permanent-magnet machines under open-circuit and short-circuit faults. IEEE Trans Power Electron 31(7):4684–4696
20. Dwari S, Parsa L (2008) An optimal control technique for multiphase pm machines under open-circuit faults. IEEE Trans Industr Electron 55(5):1988–1995
21. Mohammadpour A, Sadeghi S, Parsa L (2014) A generalized fault-tolerant control strategy for five-phase PM motor drives considering star, pentagon, and pentacle connections of stator windings. IEEE Trans Industr Electron 61(1):63–75
22. Mohammadpour A, Parsa L (2015) Global fault-tolerant control technique for multiphase permanent-magnet machines. IEEE Trans Ind Appl 51(1):178–186
23. Bianchi N, Bolognani S, Pre MD (2007) Strategies for the fault-tolerant current control of a five-phase permanent-magnet motor. IEEE Trans Ind Appl 43(4):960–970
24. Dwari S, Parsa L (2011) Fault-tolerant control of five-phase permanent-magnet motors with trapezoidal back EMF. IEEE Trans Industr Electron 58(2):476–485
25. Mohammadpour A, Parsa L (2013) A unified fault-tolerant current control approach for five-phase PM motors with trapezoidal back EMF under different stator winding connections. IEEE Trans Power Electron 28(7):3517–3527
26. Zhao P, Yang G, Li Y (2011) Fault-tolerant control strategy for five-phase permanent magnetic synchronous motor under single phase open-circuit fault condition. Proc CSEE 31(24):68–76
27. Gao H, Yang G (2016) Modeling and control of five-phase permanent magnet synchronous motor with one phase open-circuit fault. Trans China Electrotech Soc 31(20):93–101
28. Tian B, An Q, Duan J et al (2017) Decoupled modeling and nonlinear speed control for five-phase PM motor under single-phase open fault. IEEE Trans Power Electron 32(7):5473–5486
29. Liu G, Gao M, Zhou H et al (2019) Flux-modification-based fault-tolerant DTC for five-phase PMSM. Proc CSEE 39(2):359–365+633

30. Liu G, Lin Z, Zhao W, Chen Q, Xu G (2018) Third harmonic current injection in fault-tolerant five-phase permanent-magnet motor drive. IEEE Trans Power Electron 33(8):6970–6979
31. Zhou H, Zhao W, Liu G, Cheng R, Xie Y (2017) Remedial field-oriented control of five-phase fault-tolerant permanent-magnet motor by using reduced-order transformation matrices. IEEE Trans Industr Electron 64(1):169–178
32. Liu G, Song C, Chen Q (2020) FCS-MPC-based fault-tolerant control of five-phase IPMSM for MTPA operation. IEEE Trans Power Electron 35(3):2882–2894
33. Tao T, Zhao W, Du Y, Cheng Y, Zhu J (2020) Simplified fault-tolerant model predictive control for a five-phase permanent-magnet motor with reduced computation burden. IEEE Trans Power Electron 35(4):3850–3858
34. Tao T, Zhao W, He Y, Cheng Y, Saeed S, Zhu J (2021) Enhanced fault-tolerant model predictive current control for a five-phase PM motor with continued modulation. IEEE Trans Power Electron 36(3):3236–3246
35. Zhou H, Xu J, Chen C, Tian X, Liu G (2021) Disturbance-observer-based direct torque control of five-phase permanent magnet motor under open-circuit and short-circuit faults. IEEE Trans Industr Electron 68(12):11907–11917
36. Zhou H, Liu G, Zhao W, Yu X, Gao M (2018) Dynamic performance improvement of five-phase permanent-magnet motor with short-circuit fault. IEEE Trans Industr Electron 65(1):145–155
37. Bolognani S, Petrella R, Prearo A et al (2011) Automatic tracking of MTPA trajectory in IPM motor drives based on AC current injection. IEEE Trans Ind Appl 47(1):105–114
38. Kim S, Yoon Y, Sul S et al (2013) Maximum torque per ampere (MTPA) control of an IPM machine based on signal injection considering inductance saturation. IEEE Trans Power Electron 28(1):488–497
39. Bolognani S, Peretti L, Zigliotto M (2011) Online MTPA control strategy for DTC synchronous-reluctance-motor drives. IEEE Trans Power Electron 26(1):20–28
40. Antonello R, Carraro M, Zigliotto M (2014) Maximum-torque-per-ampere operation of anisotropic synchronous permanent-magnet motors based on extremum seeking control. IEEE Trans Ind Electron 61(9):5086–5093
41. Bedetti N, Calligaro S, Olsen C et al (2017) Automatic MTPA tracking in IPMSM drives: loop dynamics, design, and auto-tuning. IEEE Trans Ind Appl 53(5):4547–4558
42. Sun T, Wang J, Chen X (2015) Maximum torque per ampere (MTPA) control for interior permanent magnet synchronous machine drives based on virtual signal injection. IEEE Trans Power Electron 30(9):5036–5045
43. Sun T, Wang J (2015) Extension of virtual-signal-injection-based MTPA control for interior permanent-magnet synchronous machine drives into the field-weakening region. IEEE Trans Ind Electron 62(11):6809–6817
44. Liu G, Wang J, Zhao W et al (2017) A novel MTPA control strategy for IPMSM drives by space vector signal injection. IEEE Trans Ind Electron 64(12):9243–9252
45. Chen Q, Gu L, Lin Z et al (2020) Extension of space-vector-signal-injection-based MTPA control into SVPWM fault-tolerant operation for five-phase IPMS. IEEE Trans Ind Electron 67(9):7321–7333
46. Chen Q, Zhao W, Liu G et al (2019) Extension of virtual-signal-injection-based MTPA control for five-phase IPMSM into fault tolerant operation. IEEE Trans Ind Electron 66(2):944–955

Modeling and Simulation of Electric Motors

Jie Mei

Abstract This chapter introduces basic procedures for designing and optimizing an electric motor for electric vehicle applications using the example of an axial flux induction motor. The analytical model for the motor is derived firstly, which can provide output values close to that provided by finite element analysis in ANSYS Maxwell with a much shorter simulation time. Then, based on the developed analytical model, the corresponding optimization and solution will be formulated to help optimize the motor to meet all design requirements and achieve excellent performance. Finally, the optimized motor design will be validated by ANSYS Maxwell.

1 Introduction

Over the past few decades, electric vehicles (EVs) have grown in proportion in the commercial market, primarily for the purpose of reducing pollution and improving the driving experience [1]. The first modern cars in the last century were powered by electric motors rather than internal combustion engines. But due to poor battery performance and cheap gasoline at the time, nearly the automobile market was dominated by combustion-engine vehicles. Growing concerns about global warming and the depletion of fossil fuels, along with dramatic improvements in battery performance, have led to a resurgence of EVs.

The core of an EV is the electric motor, which converts electrical energy into mechanical energy to drive the wheels [2]. Electric motors have the advantage of being more powerful, more energy-efficient, and more compact than classic internal combustion engines. To date, many different types of motors have been applied to EVs, such as induction motors (IM) [3–5], interior permanent magnet synchronous motors (IPM) [6–10], switched reluctance motors (SRM) [11–13], and switched flux motors (SFM) [14–16]. To further meet the power demands of modern EVs, more efficient and powerful electric motors need to be developed.

J. Mei (✉)
Research Laboratory of Electronics, Massachusetts Institute of Technology, Cambridge, USA
e-mail: jameymei.mit@gmail.com

C. H. T. Lee (ed.), *Emerging Technologies for Electric and Hybrid Vehicles*,
Green Energy and Technology, https://doi.org/10.1007/978-981-99-3060-9_6

Designing a motor that satisfies multiple objectives simultaneously is a nonlinear and complex optimization problem, subject to special design specifications. Design variables are coupled and changing one variable can cause big changes in the output values and violate design constraints. Traditionally, electric machines are designed in software such as ANSYS Maxwell through finite element analysis (FEA), where design constraints and initial specifications are verified based on input design variables. However, this FEA-based design approach is very inefficient and time-consuming, as simply simulating a potential design candidate often takes hours, especially considering the fact that special-structure motors such as axial-structure motors are difficult to perform 2D simulations thus usually done in 3D FEA, which results in even longer simulation times.

Considering those problems, the design process of modern EV motors cannot completely rely on FEA in certain software but requires the assistance of analytical models with high accuracy. The main process has the following three steps: (1) Derive the analytical model of the motor to be designed, which can provide results close to FEA in ANSYS Maxwell in a much shorter simulation time. (2) Based on the analysis model, formulate the corresponding optimization model and solution method to optimize the design to meet all design requirements while achieving excellent motor performance. (3) Use FEA in ANSYS Maxwell again to validate the final design. This chapter takes axial flux induction motor (AFIM) for EV applications as an example and describes in detail how to design and optimize an electric motor from the above three steps.

The rest of this chapter is organized as follows: in Sect. 2, the derived analytical model for AFIM is presented. Section 3 introduces the corresponding optimization formulation for AFIM and a solution method based on a genetic algorithm (GA) for finding the optimal design variables given design specifications. FEA results of flux density and motor performance in ANSYS Maxwell of the final selected AFIM design are presented in Sect. 4. Finally, Sect. 5 summarizes the chapter.

2 Analytical Model of AFIM

Among all the prevalent motors of EV applications, AFIM has attracted more attention for its special features, which include better power-to-weight ratio, compact structure, higher efficiency, especially in multi-pole motors, high utilization of materials, and better ventilation and cooling [17, 18]. In this section, the steps to derive the AFIM equivalent analytical model are detailed. Figure 1 shows a two-layer distributed-winding and double-stator-double-rotor AFIM to be designed and optimized in this work where the three-phase windings are plotted in red, blue, and yellow, respectively, the two rotors are drawn in light grey in the middle, and the two stators on two sides are in dark grey. Based on the analytical model, further motor analysis and optimal motor parameter selection can be greatly facilitated.

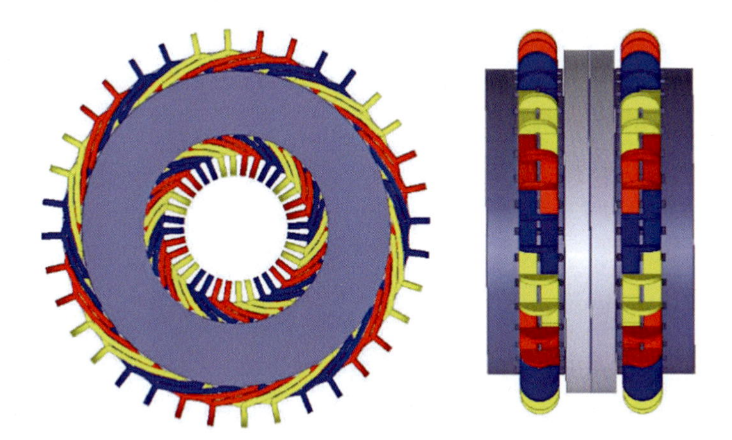

Fig. 1 Appearance of the AFIM from different angles

A. *Stator Synchronous Inductance*

Assume balanced currents are in both sets of windings as Eq. (1)–(3) below.

$$i_a = \mathrm{Re}\left(\underline{I}_S e^{j\omega t}\right) \tag{1}$$

$$i_b = \mathrm{Re}\left(\underline{I}_S e^{j(\omega t - 2\pi/3)}\right) \tag{2}$$

$$i_c = \mathrm{Re}\left(\underline{I}_S e^{j(\omega t + 2\pi/3)}\right) \tag{3}$$

For each set of windings, the fundamental magnetic flux density of its three-phase current can be expressed as:

$$B_S = \mathrm{Re}\left(\underline{B}_S e^{j(p\theta - \omega t)}\right) \tag{4}$$

$$\underline{B}_S = j\frac{3}{2}\frac{\mu_o}{g}\frac{4}{\pi}\frac{N_S \underline{I}_S}{2p} \tag{5}$$

Then the total linked flux for N_S full-pitched turns is:

$$\lambda = N_S \int_{r_i}^{r_o} \int_0^{\pi/p} B_S r d\theta dr = L_S i_a \tag{6}$$

From Eq. (6), the fundamental stator apparent self-inductance can be expressed as:

$$L_S = \frac{3}{2}\frac{4}{\pi}\frac{\mu_o N_S^2 k_S^2}{p^2 g}\frac{r_o^2 - r_i^2}{2} \quad (7)$$

Note that Eq. (7) contains k_S, which is the stator winding factor and is the product of the pitch factor k_p and the breadth factor k_b as expressed in Eq. (8) and Eqs. (9):

$$k_{pn} = \sin(n\frac{\pi}{2}) \cdot \sin(n\frac{\alpha p}{2}) = \sin(n\frac{\pi}{2}) \cdot \sin(n\frac{\beta\pi}{2}) \quad (8)$$

$$k_{bn} = \frac{\sin(\frac{np\gamma m}{2})}{m\sin(\frac{np\gamma}{2})} \quad (9)$$

where the shorted pitch angle α is expressed as $\alpha = \beta\pi/p$.

B. *Equivalent Rotor Current*

The rotor slots and rings are made of aluminum with steel inserted between the slots. Since the rotor is symmetric and each rotor slot has the same current with phase delayed according to the slot number k, the rotor current density can be expressed as the sum of impulses in Eq. (10):

$$K_r = \mathrm{Re}\left(\sum_{k=0}^{N_R-1} \frac{1}{r}\underline{I}e^{j\left(\omega_r - \frac{2\pi pk}{N_R}\right)}\delta\left(\theta\prime - \frac{2\pi k}{N_R}\right)\right) \quad (10)$$

which can also be expressed as a Fourier series as follows:

$$K_r = \mathrm{Re}\left(\sum_{n=-\infty}^{+\infty} \underline{K}_n e^{j(\omega_r - np\phi\prime)}\right) \quad (11)$$

$$\underline{K}_n = \frac{\underline{I}}{2\pi r}\left(\sum_{k=0}^{N_R-1} e^{j(n-1)\frac{2\pi pk}{N_R}}\right) \quad (12)$$

If $(n - 1)\cdot p/N_R$ is an integer, the summation part in the parentheses of Eq. (12) is evaluated to be N_R, otherwise zero. Integers of -1, 0, and 1 are chosen for the maximum magnetic field, which produces harmonics order of $n = 1$, and:

$$n_+ = \frac{N_R}{p} + 1, \quad n_- = \frac{N_R}{p} + 1 \quad (13)$$

Each of these three space harmonics will have surface currents of the form:

$$K_{rn} = \mathrm{Re}\left(\frac{N_R \underline{I}}{2\pi r}e^{j(\omega_r - np\phi\prime)}\right) \quad (14)$$

which results in a magnetic flux density of:

$$B_{zn} = \text{Re}\left(j\frac{\mu_o N_R \underline{I}}{2\pi npg} e^{j(\omega_r - np\phi\prime)} \right) \tag{15}$$

Based on the following angular transformation:

$$\theta\prime = \theta - \omega_m t \tag{16}$$

it shows that:

$$\omega_r t - p\theta\prime = \omega t - p\theta \tag{17}$$

Similar to Eq. (6), at $n = 1$, the linked flux to the armature winding due to this flux can be calculated as:

$$\lambda_{ag} = N_S \int_{r_i}^{r_o} \int_{0}^{\pi/p} B_{z1} r d\theta dr = \text{Re}\left(\underline{\Lambda}_{ag} e^{i\omega t}\right) \tag{18}$$

$$\underline{\Lambda}_{ag} = \frac{\mu_o N_R N_S k_S \left(r_o^2 - r_i^2\right)}{2\pi p^2 g} \underline{I} \tag{19}$$

Then the total stator flux per phase can be expressed as:

$$\underline{\Lambda}_a = (2L_S + L_l)\underline{I}_S + \frac{\mu_o N_R N_S k_S \left(r_o^2 - r_i^2\right)}{2\pi p^2 g} 2\underline{I} \tag{20}$$

Equation (20) motivates a definition of an equivalent rotor current $\underline{I}_R$ from the space fundamental of rotor surface current density, so the magnetic flux of different current sources can be calculated using the same inductance. From Eqs. (19) and (20) the link between $\underline{I}$ and $\underline{I}_R$ can be generated:

$$\underline{I}_R = \frac{N_R}{6N_S k_S} \underline{I} \tag{21}$$

Then Eq. (20) can be rewritten as:

$$\underline{\Lambda}_a = (2L_S + L_l)\underline{I}_S + 2L_S \underline{I}_R \tag{22}$$

C. *Equivalent Analytical Model*

According to Faraday's law, the induced electric field of the air-gap flux components in Eq. (15) can be expressed as:

$$E_{rn} = \text{Re}\left(\underline{E}_{rn} e^{j(\omega_r - np\phi\prime)}\right) \tag{23}$$

$$\underline{E}_{rn} = -j\frac{\mu_o N_R \omega_r r \underline{I}}{2\pi n^2 p^2 g} \tag{24}$$

For unit rotor slot impedance $\underline{Z}_{slot} = R_{slot} + j\omega_r L_{slot}$, the rotor slot inductance per unit length can be calculated as Eq. (25):

$$L_{slot} = \mu_o \frac{h_r}{3w_r} \tag{25}$$

The rotor slot resistance per unit length can be calculated as Eq. (26).

$$R_{slot} = \frac{1}{w_r h_r \sigma_r} \tag{26}$$

The induced voltage will push current through the slot and can be expressed as:

$$\underline{E}_{r1} + \underline{E}_{rn-} + \underline{E}_{rn+} = \underline{Z}_{slot}\underline{I} \tag{27}$$

Based on Eq. (27), the spatial fundamental electric field can be expressed as Eq. (28), where the second part comes from zigzag leakage component and $\xi(N_R)$ is a constant function of N_R.

$$\underline{E}_{r1} = \underline{Z}_{slot}\underline{I} + j\frac{\mu_o N_R \omega_r r}{2\pi g}\xi(N_R)\underline{I} \tag{28}$$

$$\xi(N_R) = \frac{1}{(N_R + p)^2} + \frac{1}{(N_R - p)^2} \tag{29}$$

The corresponding magnetic flux density generated by the fundamental space electric field can be expressed as:

$$\underline{B}_{z1} = \left(\underline{Z}_{slot} + j\frac{\mu_o N_R \omega_r r}{2\pi g}\xi(N_R)\right)\frac{\underline{I}p}{\omega_r r} \tag{30}$$

Similar to Eq. (6) and in combination with Eq. (21), the linked flux to the armature winding due to this flux can be calculated by:

$$\lambda_{ag} = N_S \int_{r_i}^{r_o} \int_0^{\pi/p} \text{Re}\left(\underline{B}_{z1} e^{i(\omega t - p\theta)}\right) r d\theta dr = \text{Re}\left(\underline{\Lambda}_{ag} e^{i\omega t}\right) \tag{31}$$

Then the air-gap voltage $\underline{V}_{ag} = j\omega\underline{\Lambda}_{ag}$ can be calculated as:

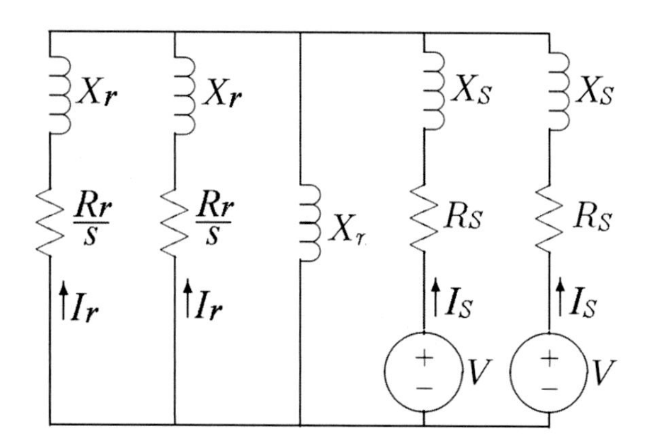

Fig. 2 Equivalent analytical model of the AFIM

$$\underline{V}_{ag} = -\underline{I}_R\left[\frac{12N_S^2k_S^2(r_o - r_i)}{N_R}\left(j\omega L_{slot} + \frac{R_2}{s}\right) + j\omega\frac{3\mu_o N_S^2k_S^2(r_o^2 - r_i^2)}{\pi g}\xi(N_R)\right] \tag{32}$$

where s is defined as per-unit slip, $s = \omega_r/\omega$. Note that the same relationship as in equation Eq. (20) can be written for the other stator. Combining Eq. (22) and Eq. (32), a per-phase equivalent circuit model can be generated, as shown in Fig. 2.

We can then make the following definitions for the components in the equivalent circuit model:

$$X_r = \omega\frac{12N_S^2k_S^2(r_o - r_i)}{N_R}L_{slot} + \omega\frac{3\mu_o N_S^2k_S^2(r_o^2 - r_i^2)}{\pi g}\xi(N_R) \tag{33}$$

$$X_m = \omega L_S \tag{34}$$

$$X_S = \omega L_l \tag{35}$$

$$R_r = \frac{12N_S^2k_S^2(r_o - r_i)}{N_R}R_{slot} \tag{36}$$

If rotor ring effects are considered, Eq. (36) can be further modified as Eq. (37).

$$R_r = \frac{12N_S^2k_S^2(r_o - r_i)}{N_R}R_{slot}\left[1 + \frac{2\pi(r_o - r_i)T_{ring}p^2}{N_r(r_o + r_i)w_r}\right] \tag{37}$$

The length of per-phase armature winding is calculated as follows:

$$l_w = 2N_S(r_o - r_i) + N_S l_{end\_o} + N_S l_{end\_i} \tag{38}$$

Given the number of strands in parallel N_{SIP} and the wire diameter d_w, the winding area A_w can be calculated:

$$A_w = \frac{\pi}{4} d_w^2 N_{SIP} \tag{39}$$

The winding resistance is expressed as:

$$R_S = \frac{l_w}{\sigma A_w} \tag{40}$$

D. *Stator Flux Leakage*

There are several main components of stator flux leakage, including belt flux leakage, zigzag flux leakage, end winding flux leakage, and slot leakage. All these components should add to the leakage impedance X_S. It is assumed that for this two-layer distributed-winding structure, the stator winding is short-pitched by N_{sp} slots. Given the slot dimensions, the stator slot leakage can be calculated as follows [19]:

$$X_{sl} = 2\omega(r_o - r_i)\mu_o(\frac{1}{3}\frac{h_s}{w_s} + \frac{d_s}{u_s})\frac{N_S^2}{p}\left(\frac{2}{m} - \frac{5}{4}\frac{N_{sp}}{m^2}\right) \tag{41}$$

The stator zigzag leakage components are due to the air-gap space harmonic orders $pn_s = N_{slots} \pm p$. Similar to Eq. (13), we can define n_{s+} and n_{s-}, and the corresponding winding factors.

$$X_{zl} = \frac{3}{2}\frac{4}{\pi}\omega\frac{\mu_o N_S^2}{g}\frac{r_o^2 - r_i^2}{2}\left(\frac{k_{n_{s+}}}{(N_{slots} + p)^2} + \frac{k_{n_{s-}}}{(N_{slots} - p)^2}\right) \tag{42}$$

Similar to zigzag leakage, the stator belt leakages are due to the harmonic orders of 5 and 7. Without rotor coupling, the belt leakages can be calculated as Eq. (43) by replacing n with 5 and 7:

$$X_{agn} = \frac{3}{2}\frac{4}{\pi}\omega\frac{\mu_o N_S^2 k_n^2}{n^2 p^2 g}\frac{r_o^2 - r_i^2}{2} \tag{43}$$

The belt harmonics link to the rotor appears to be parallel to the components of the rotor impedance. Again, ignoring rotor resistance, these components can be written as:

$$X_{r,n} = \frac{12\omega N_S^2 k_n^2 (r_o - r_i) L_{slot}}{N_R} + \frac{3\omega\mu_o N_S^2 k_n^2 (r_o^2 - r_i^2)}{\pi g}\xi(N_R) \tag{44}$$

$$\xi(N_R) = \frac{1}{(N_R + np)^2} + \frac{1}{(N_R - np)^2} \tag{45}$$

The belt leakage of the fifth and seventh harmonics can then be calculated as $X_{bln} = X_{agn} // X_{r,n}$. Corrected based on Alger's method in [20], the stator end winding leakage X_{el} can be calculated as:

$$X_{el} = \omega \frac{105\mu_o N_S^2}{8\pi^2 p^2} \frac{r_o - r_i}{2} (\alpha - 0.3) \tag{46}$$

Therefore, the total leakage impedance X_S due to flux leakages can be generated as:

$$X_S = X_{sl} + X_{el} + X_{bl} + X_{zl} \tag{47}$$

E. *Core Losses*

In this part, a simple estimation method will be provided to model core losses due to eddy current and hysteresis effects in the core. The loss density per unit weight in saturated steel can be modeled as:

$$P_d = P_B \left(\frac{\omega}{\omega_B}\right)^{\varepsilon_f} \left(\frac{B}{B_B}\right)^{\varepsilon_b} \tag{48}$$

where P_B is the known loss per unit weight of the selected steel at magnetic flux density B_B and frequency ω_B. The exponential coefficients ε_f and ε_b are chosen based on engineering experience with the material. In this study, ε_f and ε_b are set to be 1.53 and 1.88, respectively.

Similarly, the excitation volts-amperes dissipated per unit weight of the core can be modeled as:

$$Q_d = \left(Q_{B1}\left(\frac{B}{B_B}\right)^{\varepsilon_{v1}} + Q_{B2}\left(\frac{B}{B_B}\right)^{\varepsilon_{v2}}\right)\frac{\omega}{\omega_B} \tag{49}$$

where Q_{B1} and Q_{B1} are the known volts-amperes dissipated per unit weight of the selected steel at magnetic flux density B_B and frequency ω_B. The exponential coefficients ε_{v1} and ε_{v2} are also chosen based on engineering experience with the material. In this study, ε_{v1} and ε_{v2} are set to be 1.7 and 16.1, respectively. Given the design dimensions of the motor, the total weight of steel can be calculated. The total core power losses P_c and volts-amperes Q_c can then be calculated by multiplying the densities P_d and Q_d by the weight of the steel. The equivalent resistance and reactance can be calculated in Eq. (50) and Eq. (51) as follows:

$$R_c = \frac{3|V|^2}{P_c} \tag{50}$$

$$X_c = \frac{3|V|^2}{Q_c} \tag{51}$$

which can be placed in parallel with the air-gap reactance element X_m in the per-phase equivalent circuit model if the core losses are considered during the design process.

3 Optimization of AFIM

3.1 Optimization Formulation

Based on the proposed analytical model and the given design specifications, the design of AFIM can be formulated as the optimization below:

$$\text{OB1: } max\{P_{\max}/P_{in}\} \tag{52}$$

$$\text{OB2}: min\{M_S + M_A + M_C\} \tag{53}$$

subject to

$$\text{C1: } \frac{Vr_o}{k_S N_s \omega (r_o^2 - r_i^2) T_{iron}} < 2.1 \tag{54}$$

$$\text{C2}: \frac{2\pi r_i p V}{k_S N_S \omega (r_o^2 - r_i^2)(2\pi r_i - N_{slot} w_s)} < 2.1 \tag{55}$$

$$\text{C3}: M_S + M_A + M_C < 27 \tag{56}$$

$$\text{C4}: N_R w_r < 2\pi r_i \tag{57}$$

$$\text{C5}: N_{\text{slots}} w_s < 2\pi r_i \tag{58}$$

$$\text{C6}: 2(T_{iron} + h_s + d_s) + h_r + g < 0.16 \tag{59}$$

$$\text{C7: } \frac{4 \cdot \left|I_{S,P\max}\right| \cdot 10^{-6}}{N_{SIP} \pi d_w^2} < 26 \tag{60}$$

$$\text{C8: } P_{\max} > 80000 \tag{61}$$

$$\text{C9: } 6\frac{p}{\omega}\left|I_{R,P\max}\right|^2 \frac{R_2}{s} > 250 \tag{62}$$

$$\text{C10: } \omega_{P\max} < 3150 \tag{63}$$

$$\text{C11: } P_{\max}/P_{in} > 82\% \tag{64}$$

As in Eqs. (52) and (53), two example designing objectives (OBs) are shown. The first objective is that motor efficiency at the maximum output mechanical power, expressed as the ratio of the maximum output mechanical power to the input electrical power, should be as high as possible. The second objective is that the total motor weight, calculated as the sum of the steel rotor and stator weight, M_S, the aluminum rotor rings and bars weight, M_A, and the copper stator winding weight, M_C, should be as small as possible. Constraints (Cs) in Eqs. (54) and (55) require that the maximum magnetic flux density in the stator tooth and back iron to be less than 2.1 T, respectively. The total motor weight is required to be less than 27 kg as in Eq. (56). Motor dimension specifications are modeled in Eq. (57)–(59), which require that enough space needs to be reserved for both rotor and stator slots at inner radius, and motor length is restricted to be less than 16 cm. Based on circuit theory and the proposed equivalent analytical model, the current in stator windings and rotor branches, and the output mechanical power and torque can be easily calculated. Constraint in Eq. (60) requires that the current density in stator winding should be less than 26 A/mm^2. The maximum output mechanical power is required to be greater than 80 kW as in Eq. (61) and the maximum torque is required to be greater than 250 N-m as in Eq. (62). When those values are reached, the corresponding rotation frequency should be less than 3,150 RPM as in Eq. (63). Finally, the efficiency at the maximum output mechanical power of each valid motor design is required to be greater than 82% as in Eq. (64).

Tables 1 and 2 show the range of continuous and discrete design variables. It should be noted that to maintain the normal operating temperature of the motor, the stator slot space factor ζ needs to be less than 0.6 so that enough space can be left for liquid cooling in the final motor product.

Table 1 Range of continuous variables

Design Variable	Min	Max
d_s	0.45 mm	0.85 mm
h_r	0.90 cm	2.10 cm
h_s	2.20 cm	3.90 cm
g	0.75 mm	1.10 mm
r_o	8.40 cm	10.20 cm
r_i	4.60 cm	6.10 cm
V	144 V	164 V
T_{ring}	0.85 cm	1.35 cm
T_{iron}	1.45 cm	3.10 cm
w_r	1.05 cm	1.45 cm
w_s	0.32 cm	0.68 cm
ω	830 rad/s	1250 rad/s
ζ	0.48	0.60
u_s	0.13 cm	0.37 cm

Table 2 Range of discrete variables

Design Variable	Range
d_w	$\{1.15, 1.29, 1.45, 1.63\}$ mm
p	$\{2, 3, 4\}$
m	$\{1, 2, 3\}$
N_{SP}	$\{1, 2\}$
N_{SIP}	$\{1, 2,\ldots, 7\}$
N_R	$\{N_{slot}, N_{slot} - 1,\ldots, N_{slot}/2\}$

3.2 Solution Based on Genetic Algorithm

This section presents a solution based on GA to the design optimization problem introduced above. GA can be classified to a larger class of evolutionary algorithms that use knowledge inspired by genetic processes that occur in nature, such as selection, mutation, and crossover, to generate solutions to optimization problems [21, 22]. It is a good tool for solving nonlinear and complex multi-objective optimization problems. Compared with other classical solution tools for nonlinear optimization problems, GA has been proved empirically and theoretically to provide powerful search capabilities in complex spaces and shows a fast convergence speed [23]. Based on these advantages, GA has been widely used to solve many nonlinear multi-objective optimization problems, for example, in economics [24, 25], transportation [26–28], and biology [29–31]. GA has also been applied in the field of motor design, such as optimizing switched reluctance motor [32], DC motor [33], interior permanent magnet motor [34], surface-mounted permanent magnet motor [35], etc.

Figure 3 summarizes the general process of GA. As shown in the flowchart, a GA starts with a group of randomly generated design candidates (individuals) whose encoded strings (chromosomes) contain information about the variables of the design problem and successive iterations (generations) occur, evolving toward better design solutions over time [36]. Each individual representing a potential design candidate and is rated by a fitness function to determine which of them to use to form a new population during competition, which is called selection. Furthermore, in each generation, selected individuals are modified to form new populations by applying operators as crossover and mutation. The new generated population is then used for the next iteration of the algorithm. The algorithm terminates when the satisfactory average fitness level or the maximum number of generations is reached [37].

In the remainder of this section, the operators and element components in the proposed GA-based approach to the AFIM optimal design problem will be introduced in detail.

A. *Chromosome Representation*

The chromosome for AFIM candidates is shown in Fig. 4, which is the same size as the total number of design variables with each gene (slot on the chromosome) represents a design variable. In detail, each chromosome consists of 20 genes that

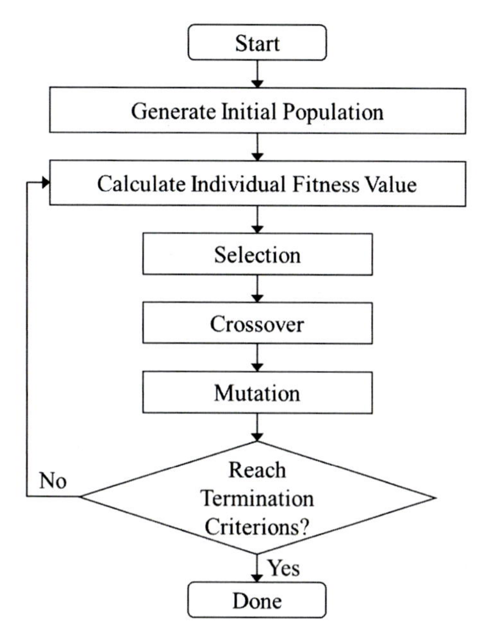

Fig. 3 General process of GA

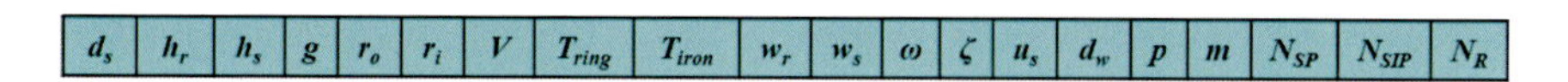

Fig. 4 Chromosome representation for the AFIM (1×20 array)

have the values of 20 design variables randomly selected from their domains. The first 14 genes of the chromosome are reserved for continuous design variables and the last 6 genes are used for discrete design variables. During evolution, the values of these genes are restricted to the corresponding domain of design variables.

B. *Fitness Value Evaluation*

A key issue in GA programming is computing chromosome fitness values to evaluate corresponding candidate designs. Chromosomes with higher fitness values will have a greater chance of being selected by the selection operator later and pass their genes to the next generation in the crossover step. Given the chromosome Λ of a candidate AFIM design, if any of the constraints in Eq. (54)–(64) are violated, the fitness value will be assigned zero, so that the gene value causing such a violation will not be passed on to the next generation. If the constraints are not violated, the fitness function can be evaluated as Eq. (65), where the motor efficiency of the first objective in Eq. (52) and the total motor weight of the second objective in Eq. (53) are denoted as $f_{OB1}(\Lambda)$ and $f_{OB2}(\Lambda)$, respectively. The weighting coefficients W_1 and W_2 of the objectives can be adjusted according to preference.

$$f_{Fit}(\Lambda) = W_1 \frac{f_{OB1\_MAX} - f_{OB1}(\Lambda)}{f_{OB1\_MAX} - f_{OB1\_MIN}} + W_2 \frac{f_{OB2}(\Lambda) - f_{OB2\_MIN}}{f_{OB2\_MAX} - f_{OB2\_MIN}} \tag{65}$$

C. *Selection*

Selection operator helps retain chromosomes with higher fitness values by assigning them a greater chance of passing genes on to the next generation, allowing the population to continue evolving toward an optimal solution. Selection process consists of three components: elitist, probability distribution, and chromosomes sampling for crossover. Elitist is used to preserve the P_e percentage chromosomes with the highest fitness value, in which copies of these chromosomes are passed directly to the next generation without involving subsequent steps. In addition, the P_e percentage chromosomes with the lowest fitness value will be removed from the current population at the same time. Through elitist, the best design candidates can always be retained from the current generation, while the worst genes are removed. After the elitist step, for each remaining chromosome Λ_i, the probability $p_s(\Lambda_i)$ of being selected later in crossover step is evaluated as:

$$p_s(\Lambda_i) = \frac{f_{Fit}(\Lambda_i)}{\sum_{j=1}^{N_Q} f_{Fit}(\Lambda_j)} \tag{66}$$

Based on the calculated probabilities, the roulette wheel sampling method [38] is applied to select chromosomes. The population is further mapped onto a virtual roulette wheel, where each chromosome Λ_i is assigned a space proportional to $p_s(\Lambda_i)$. By repeatedly spinning the roulette wheel, chromosomes are selected for crossover until the number of all available positions is reached, which is $(1\text{-}P_e)\cdot N_Q$. Through the selection process, chromosomes with high fitness values tend to generate more offspring in the next generation than those with below-average fitness values.

D. *Crossover*

Design candidates can share their information through the crossover process, in which chromosomes selected from the selection operator will be randomly paired as parental chromosomes and their genetic parts will be exchanged to produce two offspring chromosomes. Two-point crossover is applied here for demonstration. Suppose $\Lambda_{p1} = (c, c, \ldots, c)$ and $\Lambda_{p2} = (c, c, \ldots, c)$ are two paired chromosomes with n genes in the crossover step. Randomly select two indexes $i \in \{1, 2, \ldots, n\text{-}1\}$ and $j \in \{i + 1, i + 2, \ldots, n\}$ and two offspring chromosomes can then be generated as Eq. (67) and Eq. (68), respectively. Figure 5 also demonstrates this process.

$$\Lambda_{o1} = \left(c_1^1, \ldots, c_{i-1}^1, c_i^2, \ldots, c_j^2, c_{j+1}^1, \ldots, c_n^1\right) \tag{67}$$

$$\Lambda_{o2} = \left(c_1^2, \ldots, c_{i-1}^2, c_i^1, \ldots, c_j^1, c_{j+1}^2, \ldots, c_n^2\right) \tag{68}$$

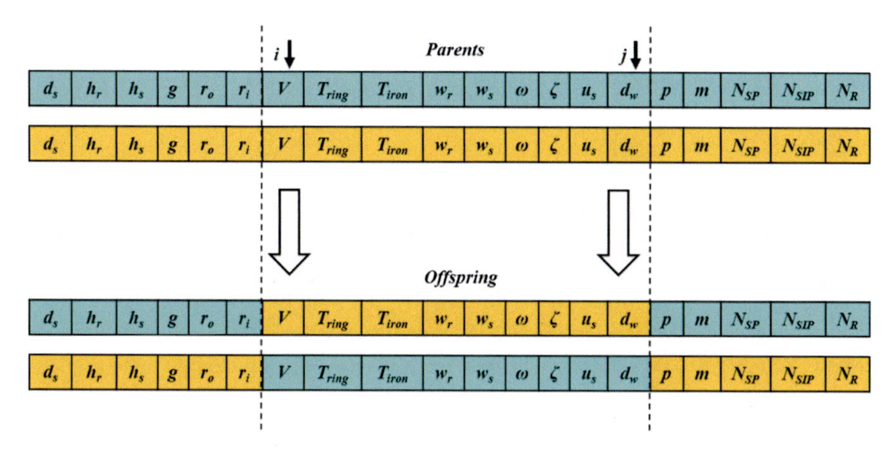

Fig. 5 Two-point crossover demonstration with random indexes i and j

E. *Mutation*

Genetic diversity can be provided to a population through mutation by restoring unexplored and lost genetic material, preventing GA from converging to suboptimal solutions, and enabling GA to search for a wider space. Each gene after the crossover step has a probability of P_m to mutate. Assuming $c_i \in [a_i, b_i]$ is a gene representing a continuous design variable, its mutated version $c_i{}'$ can be generated as Eq. (69) below:

$$c_i\prime = \begin{cases} c_i - \Delta(t, b_i - c_i) \ if \tau = 0 \\ c_i - \Delta(t, c_i - a_i) \ if \tau = 1 \end{cases} \tag{69}$$

$$\Delta(t, y) = y\left(1 - \delta^{\left(1 - \frac{t}{G_{max}}\right)^{\nu}}\right) \tag{70}$$

where δ is a random number in the interval [0, 1], τ is either one or zero with equal probability, and ν is a user-defined variable that can adjust the dependence on the number of generations.

For discrete design variables, random values from corresponding discrete domains will be assigned. Populations that have undergone the process of crossover and mutation, combined with copies of the elitist design candidates, together constitute a new generation with the same population size. The algorithm terminates when the satisfactory average fitness level or the maximum generation number G_{max} is reached.

4 Simulation Results

In this simulation, the AFIM design method consisting of the equivalent analytical model and GA-based optimization is firstly run to show the corresponding optimization process and the improvement of computation efficiency. In the second part of this section, the results from ANSYS Maxwell and the analytical model are compared and analyzed to validate the accuracy of the proposed model.

A. *GA Optimization Process*

Table 3 shows the parameters for GA. The weight coefficients of objectives are set to be 0.5 and 0.5. The GA starts with a randomly generated design that meets all design constraints with a population size of 500. The algorithm terminates whenever the motor efficiency and weight of the top 5 designs with the highest fitness value in current generation has not improved comparing with those in the last generation or the maximum generation number $G_{max} = 60$ is reached.

The proposed GA is terminated at the 37th generation. In Fig. 6, the motor efficiencies and weights at the maximum output power for the top 5 design candidates of different generations are plotted as hexagrams in turquoise (1st generation), green (5th generation), yellow (10th generation), orange (15th generation), red (20th generation), pink (25th generation), purple (30th generation), light blue (35th generation), and dark blue (37th generation) to show the motor optimization process. It can be seen from the figure that the design candidates are optimizing themselves toward being more efficient and lighter. At the 37th generation, all the top 5 design candidates can achieve efficiency over 93% at the maximum output power, which is much higher than that of the initial design. Compared with traditional FEA-based motor design methods such as in ANSYS Maxwell, the simulation time is greatly reduced. The FEA 3D simulation typically takes an hour to test one possible AFIM design at a fixed rotation speed, while the proposed method takes only 30 s on average to terminate and output the optimized AFIM designs. The AFIM design with the highest fitness value in the 37th generation was selected for further FEA in ANSYS Maxwell to verify the accuracy of the analytical model, where the design variables are shown in Table 4.

B. *Motor Performance and Accuracy of Analytical Model*

Table 3 Parameters for GA

Parameter	Value
W_1	0.5
W_2	0.5
N_Q	500
G_{max}	60
ν	5
P_e	0.20
P_m	0.05

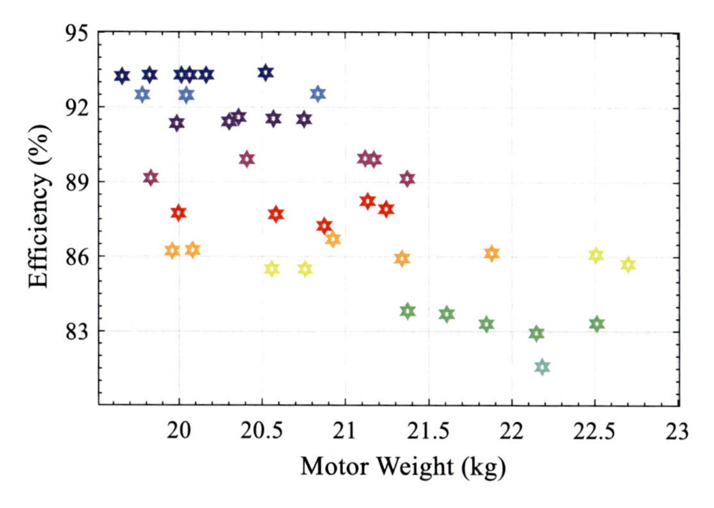

Fig. 6 Optimization path of the GA solution

Table 4 Parameters for the selected motor design

Design variable	Value
d_s	0.72 mm
h_r	1.48 cm
h_s	3.57 cm
g	0.87 mm
r_o	9.96 cm
r_i	6.06 cm
V	159 V
T_{ring}	1.22 cm
T_{iron}	2.10 cm
w_r	1.19 cm
w_{st}	0.46 cm
ω	1016 rad/s
ζ	0.584
u_s	0.21 mm
d_w	1.64 mm
p	3
m	2
N_{SP}	1
N_{SIP}	4
N_R	20

The output mechanical power, motor efficiency, and output torque at different rotor speeds of the selected AFIM based on the analytical model are drawn with green lines, as shown in Fig. 7, Fig. 8, and Fig. 9. The selected AFIM is also simulated in ANSYS Maxwell by inputting the parameters in Table 4, and the output mechanical power, motor efficiency and torque is obtained at selected rotor speeds (per 400 RPM from 400 to 3,200 RPM plus the rated speed 3,051 RPM) and indicated by red dots in the same figures. At these selected speeds, the results of the proposed analytical model and ANSYS Maxwell and their deviations are also indicated in green, red, and black fonts, respectively. The maximum output power given by the analytical model is 86.5 kW, which corresponds to an efficiency and torque of 93.5% and 275.1 N-m, respectively. At the same rotation speed 3,051 RPM, ANSYS Maxwell outputs 83.7 kW of mechanical power, 90.1% efficiency, and 265.2 N-m of torque. All these values meet the initial design specifications in the optimization formulation. The mean absolute percentage error (MAPE) of these three output values as in Fig. 8, Fig. 9, and Fig. 10 are 2.84%, 3.10%, and 3.07%, respectively suggesting that the proposed analytical model is accurate.

Due to the narrow cross-sectional areas, stator back irons and tooth tend to have the highest magnetic flux density. Therefore, in order to verify whether the final selected AFIM design can meet the magnetic flux density requirements, the surface magnetic flux density at the maximum output mechanical power of one stator core is plotted as in Fig. 10. It can be seen that the magnetic flux densities in the back iron and stator tooth can meet the corresponding design requirements in constraints Eq. (54) and Eq. (55). Part of the stator slot depressions used to fix coil windings may have high magnetic flux density, as marked in yellow. However, considering that the places with high magnetic flux density are very small and only in the depression of

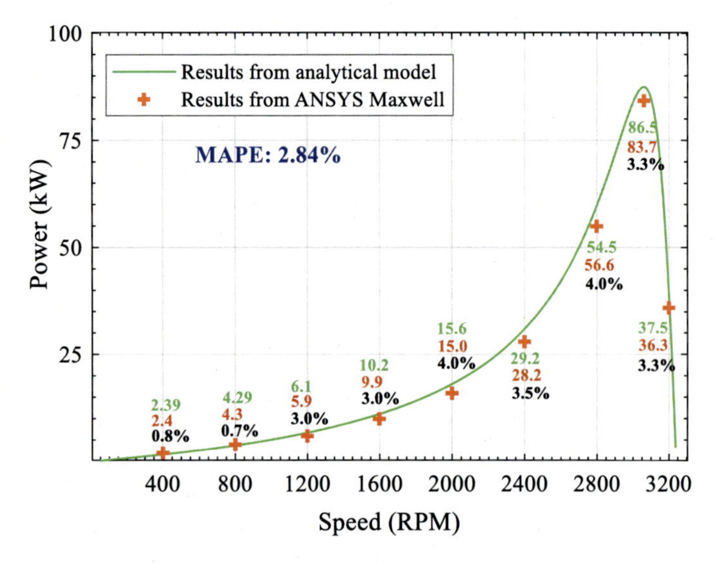

Fig. 7 Output mechanical power of the optimized AFIM design

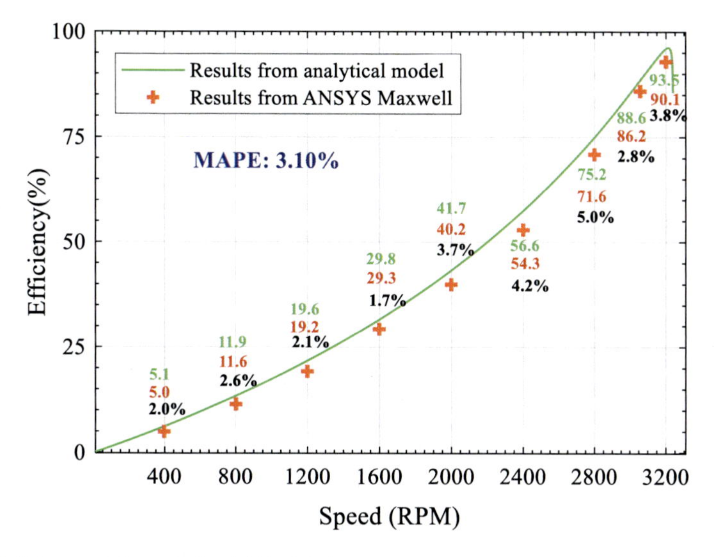

Fig. 8 Efficiency of the optimized AFIM design

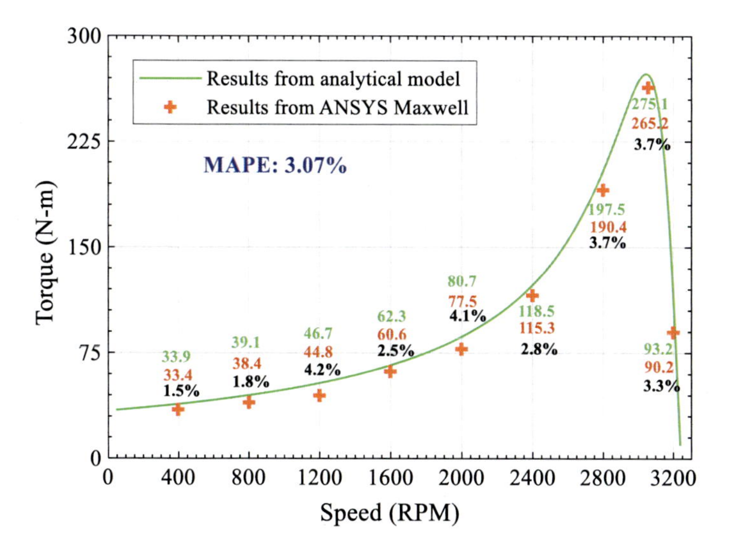

Fig. 9 Torque of the optimized AFIM design

stator tooth not in the tooth and back iron, they have little effects on the operation of the motor. Therefore, the overall magnetic flux density of the selected AFIM design can also meet the initial design specifications.

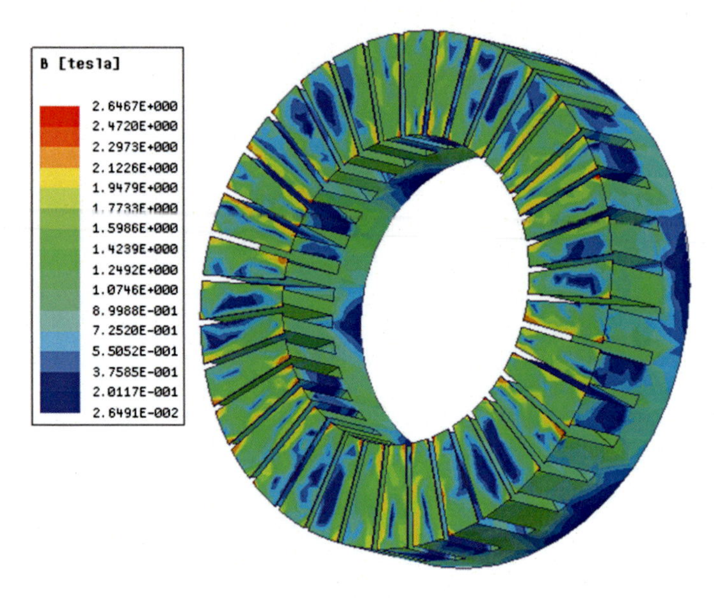

Fig. 10 Surface magnetic flux density magnitude of one stator

5 Conclusions

This chapter takes AFIM for EV applications as an example and describes in detail how to design and optimize an electric motor. The analytical model of AFIM was firstly derived, which can provide results close to FEA in ANSYS Maxwell in a much shorter simulation time. Based on the analytical model, the corresponding optimization model was formulated considering all design requirements and further solved by GA. Finally, the optimized design was validated in ANSYS Maxwell. It is worth mentioning that motor design is a very complex process, and other factors such as heat dissipation and cost also need to be considered to finally obtain a satisfactory motor design. The content of this chapter can be treated as a foundation for this field.

Nomenclature

T_{iron} Thickness of stator back iron
t Current generation number
u_{s} Width of stator slot depression
V Magnitude of input voltage
w_{r} Width of rotor bar
w_{s} Width of stator slot
W_1 Linear weight coefficient of the first objective
W_2 Linear weight coefficient of the second objective

X_{S} Leakage impedance of analytical model
X_{m} Magnetic reactance of analytical model
X_1 Stator leakage reactance of analytical model
X_2 Rotor leakage reactance of analytical model
X_{el} Stator end winding leakage of analytical model
X_{sl} Stator slot leakage of analytical model
X_{bl} Stator belt leakage of analytical model
X_{zl} Stator zigzag leakage of analytical model
θ Stator angular position
θ' Rotor angular position
λ Linked flux
μ_{o} Permeability constant
σ_{r} Electrical conductivity of rotor bar
σ Electrical conductivity of winding
ζ Stator slot space factor
ω Frequency of input voltage
ω_{r} Rotor current frequency
ω_{m} Rotor mechanical frequency
α Short pitch angle
β Ratio number
Λ Chromosome
γ Angular angle between two stator slots

References

1. Zhu J, Cheng K, Xue X, Zou Y (2017) Design of a new enhanced torque in-wheel switched reluctance motor with divided teeth for electric vehicles. IEEE Trans Magn 53(11)
2. Zhang Y, Cao W, Mcloone S, Morrow J (2016) Design and flux-weaking control of an interior permanent magnet synchronous motor for electric vehicles. IEEE Trans Appl Supercond 26(7)
3. Pellegrino G, Vagati A, Boazzo B, Guglielmi P (2012) Comparison of induction and PM synchronous motor drives for EV application including design examples. IEEE Trans Ind Appl 48(6):2322–2332
4. Kim SH, Sul SK (1997) Voltage control strategy for maximum torque operation of an induction machine in the field weakening region. IEEE Trans Industr Electron 44(4):512–518
5. Seok J, Sul S (2001) Induction motor parameter tuning for high performance drives. IEEE Trans Ind Appl 37(1):35–41
6. Mun JM et al (2017) Design characteristics of IPMSM with wide constant power speed range for EV traction. IEEE Trans Magn 53(6):1–4
7. Torregrossa D, Peyraut F, Fahimi B, Boua JM, Miraoui A (2011) Multi-physics finite-element modeling for vibration and acoustic analysis of permanent magnet synchronous machine. IEEE Trans Energy Convers 26(2):490–500
8. Ma C, Liu Q, Wang D, Li Q, Wang L (2016) A novel black and white box method for diagnosis and reduction of abnormal noise of hub permanent magnet synchronous motors for electric vehicles. IEEE Trans Industr Electron 63(2):1153–1167

9. Islam MS, Islam R, Sebastian T (2014) Noise and vibration characteristics of permanent-magnet synchronous motors using electromagnetic and structural analyses. IEEE Trans Ind Appl 50(5):3214–3222
10. Zhang Y, Cao W, Mcloone S, Morrow J (2016) Design and flux-weakening control of an interior permanent magnet synchronous motor for electric vehicles. IEEE Trans Appl Supercond 26(7)
11. Vijayakumar K, Karthikeyan R, Paramasivam S, Arumugam R, Srinivas KN (2008) Switched reluctance motor modeling, design, simulation, and analysis: A comprehensive review. IEEE Trans Magn 44(12):4605–4617
12. Xue XD, Cheng KWE, Ng TW, Cheung NC (2010) Multi objective optimization design of in-wheel switched reluctance motors in electric vehicles. IEEE Trans Industr Electron 57(9):2980–2987
13. Yang Z, Shang F, Brown I, Krishnamurthy M (2015) Comparative study of interior permanent magnet, induction, and switched reluctance motor drives for EV and HEV applications. IEEE Trans Transp Electrification 1(3):245–254
14. Xiang Z, Zhu X, Quan L, Du Y, Zhang C, Fan D (2016) Multilevel design optimization and operation of a brushless double mechanical ports flux-switching permanent magnet motor. IEEE Trans Industr Electron 63(10):6042–6054
15. Kim D, Hwang H, Bae S, Lee C (2016) Analysis and design of a double-stator flux-switching permanent magnet machine using ferrite magnet in hybrid electric vehicles. IEEE Trans Magn 52(7), Art. No. 8106604
16. Li H, Zhu H (2016) Design of bearingless flux-switching permanent magnet motor. IEEE Trans Appl Supercond 26(4)
17. Mei J, Lee CHT, Kirtley JL (2020) Design of axial flux induction motor with reduced back iron for electric vehicles. IEEE Trans Veh Technol 69(1):293–301
18. Mei J, Zuo Y, Lee CHT, Kirtley JL (2020) Modeling and optimizing method for axial flux induction motor of electric vehicles. IEEE Trans Veh Technol 69(11):12822–12831
19. Fitzgerald AE, Kingsley C, Umans SD (2003) Electric machinery, 6th edn. McGraw Hill
20. Alger PL (1969) Induction machines. Gordon and Breach
21. Deb K, Pratap A, Agrawal S, Meyarivan T (2002) A fast and elitist multiobjective genetic algorithm: NSGA-II. IEEE Trans Evol Comput 6(2):182–197
22. Panda D, Ramteke M (2019) Preventive crude oil scheduling under demand uncertainty using structure adapted genetic algorithm. Appl Energy 235:68–82
23. Golberg D (1989) Getrelic algorithms. Addison-Wesley
24. Zou D, Li S, Kong X, Ouyang H, Li Z (2019) Solving the combined heat and power economic dispatch problems by an improved genetic algorithm and a new constraint handling strategy. Appl Energy 237:646–670
25. Nemati M, Braun M, Tenbohlen S (2018) Optimization of unit commitment and economic dispatch in microgrids based on genetic algorithm and mixed integer linear programming. Appl Energy 210:944–963
26. Ahn CW, Ramakrishna RS (2002) A genetic algorithm for shortest path routing problem and the sizing of populations. IEEE Trans Evol Comput 6(6):566–579
27. Wang K, Shen Z (2012) A GPU-based parallel genetic algorithm for generating daily activity plans. IEEE Trans Intell Transp Syst 13(3):1474–1480
28. Jiau MK, Huang SC (2015) Services-oriented computing using the compact genetic algorithm for solving the carpool services problem. IEEE Trans Intell Transp Syst 16(5):2711–2722
29. Shin SY, Lee IH, Kim D, Zhang BT (2005) Multiobjective evolutionary optimization of DNA sequences for reliable DNA computing. IEEE Trans Evol Comput 9:143–158
30. Hu Z, Chan CW, Huang GH (2007) Multi-objective optimization for process control of the in-situ bioremediation system under uncertainty. Eng Appl Artif Intell 20(2):225–237
31. Bates Congdon C, Aman JC, Nava GM, Gaskins HR, Mattingly CJ (2008) An evaluation of information content as a metric for the inference of putative conserved non-coding regions in DNA sequences using a genetic algorithms approach. IEEE/ACM Trans Comput Biol 5(1):1–14
32. Mirzaeian B, Moallem M, Tahani V, Lucas C (2002) Multiobjective optimization method based on a genetic algorithm for switched reluctance motor design. IEEE Trans Magn 38(3):1524–1527

33. Ishikawa T, Yonetake K, Kurita N (2011) An optimal material distribution design of brushless DC motor by genetic algorithm considering a cluster of material. IEEE Trans Magn 47(5):1310–1313
34. Okamoto Y, Tominaga Y, Wakao S, Sato S (2014) Topology optimization of rotor core combined with identification of current phase angle in IPM motor using multistep genetic algorithm. IEEE Trans Magn 50(2):725–728
35. Hwang C, Lyu L, Liu C, Li P (2008) Optimal design of an SPM motor using genetic algorithms and taguchi method. IEEE Trans Magn 44(11):4325–4328
36. Fonseca CM, Fleming PJ (1998) Multiobjective optimization and multiple constraint handling with evolutionary algorithms—Part I: A unified formulation. IEEE Trans Syst, Man, Cybern Part A: Syst Hum 28(1):26–37
37. Wieczorek M, Lewandowski M (2017) A mathematical representation of an energy management strategy for hybrid energy storage system in electric vehicle and real time optimization using a genetic algorithm. Appl Energy 192:222–233
38. Gupta N, Shekhar R, Kalra PK (2012) Congestion management-based roulette wheel simulation for optimal capacity selection: probabilistic transmission expansion planning. Int J Electr Power Energy Syst 43(1):1259–1266

Advanced Electrochemical Energy Sources for Electric and Hybrid Vehicles

Rodney Chua, Yi Cai, William Manalastas Jr, Ernest Tang Jun Jie, Deepika Ranganathan, Eldho Edison, Tan Fu Xing Ivan, and Madhavi Srinivasan

Abstract With the ever-growing concerns on the impact of climate change, the world has stepped up efforts to reduce carbon emission by implementing policies in several key areas. One of the main strategies is to phase out the internal combustion engine vehicles (ICEVs) that have contributed significantly to greenhouse gases and replace them with electric and hybrid vehicles (EVs and HEVs) that run on cleaner energy alternatives. This chapter focuses on three different electrochemical energy sources employed in EVs and HEVs: batteries, supercapacitors, and fuel cells. We will discuss the capabilities and development of the state-of-art electrochemical energy sources, outlining their key challenges and providing the outlook toward realizing a safer, greener, and efficient urban transportation system.

1 Background

Greenhouse gases are found to be responsible for the global temperature rise, which leads to climate change [1, 2]. A significant fraction of the greenhouse gas emissions are composed of carbon dioxide (CO_2), contributed by the transport sector due to the burning of fossil fuels in the operation of internal combustion engine vehicles; ICEVs [3–5]. In a united effort across the world, the ICEVs are in the process of being replaced with cleaner and sustainable electric and hybrid electric vehicles (EVs and HEVs) [6–8]. In this chapter, we focus mainly on three electrochemical energy sources that are employed to power the EVs and HEVs: (1) batteries, (2) supercapacitors, and (3) fuel cells. Figure 1 presents the typical configurations of the battery (Fig. 1a), supercapacitor (Fig. 1b), and fuel cell (Fig. 1c) system.

R. Chua · W. Manalastas Jr · E. T. J. Jie · D. Ranganathan · E. Edison · T. F. X. Ivan · M. Srinivasan (✉)
School of Materials Science and Engineering, Nanyang Technological University, 50 Nanyang Avenue, Block No. 4.1, Singapore 639798, Singapore
e-mail: madhavi@ntu.edu.sg

Y. Cai · D. Ranganathan · M. Srinivasan
Energy Research Institute at Nanyang Technological University, Research Techno Plaza, 50 Nanyang Drive, Singapore 637553, Singapore

C. H. T. Lee (ed.), *Emerging Technologies for Electric and Hybrid Vehicles*, Green Energy and Technology, https://doi.org/10.1007/978-981-99-3060-9_7

Although these systems share "electrochemical similarities" where the electrochemical processes occur at the interface between the electrode and electrolyte, they have different charge-storage and conversion mechanisms, kinetics, and thermodynamics reactions that enable distinct electrochemical attributes shown in Fig. 2 [9]. Among the system, electrochemical capacitors (e.g., capacitors/supercapacitors) exhibit the highest specific power (refers to the rate capability), whereas fuel cells possess the highest specific energy (refers to the amount of energy content). To improve the overall performance (specific power/energy, operating life, and efficiencies) for EVs and HEVs applications, one of the critical approaches adopted was to hybridize the different electrochemical energy systems [8, 10–12]. However, to achieve a comparable driving range (~700 km per 50 L tank) and cost as ICEVs, there are still several milestones that need to be achieved [9]. In the following, we will further discuss the state-of-art and critical challenges of each electrochemical energy source in EVs and HEVs applications and provide an outlook for a safer, greener, and sustainable land transport system and beyond.

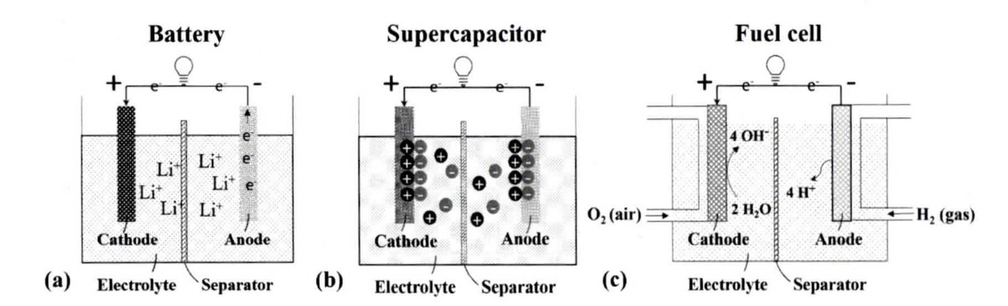

Fig. 1 Configurations of **a** battery, **b** supercapacitor, and **c** fuel cell

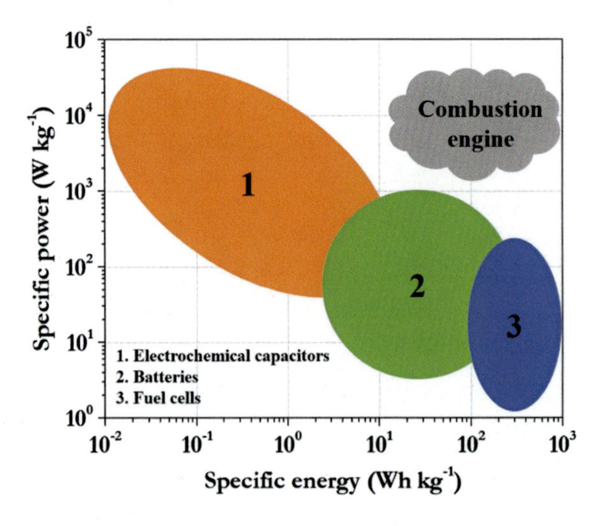

Fig. 2 Ragone plot (simplified version) of the different electrochemical energy systems in comparison with the combustion engine

2 Batteries

Today's emerging EV market calls for lower cost, higher energy density, and sustainable automotive batteries to suit the typical driving patterns of twenty-first-century consumers [13]. To achieve a competitive edge against the ICEVs, the United States Department of Energy (DOE) and the United States Advanced Battery Consortium (USABC) had proposed energy density targets of 350 Wh kg^{-1} and 750 Wh L^{-1} at the cell level for automotive batteries to enable at least 300 miles (~482.8 km) of driving range per charge and reduction in battery cost to US$125 per kWh [13, 14]. Lithium-ion batteries (LIBs), having the highest energy density (~150 Wh kg^{-1}) as compared to their counterparts, such as lead-acid batteries (~60 Wh kg^{-1}) and Ni-MH batteries (~110 Wh kg^{-1}), is the most promising candidate for energy storage application in EVs and HEVs [15]. As illustrated in Table 1, there are various types of lithium-ion battery chemistries based on different active electrode materials that enable unique attributes towards energy storage in EV applications, as discussed in the following [16–21].

2.1 *Battery Chemistries for Long-Driving Range Applications*

Layered nickel-rich oxides such as lithium nickel manganese cobalt oxides (NMC) and lithium nickel cobalt aluminum oxides (NCA) are commonly incorporated into

Table 1 Overview of the lithium-ion based batteries with their unique attributes.

Li-ion based batteries	Cell energy density	Average voltage	Cycle life	Working temperature	Thermal safety	Cost
NMC/G		2V 4V				
NCA/G						
LFP/G						
LFP/LTO						
LMO/G						

NMC—lithium nickel manganese cobalt oxides; NCA—lithium nickel cobalt aluminum oxides; LFP—lithium iron phosphate; G—graphite; LTO—lithium titanate oxides and LMO—lithium manganese oxides long-range electric vehicles (Fig. 3) [13, 14, 17, 22, 23]. Owing to the utilization of nickel redox couples (Ni^{2+}/Ni^{3+} and Ni^{3+}/Ni^{4+}), these layered nickel-rich oxides have high operating cell voltages that enable superior cell-level specific energy (NMC/graphite (G): ~200 Wh kg^{-1}; NCA/G: ~250 Wh kg^{-1}), enabling long-driving range applications [16, 18, 21]. For instance, Tesla's EVs based on NCA chemistry ($LiNi_{0.85}Co_{0.10}Al_{0.05}O_2$) are capable of a driving range close to 500 km in a single charge [24]. Although there is a significant breakthrough in the performance to bring us one step closer toward the total replacement of ICEVs, there are still challenges in terms of cost and sustainability. The utilization of cobalt in these chemistries is associated with the high cost of battery packs due to the limited global reserves of the cobalt (~7.1 million tons) and poor geographical distribution (~51% reserves in DRC) [16]. Moreover, even the "low-cobalt" battery chemistry is still estimated to contain around 6.6 kg of cobalt in a 75 kWh EV battery, posing environmental concerns on the toxicity brought by cobalt mining activities and the end-of-life cycle of these cobalt-containing batteries [25, 26].

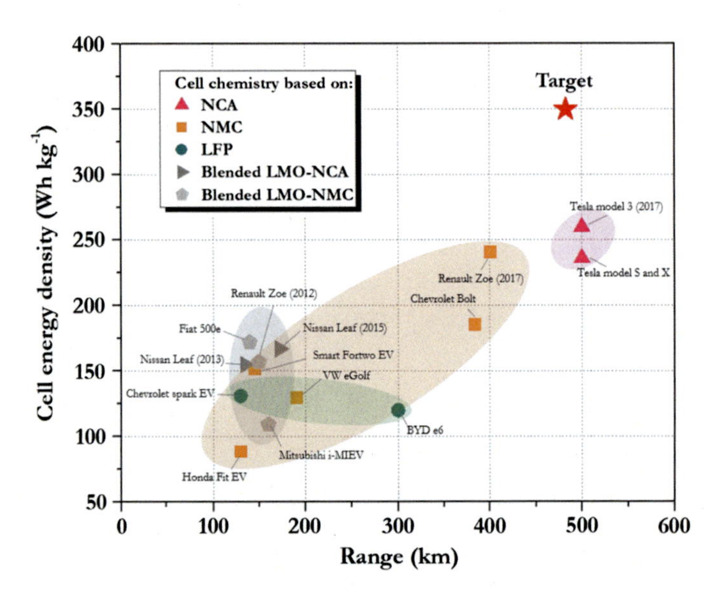

Fig. 3 Performance of EVs based on different cell chemistries. Data are plotted based on Ref. [13, 14, 17, 22, 23]

2.2 Battery Chemistries for Economical, Durable, and Short-Driving Range Applications

LIBs based on olivine lithium iron phosphates ($LiFePO_4$; LFP) and spinel lithium manganese oxides ($LiMn_2O_4$; LMO) cathodes materials have lower cell energy densities due to lower transition metal redox couples potential and theoretical specific capacities (LFP: 170 mA h g^{-1}; LMO: 148 mA h g^{-1}) [19, 20]. Nevertheless, both LFP and LMO could offer a more economical alternative toward mid-end EVs due to their lower cost (in terms of price per kWh) [16]. Based on reports, major car manufacturers such as Volkswagen and Tesla are transiting toward LFP-based chemistry, as LFP batteries have recently recorded the lowest reported price of $80 per kWh [27, 28]. Both LFP and LMO-based chemistries generally have a better thermal safety feature (with thermal runaway temperature beyond 250 °C) than their lithium nickel rich oxides counterparts (NCM: ~210 °C/NCA: ~150 °C) [16, 17]. For a harsher operating environment, LFP can be paired with lithium titanate oxides (LTO) anode could potentially enable EV to operate in a much colder working temperature down to −40 °C [16]. To boost the performances (in terms of safety and energy densities), blended chemistries (LMO-NCA and LMO-NMC) strategies have been widely adopted into several EVs (Fig. 3) [17, 22, 23]. Current research efforts are devoted to enhancing the electrochemical performance of LFP and LMO through surface modification, morphological and structural engineering and bulk substitution, as well as electrolyte engineering [18].

2.3 Toward Safer Battery Chemistry: Solid-State Battery

To date, LIBs are widely applied for EVs and HEVs. However, their energy density still needs further improvement to meet the increasing marketing demand. Meanwhile, the safety issue is one of the essential concerns when using conventional LIBs. Therefore, solid-state batteries (SSBs) have attracted much attention because they may solve the issues hindering widespread EV or HEV adoption. Conventional batteries used in EV or HEV typically contain organic liquid or gel electrolytes which are flammable and can be frozen in extreme conditions. In addition, the electrolyte can participate in a rapid exothermic reaction when an improper operation occurs, such as mechanical or electrical abuse, leading to fire or explosion. Therefore, additional safety and cooling monitoring systems are usually required to couple with the conventional batteries in EV and HEV, which are costly and complicated. In comparison, SSBs possess inherent safety since they are solid-state cells using inorganic solid electrolytes instead of flammable carbonate-based organic liquid electrolytes, eliminating the fire risk showing great potential for EV and HEV applications, as shown in Fig. 4.

Solid-state electrolytes are divided into inorganic and polymer-based electrolytes. The inorganic-based solid electrolytes usually are oxides, sulfides, or phosphates.

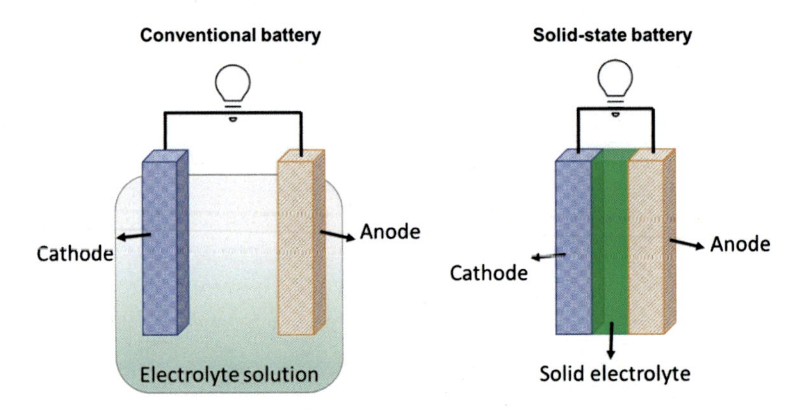

Fig. 4 Scheme diagram comparison of conventional battery and solid-state battery

The ions diffuse through the lattice, with a relatively high ionic conductivity of a few mS cm^{-2} at room temperature [29]. Polymer-based solid electrolytes conduct ions via the polymer chains. Compared to inorganic-based, they own lower ionic conductivity but higher elasticity and plasticity, enhancing the interface stability and more flexibility that can adapt to volume changes during operation. Moreover, they are easy to process suitable for large-scale manufacturing processes.

Another advantage of exploring SSBs for EVs and HEVs is the higher energy density than currently used LIBs. In EV and HEV markets, graphite-based anode materials are the common choice for LIBs. The relatively low theoretical capacity (372 mA h g^{-1}) of graphite limits the full cell energy density. Lately, the best-in-class LIBs can achieve energy densities of 250 to 300 Wh kg^{-1}. In comparison, SSBs are believed to achieve higher energy densities because the solid electrolyte may enable lithium metal as anode with a higher theoretical specific capacity of 3860 mA h g^{-1}. Specifically, the volumetric energy densities can be increased because SSBs could realize bipolar electrode stacking cell design so that the thickness of the overall current collector could be reduced [30].

Despite the promising advantages, there are many constraints that impede the usage of SSBs for EVs and HEVs. One of the main challenges is the poor ionic conductivity compared to liquid counterparts. Therefore, the operation of some solid electrolytes needs extreme conditions, such as high temperature. Cycle stability is also another issue. The current SSBs have hardly fulfilled the standard requirement (800 cycles while retaining 80% capacity). Therefore, more attention should also be paid to solving these problems for realizing the successful commercialization of SSBs.

Beyond academic research, many battery companies and car manufacturers are working on commercializing SSBs, such as QuantumScape Corp., Factorial, etc. [31–34]. For example, the Factorial company introduced the first 40 Ah solid-state battery using "Factorial Electrolyte System Technology" (FEST) for EVs, increasing the electric cars driving range up to 50%. The current solid-state battery can reach an energy density of 350 Wh kg^{-1} and 770 Wh L^{-1}. Thereafter, many car manufacturers such as Hyundai Motor Company, Kia Corporation, Stellantis, and Mercedes-Benz are interested in the new battery technology and have announced joint development agreements with Factorial to integrate cell, model and system for future vehicles using the advanced technology of Factorial company. Furthermore, as SSBs manufacturing matures, costs could potentially reduce. Therefore, the marketing and adoption of SSBs in EVs and HEVs would be expected in the foreseeable future.

2.4 *Toward Sustainable Battery Chemistries: Beyond Lithium-Ion Technologies*

As the EV market with a strong reliance on lithium-ion technologies continues to proliferate, there is a need to explore a sustainable energy solution for the long term. As depicted in Fig. 5, abundant metals such as sodium, potassium, magnesium, and calcium have emerged as today's promising candidates for beyond lithium-ion battery technologies [35–41]. One of the critical advantages of transitioning beyond lithium-ion technologies is the lower battery manufacturing cost due to the cheaper raw materials. Also, aluminium foil can be employed as the current collector for both cathodes and anodes for the non-lithium containing battery, further lowering the material and manufacturing costs for EV batteries [37]. In these regards, it can potentially bring cost-savings to the consumers, thus encouraging widespread adoption of EVs. For alternative battery chemistries based on sodium (Na^+) and potassium (K^+) ions, due to their similar chemical properties as lithium, this enables more accessible adaptation from lithium-based battery systems from battery design to manufacturing and end-products commercialization processes [42]. With smaller Stokes radii of sodium and potassium than lithium in the electrolyte solvents such as propylene carbonate, they have lower desolvation energy, which can enable faster ion transport kinetics beneficial in boosting the power performance of these monovalent Na^+/K^+ ions-based batteries [37]. On the other hand, battery chemistries based on multivalent ions such as magnesium (Mg^{2+}), calcium (Ca^{2+}), and aluminum (Al^{3+}) could offer higher energy charge storage capability of the batteries in terms of volumetric and specific capacities owing to the higher number of electron transfer per charge in these multivalent-ions based system (Fig. 6) [39, 43–47].

For beyond lithium-ion technologies to enter the EV market, there are still several milestones to be achieved with holistic assessment and validation to be made in terms of feasibility (theoretical modeling and experimental), compatibility (cathode, anode, and electrolyte), and reliability (safety, cycle life, and durability test) [39, 47, 48]. Nevertheless, the number of research efforts devoted to these next-generation technologies has been growing rapidly recently, given the limitations of existing LIBs towards powering a safer and sustainable future [38, 39].

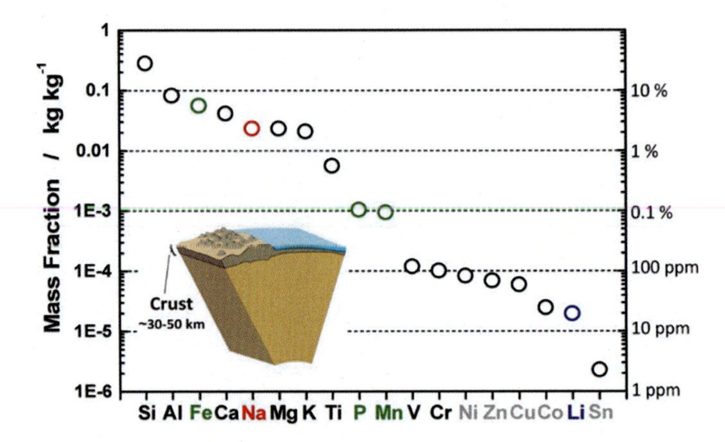

Fig. 5 Elemental abundance in the Earth's crust. Reprinted with permission from [35]. Copyright 2014 American Chemical Society

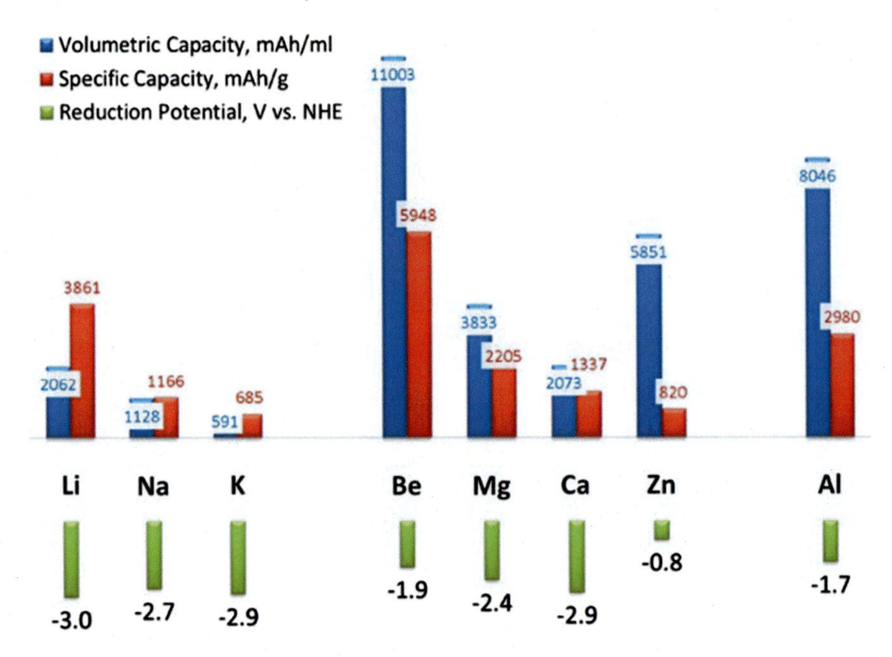

Fig. 6 Capacities and reduction potentials for various alternative metal anodes in comparison with traditionally used lithium. Reprinted with permission from [47]. Copyright 2014 American Chemical Society

3 Supercapacitors

Electrochemical capacitors, owing to their high-power capabilities, play a vital role in improving the peak power performance of EVs and HEVs. Incorporating electrochemical capacitors alongside primary energy storage systems such as batteries and fuel cells could improve both the efficiency and the cycle life of the energy storage unit. Additionally, regenerative braking energy storage into the capacitor system could further improve the cycle life of the primary energy storage systems. Further research into reducing the $ Wh^{-1} by developing new materials and novel vehicle design features incorporating capacitors within the functional components would lead to dramatic space and energy savings in EVs and HEVs. The power capability of electric drivetrains has increased dramatically in recent years; however, a major problem is the design of efficient energy storage systems that can support the peak power requirements. Electrochemical capacitors (EC) could mitigate the poor power performance of batteries and fuel cells that are employed as primary energy storage systems in EVs and HEVs. EC offers several advantages, including high efficiency, better charge acceptance, and longer cycle life [49]. Although an electrochemical capacitor (EC) is configured similar to a battery (e.g., two electrodes separated by a separator and soaked in the electrolyte), EC stores energy through charge separation in the double layer formed in the micropores of a very high surface area (~500–2000 m^2g^{-1}) electrode material (Fig. 7) [50–52]. Additionally, hybrid electrochemical capacitors (HEC) could be fabricated with one electrode composed of battery-like material and the other with double-layer capacitance. HECs can provide significantly higher energy densities than double-layer capacitors. The readers are encouraged to refer to sources [53, 54] for an in-depth discussion on the electrochemistry of batteries and ECs. To meet the peak power requirements of the energy storage system, ECs and HECs potentially form an integral part of the pulse power unit used in conjunction with a battery in EVs and HEVs [50–52].

3.1 Application of Capacitors in Electric and Hybrid Electric Vehicles

A series of ECs and HECs were previously tested and reported by Burke et al. from the University of California-Davis [55]. In the case of commercial ECs based on carbon/carbon electrodes, the energy densities of these ECs are typically around 4–5 Wh kg^{-1} with a specific power of 1000 W kg^{-1} [55]. With electrode optimization (to enhance specific capacitance of carbon) and electrolyte modifications (to expand the cell voltage), the ECs from Skeleton Technologies can exhibit high power capability (1730 W kg^{-1}) and energy density (9 Wh kg^{1}) [55]. To boost the energy density, it can be achieved by utilizing carbon and metal oxide composite electrodes such as the HECs from Yunasko [55]. Thus, the HEVs, which are sized by the energy storage requirements, could fully utilize the high energy and power capabilities of HECs.

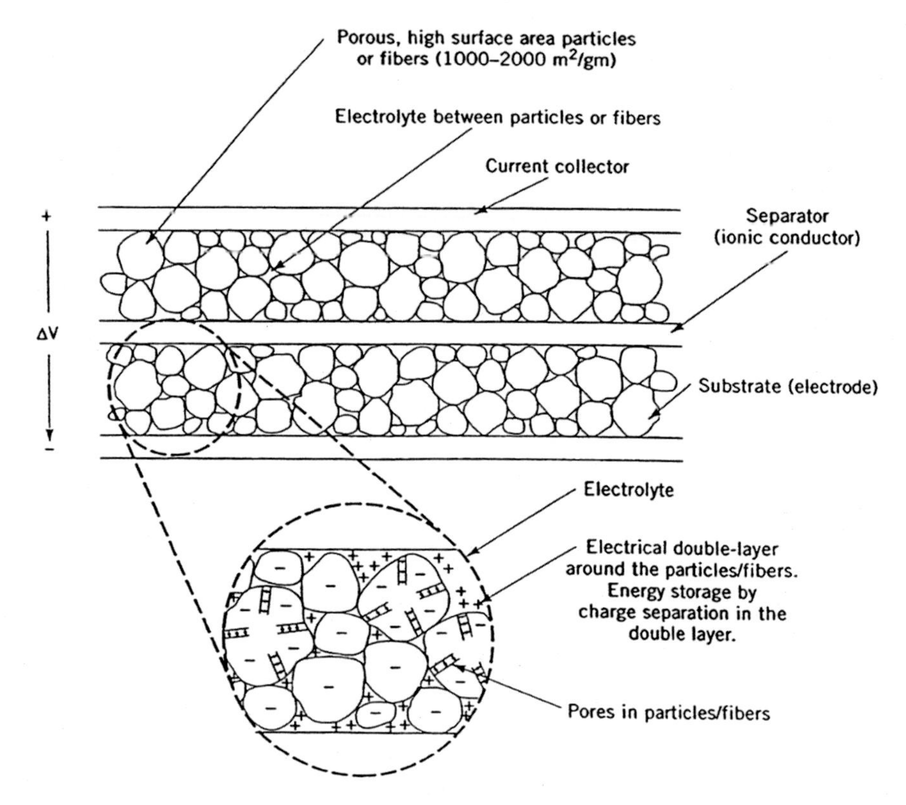

Fig. 7 Schematic of an electrochemical capacitor. Reprinted with permission from [51]. Copyright 2000, Elsevier

During the design of the powertrain, often lithium-ion batteries are chosen due to their higher energy density. However, this choice leads to oversized batteries to attain the required power performance and leads to reduced efficiency of the vehicle. Burke et al. also conducted a simulation test on light-duty vehicles which are powered by mild hybrid and fuel cell drivelines [55]. Although the energy density of the ECs is typically ten times smaller than batteries, its efficiency is higher than that of the battery, and this could significantly influence the overall energy efficiency of the vehicle. Thus, employing ECs or HECs in the electric drivetrains is highly efficient to meet the maximum power required by large motors, especially for fuel-cell vehicles. Additionally, ECs exhibit energy regenerative characteristics and could efficiently absorb the braking energy with an efficiency of up to 88% [56]. Thus, hybridizing ECs with other energy storage systems such as battery/fuel cells could lead to longer cycle life and improve the unit's overall energy efficiency [57].

4 Fuel Cell

The general concept of a fuel cell stack is the controlled release of stored chemical energy in the form of electrical power instead of heat. This is analogous to the internal combustion engine system in that it combines fuel and oxygen, but instead of combustion moving pistons, the energy is extracted directly as an electromotive impulse which can be used to drive vehicular motor powertrains.

The thermodynamic reaction of hydrogen and oxygen packs about 33,600 Wh kg^{-1} of energy (on a hydrogen mass basis, assuming the oxygen is derived from ambient air), which is about two orders of magnitude higher than conventional battery chemistries. To illustrate, 10 g of hydrogen would be the equivalent of about 1 kg of battery cell on a one-time energy discharge basis. This is how much more energy-dense fuel-cell concepts can potentially be, and hence the attractiveness. Further, fuel cells allow for semi-permanent continuous operation, unlike batteries and supercapacitors, which need to be periodically recharged. Lastly, the fuel cell infrastructure network is very much analogous to the fossil-fuel supply chain already deployed ubiquitously; hence social engineering can be expected to be quite straightforward for mass transport adoption, and pre-existing gas stations can be earmarked for direct conversion as hydrogen refueling stations [58].

4.1 Fuel Cell Technologies: Considerations for Transport Applications

There are several types of fuel cells, as shown in Fig. 8. In general, fuel cell stacks can use cryogenic hydrogen, compressed hydrogen, metal hydrides, ammonia, hydrocarbons, carbohydrates, or alcohols as fuels.

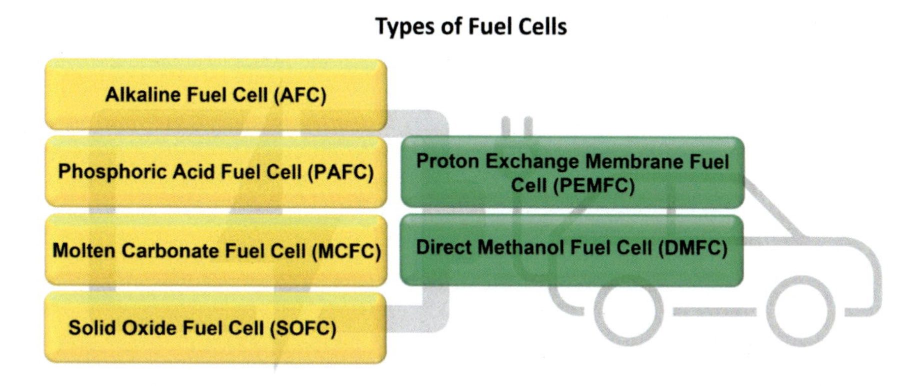

Fig. 8 An overview of fuel cell technologies

The most popular implementation is the Proton Exchange Membrane Fuel Cell (PEMFC) due to its relatively low-temperature operation (typically $\leq$ 100 °C) and the simplicity of mass exchange (hydrogen + oxygen $\rightarrow$ water), assuming high-purity inputs. The electrochemical potential across each cell is theoretically 1.23 V, subject to activation, concentration, and ohmic losses. Series stacking allows for higher voltage output, but lower current operation typically enables higher device efficiencies. However, there is an essential economic trade-off, as state-of-the-art PEMFCs typically use expensive Nafion® perfluorinated membranes and precious metal catalysts (Pt/Ir loadings typically on the order of 0.1–0.5 mg cm^{-2}) as critical components [59]. Therefore, there is great interest in operating at the maximum power point (best trade-off between efficiency and cell material) to recoup fabrication investments.

The architecture of a PEMFC involves anode/cathode bipolar plates (ABP, CBP), gas diffusion layers (AGDL/CGDL), and carbon catalyst support (ACC/CCC). The electrolyte, the coated catalyst layer, and the gas-diffusion layers are collectively known as the membrane electrode assembly (MEA). The anode generates positively charged protons from the hydrogen fuel. The protons traverse the electrolyte via a Grotthuss mechanism of ion-hopping across the Nafion® backbone. The cathode then catalyzes the hydrogen–oxygen recombination to produce water, which occurs with a relatively higher activation penalty (compared to the anode). The electrolyte is a solid membrane that must be constantly wet with controlled humidity to achieve optimum ion/mass transfer. The critical problem with PEMFCs is catalyst poisoning due to the carbon monoxide derived from reformed gas by-products or the ambient atmosphere, which permanently decimates cell efficiency [60, 61]. Hence, auxiliary components such as carbon monoxide-elimination systems, fluid circulation pumps, and electrical control devices are mandatory.

Direct methanol fuel cells (DMFCs) use the same setup as PEMFCs but utilize a liquid fuel (methanol), and hence the room-temperature liquid phase can be stored much more easily compared to gaseous hydrogen. Again, it can use a Nafion® membrane electrolyte and Pt/Ir catalysts at about the same temperature range as PEMFCs. The main problem is the low electrochemical performance as compared to hydrogen due to catalytic complications arising from the bulkier functional groups of methanol [62].

On the other hand, alkaline fuel cells (AFCs) operate in basic conditions using high-concentration potassium hydroxide as a hydroxyl-ion (OH^-) carrier. Here, the system has ion transport occurring from the cathode to the anode instead, allows for a broader range of operating temperatures, and enjoys lower activation penalties for oxygen splitting. The critical problem, however, is carbon dioxide poisoning [63]. The use of corrosive liquids at elevated temperatures also degrades the cell housing significantly.

Phosphoric acid fuel cells (PAFCs) are a cheap version of PEMFCs because they replace the perfluorinated polymer with a viscous phosphoric acid liquid wetting a silicon carbide support matrix. The key disadvantage is the requirement to keep the module temperature constantly above 42 °C, else solidification of the electrolyte will cause mechanical rupture of the cell assembly. The operating temperature is higher at 150 °C-200°C as well, and the system suffers from carbon dioxide poisoning [64].

Molten carbonate fuel cells (MCFCs) circumvent the problem of carbon monoxide/carbon dioxide poisoning by operating at higher temperatures (500–800 °C). Here, water is not the ion-carrier but $CO_3{}^{2-}$ ions. The advantage of the system is that it also allows the use of hydrocarbons to be directly used as fuel, besides hydrogen. However, having to maintain ultrahigh temperatures is not particularly conducive to maintaining vehicle structural health and energy efficiency [65].

$$\text{Anode: } H_2 + CO_3^{-2} \rightarrow H_2O + CO_2 + 2e^-$$
$$\text{Cathode: } 1/2O_2 + CO_2 + 2e^- \rightarrow CO_3^{2-}$$

Solid oxide fuel cells (SOFCs) utilize a ceramic membrane operating at a very high temperature (800–1200 °C) consisting of the cathode catalyst ($La_{1-x}Sr_xMnO_{3-\delta}$), electrolyte (yttrium-stabilized zirconia) and the anode catalyst (Ni-doped yttrium-stabilized zirconia). Much efforts are underway to bring down the operating temperature to intermediate 600–800 °C levels, but in general, the key advantages for SOFCs are the ruggedness to feed fuels, the ability to use low-cost catalyst elements and reduced poisoning concerns (except for oxygen-starved periods resulting in coking) [66]. Other main problems are the extreme brittleness which may make SOFCs problematic for vehicular ground applications, and the higher thermal cycling stresses.

Although there are numerous fuel cell subcategories as presented in Fig. 8 however, given that most FCEVs use the Proton Exchange Membrane Fuel Cell (PEMFC), the remainder of this section will focus on PEMFC-based vehicles. The other types of fuel cells are more suited for stationary energy storage/power generation at the current technological level [67].

4.2 *Fuel Cell Electric Vehicles (FCEVs)*

The automobile magnate Henry Ford was responsible for the mass proliferation of the internal combustion engine vehicles (ICEVs). The recent renaissance of electric drivetrains is pushed by a combination of government policies, technological advancement, economic-cost scaling, predicted eventual depletion of fossil fuel reserves and a greater emphasis on a renewable, non-polluting circular economy. Transitioning the passenger/cargo transport system to electric is viewed as the ultimate urgency of this decade due to catastrophic global destruction brought by climate change.

Presently, EVs and HEVs lead the way due to cost practicality and technological maturity, but fuel cell electric vehicles (FCEVs) are the holy grail if practical economic and engineering concerns can be addressed. Most commercially available FCEVs are actually a hybrid combination of a fuel cell, and battery pack (Fig. 9)

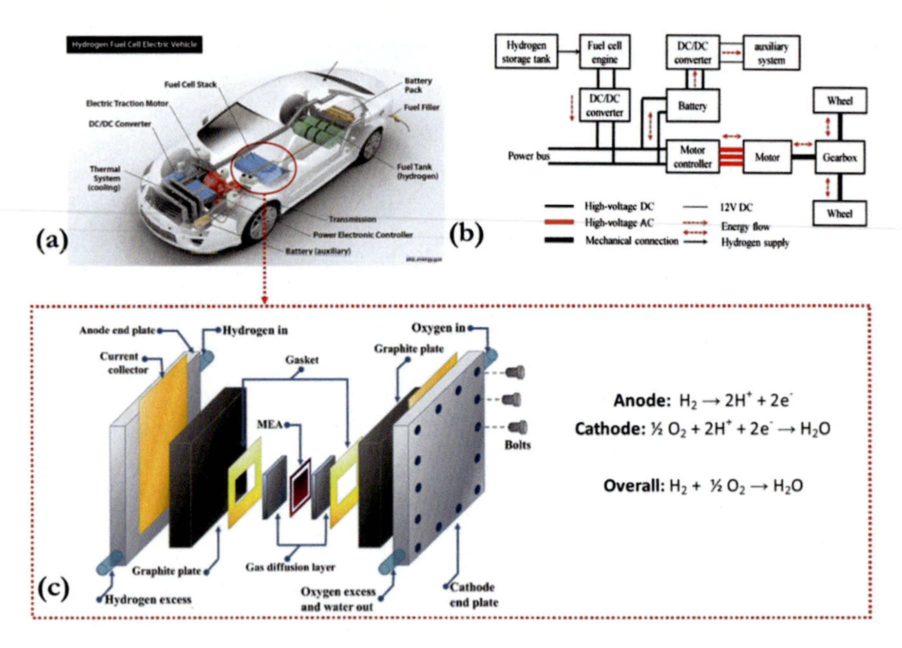

Fig. 9 **a** Hydrogen fuel cell electric vehicle. Adapted with permission from [68] Copyright 2021 U.S. Department of Energy Alternative Fuels Data Center. **b** FCEV powertrain configuration. Adapted with permission [69]. Copyright 2018 Elsevier. **c** Components of a single PEMFC. Adapted with permission from [70]. Copyright 2018 Elsevier.

[68–70] and the vehicles using this technology are more precisely termed as "hybrid fuel cell electric vehicles". Here, we will be referring to them as FCEVs.

4.3 Current Standing of FCEVs: Technical and Commercial Aspects

The comparisons of ICEV, HEV, EV, and FCEV are outlined in Table 2 [65, 67, 71, 72]. One of the most concerning factors to drive FCEVs into the market is cost. To build an infrastructure to support FCEVs, a large amount of capital is needed to develop, operate, and maintain hydrogen production facilities, refueling stations, and a logistics system for delivering hydrogen from production to refueling stations as indicated in Fig. 10 [73]. Also, the high cost of FCEVs production goes into the fuel cell stack due to the utilization of the precious metal catalyst, which is the most expensive component in each fuel cell. Using platinum as both the electrodes while conferring the highest performance will significantly drive up the cost price. Additionally, degradation mechanisms of the Pt electrodes such as oxidation, migration, loss of active surface area, and disintegration of the carbon support necessitate regular replacement of the Pt electrode assembly, thus leading to astronomical refurbishment costs [74]. As such, new materials or strategies to mitigate the degradation of Pt electrodes will be a significant way forward to ensure that FCEVs are made affordable for the masses. Currently, based from the IEA annual report in the year 2021, there are only 34,804 FCEVs registered worldwide, which is a stark contrast to the EV count with over 2 million units sold in the year 2020 alone [75]. Nevertheless, various types of FCEVs are continually being launched in the market, and their general technical specifications can be found in (Tables 3, 4, and 5). Regarding global FCEV distribution, Asia occupies the largest market share with 65%, followed by North America with 27% and the remaining in Europe [75]. The strong impetus to pursue FCEVs in Asia (especially in China and Japan), is due to the various government initiatives complemented by the existing and future hydrogen infrastructure plans [75, 76]. For instance, the refueling infrastructures for hydrogen and petroleum are similar, thus allowing consumers to transit easily to hydrogen refueling infrastructures. Other advantages include rapid refueling time as compared to EVs, modularity of components, reduced harmful emissions, and higher energy density [67]. Between 1990 and 2020, there were massive investments in fuel cell technology for electric vehicle applications (e.g., Daimler Chrysler's' much-publicized NECAR 4, Nissan's X-trail, Ford's Fusion Hydrogen 999, Toyota Mirai, etc.). However, there is an increasing epiphany for a predominantly battery-aided green transport transformation. Of the major automobile companies, Toyota is the lone stalwart warrior betting heaviest on hydrogen technology as part of Japan's hydrogen plan [76, 77].

Table 2 Comparisons ICEV, HEV, EV and FCEV

	ICEV	HEV	EV	FCEV
Time taken for refueling/recharging	<5 min	<5 min	<1 h	<10 s
Cost of refueling 1L/kWh/kg (price in USD)	~$2.00/L	~$2.00/L $0.14/Wh[a]	$0.14/Wh[a]	~$10/g[b]
Cost				
Maintenance				
Emission during operation	CO_2, NO_x, SO_x, H_2O	CO_2, NO_x, SO_x, H_2O	Nil	H_2O
Stationary engine noise level				

[a] Average electricity price in United States of America
[b] Green hydrogen pricing

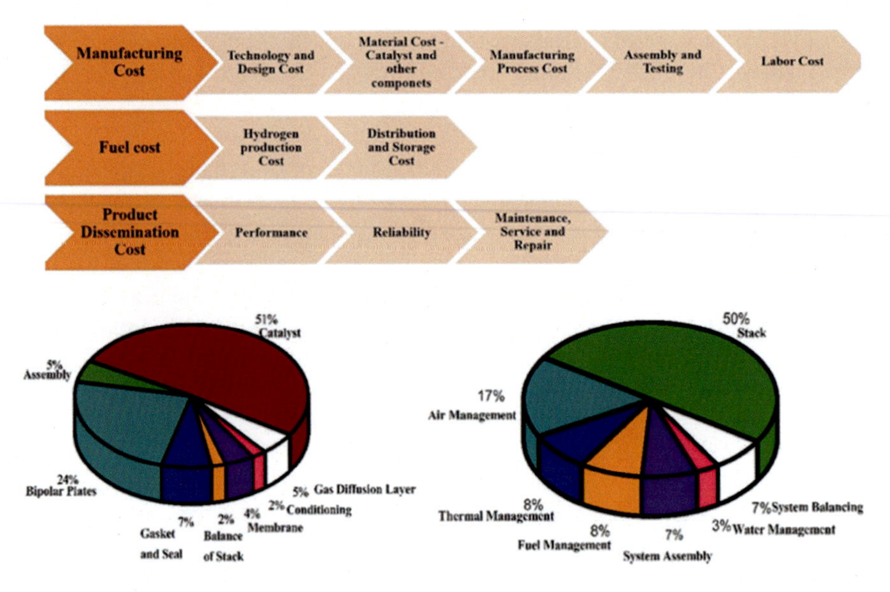

Fig. 10 Overview of cost for FCEVs implementation. Reproduced with permission [73]. Copyright 2021 John Wiley and Sons

5 Summary and Outlook

This chapter has provided an insight into the state-of-art electrochemical sources (batteries, supercapacitors, and fuel cells) employed in EVs and HEVs. Although lithium-ion battery technologies have dominated the EVs and HEVs market, its long-term future is daunted by its limited resources and alternative energy storage chemistries or "beyond lithium-ion technologies" must be explored. Furthermore, the safety of LIBs needs to be improved to tackle issues including accidental mechanical damage and thermal runaway. A solid-state battery is a promising solution due to its non-usage of flammable liquid organic electrolyte that eliminates the fire hazard. However, the practical application of solid-state batteries is still hindered due to sluggish kinetics, low conductivity, and poor cycle life. Nevertheless, there are several joint collaborations between battery companies and car manufacturers investing in a safer energy storage future. Electrochemical capacitors were found to possess significantly higher power performance than batteries/fuel cells, and hence, the utilization of ECs and HECs in the drivetrain would significantly improve the energy efficiency of the EVs and HEVs. Increasing the energy density of ECs is expected to reduce the cost ($\$ Wh^{-1}$), and the higher overall efficiency of the energy storage unit would make the implementation of ECs and HECs plausible. Additionally, research into developing novel design aspects integrating the ECs into the structure of EVs could further improve the design and minimize the space requirements. On the other hand, vehicles powered by fuel cell systems will ultimately require an ancillary battery system

Table 3 Technical specifications of various fuel cell electric vehicles (light passenger vehicles)

Light passenger vehicles		Fuel cell stack	Tank system		Battery			Electric motor		Performance		Driving range	Ref.
Vehicle company	Vehicle model	Max power (kW)	Working pressure (bars)	Fuel weight capacity (kg)	Type	Voltage (V)	Capacity (kW h)	Power output (kW)	Max torque (Nm)	Acceleration [0 to 100 km/h] (s)	Top speed (km/h)	WLTP[b] NEDC[c] (km)	
Toyota	Mirai	128	700	5.6	Lithium-ion	310.8	1.24	134	300	9.2	175	650[b]	[75]
Hyundai	Nexo	95	–	6.33	Lithium-ion polymer	240	1.56	120	395	9.2	179	666[b] 756[c]	[75]
Honda	Clarity Fuel Cell	103	700	5.46	Lithium-ion	346	–	130	300	9	165	650[c]	[75]
SAIC	EUNIQ 7	130	700	-	Lithium-ion	–	–	150	–	–	–	605[c]	[78, 79]
Mercedes Benz	GLC F-cell	–	700	4.4	Lithium-ion	–	13.5	155	365	–	160	478[c]	[75]
Hyundai	ix35 (2013)	100	700	5.64	Lithium-ion polymer	–	–	100	300	12.5	160	594[c]	[80]
Ford	Focus (FCEV)	–	344	–	Ni-MH	–	–	65	230	–	128	320[c]	[81]
Hyundai	Vision FKa	95	–	4	–	–	60	500	–	3.9	260	>600[c]	[82]
BMW	I Hydrogen Nexta	–	700	6	–	–	–	125	–	–	–	–	[83]

[a] Test bed vehicles
[b] WLTP—Worldwide Harmonised Light Vehicle Test Procedure
[c] NEDC—New European Driving Cycle

Table 4 Technical specifications of various fuel cell electric vehicles (trucks/heavy mover)

Trucks/heavy movers		Fuel Cell stack	Tank system		Battery		Electric motor		Performance	Driving Range	Ref.
Vehicle company	Vehicle model	Max power (kW)	Working pressure (bars)	Fuel weight capacity (kg)	Voltage (V)	Capacity (kWh)	Power output (kW)	Max torque (Nm)	Top speed (km/hr)	WLTP[a] NEDC[b](km)	
Hyzon	Class 8 Fuel cell electric truck (FCET 6)	100	350	40	600	55	320	1799	104	563	[84]
Hyzon	Class 8 Fuel cell electric truck (FCET 8)	120	350	50–70	700	110	320	1599	88	603–804	[84]
Hyundai	Xcient Fuel Cell	190	350	32.09	661	73.2	350	3400	85	~400	[85, 86]
Toyota	Project Portal Beta	226	700	40	–	12	499	1796	–	482	[87, 88]

[a] WLTP—Worldwide Harmonised Light Vehicle Test Procedure
[b] NEDC—New European Driving Cycle

Table 5 Technical specifications of various fuel cell electric vehicles (bus/others)

Bus/others		Fuel cell stack	Tank system		Battery		Electric motor		Performance	Driving range	Ref.
Vehicle company	Vehicle model	Max power (kW)	Working pressure (bars)	Fuel weight capacity (kg)	Type	Capacity (kWh)	Power output (kW)	Max torque (Nm)	Top speed (km/hr)	WLTP[a] NEDC[b] (km)	
Hyzon	Passenger Coach	80	350	35	–	141	195	–	–	402[a]	[84]
SAIC	Maxus FCEV80	30	350	4.4	Lithium-ion (LFP)	3	100	350	100	305[b]	[89]
Renault	Kangoo Z.E Hydrogen	5	350/700	–	–	33	44	–	–	350[b]	[90]

[a] WLTP—Worldwide Harmonised Light Vehicle Test Procedure
[b] NEDC—New European Driving Cycle

anyway to support resolving cold-start issues and sluggish dynamic response (acceleration/deceleration). Of note, fuel cells operate at elevated temperatures; hence energy must be expended just for heating, and thus wastes a certain percentage of the fuel efficiency. Further, the chemical engineering of hydrogen production is usually limited to lower than 80% efficiency; hence there are energy losses from the onset, which makes energy regeneration based on renewables an absolute must to make any environmental/economic impact for mass adoption of fuel cell vehicle concepts. Many efforts must also be directed toward reducing precious metal catalyst loadings to lower the overall fuel cell costs. In particular, the heavy-push narrative for fuel cell adoption (over batteries) is based on:

1. Ultra-heavy-duty transport crafts such as trucks/aeroplanes/ships/submarines/space shuttles were to meet the energy requirements, battery module scale-up absurdly compounds the weight and volume of the energy core, as compared to a fuel cell system.
2. Off-the-grid refueling stations where the energy vector (hydrogen fuel) can be stored with a minimal real-estate footprint or transported in extremely huge mega-joule chunks (as opposed to electrical grid transmission where multiple step-down/step-up transformer installations add complexity to where battery banks can be located)
3. Waste heat recovery flows wherein the thermal energy losses can be upcycled to power chemical industry, bio farming, and habitat requirements.

Among all the fuel-cell chemistries available, PEMFCs may continue to dominate retail vehicular applications if there is sufficient economic incentive to do so, primarily due to engineering considerations of lower operating temperature, easier structural maintenance, less mechanical fatigue, and lower probability of catastrophic failure. However, the process infrastructure costs for producing, storing, transporting, and utilizing the "fuel-cell fuels" are still about an order of magnitude higher compared to full battery deployment. This is on top of land-area footprint considerations, safety, and concerns about where the fuel hydrogen/methanol/hydrocarbons ultimately derive from.

Finally, phasing out ICEVs by replacing them with greener EVs and HEVs remains one of the essential strategies to tackle the greenhouse gas emission associated with climate change. Therefore, to drive toward widespread market adoption of EVs and HEVs, a combination of government policies with infrastructures must be in place and joint research-industrial collaborations devoted with holistic considerations in terms of sustainability, safety, performance, and cost.

References

1. Lacis AA, Schmidt GA, Rind D, Ruedy RA (2010) Atmospheric CO_2: Principal control knob governing Earth's temperature. Science 330(6002):356–359
2. Ritchie H, Roser M (2020) CO_2 and greenhouse gas emissions. Our World in Data

3. Tie SF, Tan CW (2013) A review of energy sources and energy management system in electric vehicles. Renew Sustain Energy Rev 20:82–102
4. Li Z, Khajepour A, Song J (2019) A comprehensive review of the key technologies for pure electric vehicles. Energy 182:824–839
5. Jorgensen K (2008) Technologies for electric, hybrid and hydrogen vehicles: electricity from renewable energy sources in transport. Util Policy 16(2):72–79
6. Aijaz I, Ahmad A (2022) Electric vehicles for environmental sustainability. In: Smart technologies for energy and environmental sustainability. Springer, pp 131–145
7. Policies to promote electric vehicle deployment. https://www.iea.org/reports/global-ev-outlook-2021/policies-to-promote-electric-vehicle-deployment
8. Chau K, Chan CC (2007) Emerging energy-efficient technologies for hybrid electric vehicles. Proc IEEE 95(4):821–835
9. Winter M, Brodd RJ (2004) What are batteries, fuel cells, and supercapacitors? Chem Rev 104(10):4245–4270
10. Kouchachvili L, Yaïci W, Entchev E (2018) Hybrid battery/supercapacitor energy storage system for the electric vehicles. J Power Sources 374:237–248
11. Tanç B, Arat HT, Baltacıoğlu E, Aydın K (2019) Overview of the next quarter century vision of hydrogen fuel cell electric vehicles. Int J Hydrogen Energy 44(20):10120–10128
12. Ezzat M, Dincer I (2018) Development, analysis and assessment of fuel cell and photovoltaic powered vehicles. Int J Hydrogen Energy 43(2):968–978
13. Tian Y et al (2020) Promises and challenges of next-generation "beyond Li-ion" batteries for electric vehicles and grid decarbonization. Chem Rev 121(3):1623–1669
14. Liu P, Ross R, Newman A (2015) Long-range, low-cost electric vehicles enabled by robust energy storage. MRS Energy & Sustain 2
15. Ding Y, Cano ZP, Yu A, Lu J, Chen Z (2019) Automotive Li-ion batteries: current status and future perspectives. Electrochem Energy Rev 2(1):1–28
16. Porzio J, Scown CD (2021) Life-cycle assessment considerations for batteries and battery materials. Adv Energy Mater 11(33):2100771
17. Lu L, Han X, Li J, Hua J, Ouyang M (2013) A review on the key issues for lithium-ion battery management in electric vehicles. J Power Sources 226:272–288
18. Radin MD et al (2017) Narrowing the gap between theoretical and practical capacities in Li-ion layered oxide cathode materials. Adv Energy Mater 7(20):1602888
19. Kim TH, Park JS, Chang SK, Choi S, Ryu JH, Song HK (2012) The current move of lithium ion batteries towards the next phase. Adv Energy Mater 2(7):860–872
20. Hu L-H, Wu F-Y, Lin C-T, Khlobystov AN, Li L-J (2013) Graphene-modified $LiFePO_4$ cathode for lithium ion battery beyond theoretical capacity. Nat Commun 4(1):1–7
21. Sun HH et al (2021) Transition metal-doped Ni-rich layered cathode materials for durable Li-ion batteries. Nat Commun 12(1):1–11
22. Blomgren GE (2016) The development and future of lithium ion batteries. J Electrochem Soc 164(1):A5019
23. Schmuch R, Wagner R, Hörpel G, Placke T, Winter M (2018) Performance and cost of materials for lithium-based rechargeable automotive batteries. Nat Energy 3(4):267–278
24. Manthiram A, Song B, Li W (2017) A perspective on nickel-rich layered oxide cathodes for lithium-ion batteries. Energy Storage Mater 6:125–139
25. Kallitsis E, Korre A, Kelsall G, Kupfersberger M, Nie Z (2020) Environmental life cycle assessment of the production in China of lithium-ion batteries with nickel-cobalt-manganese cathodes utilising novel electrode chemistries. J Clean Prod 254:120067
26. Sharma SS, Manthiram A (2020) Towards more environmentally and socially responsible batteries. Energy Environ Sci 13(11):4087–4097
27. (2020) Battery Pack Prices Cited Below $100/kWh for the First Time in 2020, While Market Average Sits at $137/kWh. https://about.bnef.com/blog/battery-pack-prices-cited-below-100-kwh-for-the-first-time-in-2020-while-market-average-sits-at-137-kwh/
28. Ryu H-H, Sun HH, Myung S-T, Yoon CS, Sun Y-K (2021) Reducing cobalt from lithium-ion batteries for the electric vehicle era. Energy Environ Sci 14(2):844–852

29. Zhao Q, Stalin S, Zhao C-Z, Archer LA (2020) Designing solid-state electrolytes for safe, energy-dense batteries. Nat Rev Mater 5(3):229–252
30. Kato Y et al (2016) High-power all-solid-state batteries using sulfide superionic conductors. Nat Energy 1(4):1–7
31. Battery Startup Comes Out of Stealth to Seize On Investor Frenzy. https://www.bloomberg.com/news/articles/2021-04-20/battery-startup-comes-out-of-stealth-to-seize-on-investor-frenzy
32. Short Seller Targets QuantumScape, Battery Startup Responds. https://insideevs.com/news/501549/short-seller-targets-quantumscape-ssb/
33. Solid-State Battery: Mercedes, Stellantis Invest In Factorial Energy. https://insideevs.com/news/551374/mercedes-stellantis-factorial-energy-batteries/
34. Nissan Announces Proprietary Solid-State Batteries: $75/kWh Pack. https://insideevs.com/news/551144/nissan-proprietary-solid-state-batteries/
35. Yabuuchi N, Kubota K, Dahbi M, Komaba S (2014) Research development on sodium-ion batteries. Chem Rev 114(23):11636–11682
36. Hosaka T, Kubota K, Hameed AS, Komaba S (2020) Research development on K-ion batteries. Chem Rev 120(14):6358–6466
37. Kubota K, Dahbi M, Hosaka T, Kumakura S, Komaba S (2018) Towards K-ion and Na-ion batteries as "beyond Li-ion." Chem Rec 18(4):459–479
38. Shah R, Mittal V, Matsil E, Rosenkranz A (2021) Magnesium-ion batteries for electric vehicles: current trends and future perspectives. Adv Mech Eng 13(3):16878140211003398
39. Gummow RJ, Vamvounis G, Kannan MB, He Y (2018) Calcium-ion batteries: current state-of-the-art and future perspectives. Adv Mater 30(39):1801702
40. Chua R et al (2020) Hydrogen-bonding interactions in hybrid aqueous/nonaqueous electrolytes enable low-cost and long-lifespan sodium-ion storage. ACS Appl Mater Interfaces 12(20):22862–22872
41. Chua R et al (2019) 1.3 V superwide potential window sponsored by Na-Mn-O plates as cathodes towards aqueous rechargeable sodium-ion batteries. Chem Eng J 370:742–748
42. El Kharbachi A, Zavorotynska O, Latroche M, Cuevas F, Yartys V, Fichtner M (2020) Exploits, advances and challenges benefiting beyond Li-ion battery technologies. J Alloy Compd 817:153261
43. Bervas M, Klein L, Amatucci G (2005) Vanadium oxide–propylene carbonate composite as a host for the intercalation of polyvalent cations. Solid State Ionics 176(37–38):2735–2747
44. Petnikota S et al (2020) An Insight into the Electrochemical Activity of Al-doped V_2O_3. J Electrochem Soc 167(10):100514
45. Cai Y et al (2020) Bronze-type vanadium dioxide holey nanobelts as high performing cathode material for aqueous aluminium-ion batteries. J Mater Chem A 8(25):12716–12722
46. Kumar S, Salim T, Verma V, Manalastas W Jr, Srinivasan M (2022) Enabling Al-metal anodes for aqueous electrochemical cells by using low-cost eutectic mixtures as artificial protective interphase. Chem Eng J 134742
47. Muldoon J, Bucur CB, Gregory T (2014) Quest for nonaqueous multivalent secondary batteries: magnesium and beyond. Chem Rev 114(23):11683–11720
48. Li Z, Fuhr O, Fichtner M, Zhao-Karger Z (2019) Towards stable and efficient electrolytes for room-temperature rechargeable calcium batteries. Energy Environ Sci 12(12):3496–3501
49. Simon P, Gogotsi Y (2020) Perspectives for electrochemical capacitors and related devices. Nat Mater 19(11):1151–1163
50. Burke A, Murphy T (1995) Material characteristics and the performance of electrochemical capacitors for electric/hybrid vehicle applications. MRS Online Proc Libr (OPL) 393
51. Burke A (2000) Ultracapacitors: why, how, and where is the technology. J Power Sources 91(1):37–50
52. Dowgiallo EJ, Hardin JE (1995) Perspective on ultracapacitors for electric vehicles. IEEE Aerosp Electron Syst Mag 10(8):26–31
53. Conway BE (1991) Transition from "supercapacitor" to "battery" behavior in electrochemical energy storage. J Electrochem Soc 138(6):1539

54. Conway BE (2013) Electrochemical supercapacitors: scientific fundamentals and technological applications. Springer Science & Business Media
55. Burke A, Liu Z, Zhao H (2014) Present and future applications of supercapacitors in electric and hybrid vehicles. In: 2014 IEEE international electric vehicle conference (IEVC). IEEE, pp 1–8
56. Zou Z, Cao J, Cao B, Chen W (2015) Evaluation strategy of regenerative braking energy for supercapacitor vehicle. ISA Trans 55:234–240
57. Naseri F, Farjah E, Ghanbari T (2016) An efficient regenerative braking system based on battery/supercapacitor for electric, hybrid, and plug-in hybrid electric vehicles with BLDC motor. IEEE Trans Veh Technol 66(5):3724–3738
58. Greene DL, Duleep G (2013) Status and prospects of the global automotive fuel cell industry and plans for deployment of fuel cell vehicles and hydrogen refueling infrastructure. Oak Ridge National Laboratory
59. Curtin DE, Lousenberg RD, Henry TJ, Tangeman PC, Tisack ME (2004) Advanced materials for improved PEMFC performance and life. J Power Sources 131(1–2):41–48
60. Baschuk J, Li X (2001) Carbon monoxide poisoning of proton exchange membrane fuel cells. Int J Energy Res 25(8):695–713
61. Becker H et al (2020) Operando characterisation of the impact of carbon monoxide on PEMFC performance using isotopic labelling and gas analysis. J Power Sources Adv 6:100036
62. Alias M, Kamarudin S, Zainoodin A, Masdar M (2020) Active direct methanol fuel cell: An overview. Int J Hydrogen Energy 45(38):19620–19641
63. Gülzow E (2004) Alkaline fuel cells. Fuel Cells 4(4):251–255
64. Hendry C, Harborne P, Brown J (2007) Niche entry as a route to mainstream innovation: learning from the phosphoric acid fuel cell in stationary power. Technol Anal & Strat Manag 19(4):403–425
65. Mekhilef S, Saidur R, Safari A (2012) Comparative study of different fuel cell technologies. Renew Sustain Energy Rev 16(1):981–989
66. Singh M, Zappa D, Comini E (2021) Solid oxide fuel cell: Decade of progress, future perspectives and challenges. Int J Hydrogen Energy 46(54):27643–27674
67. Sharaf OZ, Orhan MF (2014) An overview of fuel cell technology: Fundamentals and applications. Renew Sustain Energy Rev 32:810–853
68. How Do Fuel Cell Electric Vehicles Work Using Hydrogen? 2021. U.S. Department of Energy Alternative Fuels Data Center. Accessed January 31, 2022. afdc.energy.gov/vehicles/how-do-fuel-cell-electric-cars-work
69. Song K, Chen H, Wen P, Zhang T, Zhang B, Zhang T (2018) A comprehensive evaluation framework to evaluate energy management strategies of fuel cell electric vehicles. Electrochim Acta 292:960–973
70. Cruz-Martínez H et al (2019) Mexican contributions for the improvement of electrocatalytic properties for the oxygen reduction reaction in PEM fuel cells. Int J Hydrogen Energy 44(24):12477–12491
71. Salleh I, Zain MZ, Raja Hamzah R (2013) Evaluation of annoyance and suitability of a back-up warning sound for electric vehicles. Int J Automot Mech Eng 8:1267–77
72. Van Vliet OP, Kruithof T, Turkenburg WC, Faaij AP (2010) Techno-economic comparison of series hybrid, plug-in hybrid, fuel cell and regular cars. J Power Sources 195(19):6570–6585
73. Chandran M, Palanisamy K, Benson D, Sundaram S (2021) A review on electric and fuel cell vehicle anatomy, technology evolution and policy drivers towards EVs and FCEVs market propagation. The Chemical Record
74. Sealy C (2008) The problem with platinum. Mater Today 11(12):65–68
75. Remzi Can Samsun LA, Rex M, Stolten D (2021) Deployment status of fuel cells in road transport: 2021 update. https://www.ieafuelcell.com/fileadmin/publications/2021 Deployment_status_of_fc_in_road_transport.pdf
76. Popov S, Baldynov O (2018) The hydrogen energy infrastructure development in Japan. In: E3S web of conferences, vol 69. EDP Sciences, p 02001

77. Buckland K, Sano N (2017, 10th January 2022). Toyota clings to hydrogen bet while electric sales soar. https://www.bloomberg.com/news/articles/2017-11-27/electric-cars-success-leaves-toyota-isolated-with-hydrogen-bet
78. S. MAXUS (8th January 2022). Know your EUNIQ7 executive edition. https://www.saicmaxus.com/config.shtml?file=UlZWT1NWRXZOeS9vb1l6bWxML25pWWc9&price=/dHVuZWQ
79. S. MAXUS (8th January 2022). Euniq7. https://www.saicmaxus.com/euniq7.shtml
80. H. M. Center. (6th January 2022). Hyundai ix35 Fuel Cell. Available: Hyundai ix35 Fuel Cell
81. E. V. F.-. sdctcher. (11th January 2022). Ford begins production of 30 hydrogen-powered focus fuel cell vehicles for evaluation fleets. https://electricvehicleforums.com/forums/off-topic-9/ford-focus-fuel-cell-press-release-1995/
82. H. M. G. Tech. (2021, 5th January 2022). Vision FK, the world's first high-performance eco-friendly hydrogen electric vehicle. https://tech.hyundaimotorgroup.com/article/vision-fk-the-worlds-first-high-performance-eco-friendly-hydrogen-electric-vehicle/
83. B. group, Everyday testing of BMW i Hydrogen NEXT with hydrogen fuel cell drive train begins, ed, 2021
84. H. Motors (5th January 2022). Hyzon fuel cell vehicle brochure. https://hyzonmotors.com/wp-content/uploads/2021/09/022583_Hyzon-Sell-Sheets-V7-Spreads.pdf
85. Hyundai (5th January 2022). XCIENT Fuel Cell. https://trucknbus.hyundai.com/global/en/products/truck/xcient-fuel-cell
86. Hyundai (2022, 5th January 2022). Hyundai Motor and H2 Energy to bring the world's first fleet of fuel cell electric trucks into commercial operation. https://www.hyundai.co.nz/hyundai-motor-and-H2-energy-to-bring-the-world-s-first-fleet-of-fuel-cell-electric-trucks-into-commercial-operation
87. Satoh N, Nakashima T, Yamamoto K (2013) Metastability of anatase: size dependent and irreversible anatase-rutile phase transition in atomic-level precise titania. Sci Rep 3:1959
88. Toyota (5th January 2022). Toyota Doubles-Down on Zero Emissions Heavy-Duty Trucks. https://global.toyota/en/newsroom/corporate/23722307.html
89. S. MAXUS (8th January 2022). Know your FCV80 ,FCV80 fuel cell bus. https://www.saicmaxus.com/config.shtml?file=UmtOV09EQXZSa05XT0RBPQ==&price=/Q29uc3VsdA
90. R. Group. (12th January 2022). Kangoo Z.E. Hydrogen and Master Z.E. Hydrogen. https://www.renaultgroup.com/en/news-on-air/news/kangoo-z-e-hydrogen-and-master-z-e-hydrogen/

Battery Management Technologies in Hybrid and Electric Vehicles

Wei Liu and K. T. Chau

Abstract Hybrid electric vehicles (HEVs) and electric vehicles (EVs) have been advocated by global governments' policies in recent decades. Besides combating the climate crisis and urban air pollution, great contributions of developing the HEVs and EVs have been identified to accelerate the process of green transportation and smart city. Battery management is one of the most crucial functions for HEVs and EVs. It can ensure safe operation and optimize the performance of EV batteries. This chapter discusses the mainstream technologies of battery management in HEVs and EVs. Wherein, battery management technologies, including battery modeling, battery state estimation, safety prognostic (such as thermal management), and fault diagnosis, are elaborated in detail. Among them, the data-driven method is most effective and promising for battery state estimation (such as for state of charge and state of temperature) and health diagnosis or prognostics with impressive accuracy. Besides, some emerging management technologies, including multi-model co-estimation, artificial intelligence, cloud computing technology, and blockchain technology, are briefed, which can play a significant role in coordinating the information and energy flows in a vehicular information and energy internet.

1 Background

Conventional mobility has been experiencing a historic transition from the era of internal-combustion-engine vehicles to another of hybrid electric vehicles (HEVs) and electric vehicles (EVs) [1]. The developments of HEVs and EVs will make great contributions to promoting the accomplishment of carbon neutrality, green mobility, and smart city. Furthermore, they have numerous advantages of improving urban air quality, alleviating energy shortages, and combating the climate crisis. Therefore,

W. Liu · K. T. Chau (✉)
Department of Electrical and Electronic Engineering, The Hong Kong Polytechnic University, Hong Kong, China
e-mail: k.t.chau@polyu.edu.hk

W. Liu
e-mail: wei.liu@polyu.edu.hk

C. H. T. Lee (ed.), *Emerging Technologies for Electric and Hybrid Vehicles*,
Green Energy and Technology, https://doi.org/10.1007/978-981-99-3060-9_8

national governments have put in efforts to advocate by making policies. Typically, the United States announced that the sales share of EVs should reach 50% by 2030. China's New Energy Vehicle Industrial Development Plan for 2021 to 2035 ("Plan 2021–2035") aims to build a green, robust, and internationally competitive auto industry [2]. Future EVs may evolve into a power and information interface, which can help users to perform energy interaction with the modern power grid by vehicle-to-grid operation and information interaction with the cloud by wireless communication, respectively [3, 4]. However, advanced battery management is essential for achieving the above functions in a vehicular information and energy internet (VIEI).

As energy storage devices, batteries, and supercapacitors are commonly used in EVs and HEVs. Compared with the battery, the supercapacitor possesses much higher specific power (W kg^{-1}) but suffers from much lower specific energy (Wh kg^{-1}), thus improving the transient power handling capability [5]. As a result, the battery serves to provide the majority of energy capacity thanks to its high energy density, while the supercapacitor usually serves 5% of energy capacity only due to its high cost and large volume. For automobile applications, the main challenges are on three key performance indicators of (1) safety issues, (2) energy density, and (3) fast-charging capability. Accordingly, lithium-ion batteries (LIBs) outperform the nickel-based batteries and lead-acid batteries, hence dominating the current battery industry for HEVs and EVs [6, 7].

Divided by electrode materials for anode and cathode, the common types of LIBs are listed in Table 8.1. Superior to nickel-based batteries and lead-acid batteries, all these LIBs can provide high energy density for HEVs and EVs [8]. Although the lithium-titanium battery suffers from relatively lower energy density than the state-of-the-art LIBs do, it has a much better fast-charging capability thanks to its good charging acceptance. Plenty of battery cells are connected in series for meeting the requirement of voltage level and in parallel for improving the energy capacity of a battery pack in EVs. Due to the manufacturing difference, the cell balancing, thermal management, and aging issue are required to be concerned during the battery charging and discharging. Optimization of charging and discharging profiles can facilitate the battery to maintain high energy capacity and long remaining useful life. Battery management technology can protect the battery from various faults and perform optimal battery performances.

Table 8.1 Common lithium-ion batteries

Cathode	Anode
• $LiCoO_2$ • $LiMn_2O_4$ • $LiFePO_4$ • $Li[Ni_xCo_yMn_z]O_2$ (such as $Li[Ni_{0.8}Co_{0.1}Mn_{0.1}]O_2$), and $Li[Ni_{0.6}Co_{0.2}Mn_{0.2}]O_2$ • $Li[Ni_xCo_yAl_z]O_2$ (such as $Li[Ni_{0.8}Co_{0.15}Al_{0.05}]O_2$)	• $Li_4Ti_5O_{12}$ • Soft carbon • Hard carbon • Graphite • Silicon/graphite

2 Battery Management System

2.1 Key Concepts

High-energy batteries will play a significant role in powering EVs. Therefore, their safety, reliability, and efficient operations are the main concerns of consumers. To achieve these goals, a battery management system (BMS) is required to monitor and manage the battery conditions. Some key concepts regarding battery management are introduced as follows.

2.1.1 State of Charge

There are many definitions of state of charge (SoC). In this work, the SoC is quantified as the percentage of the reserve capacity to the rated capacity under the same specified condition. For example, the 100% or 0% SoC indicates that the battery is fully charged or discharged, respectively. Besides the safe and reliable operation, the SoC aids to determine the optimal strategies for charging and discharging. Furthermore, accurate estimation of SoC can be used to predict the remaining useful life (RUL).

2.1.2 State of Health

Importantly, the state of health (SoH) signifies a figure of merit of battery conditions compared to its ideal conditions. It can be derived by many parameters, such as internal resistance, AC impedance, battery capacity, power density, and discharge rate. The most common definition of SoH is the ratio of the maximum available capacity to the rated value of the battery. The 100% SoH means a fresh battery. While a battery with less than 80% SoH will be out of use due to the capacity of below 80% of the rated capacity. Another way to estimate SoH is to compare the internal resistance with its initial value. If the internal resistance grows to 1.3 times its initial value, the battery should also be retired.

2.1.3 State of Temperature

State of temperature (SoT) is a crucial factor for the thermal management of batteries. High operation temperature will reduce both the cycle life and the performance of batteries. Worse still, the battery will cause fire and explode due to thermal runaway. On the other hand, low operation temperature may disable the battery charge, and thus battery preheating is necessary. Although the battery surface temperature can be readily measured by thermal sensors, the SoT is more relevant to the internal temperature that directly influences the electrochemical conditions.

2.1.4 State of Function

State of function (SoF) is to describe how a battery can meet the actual demand. It relates to many factors, such as SoC, SoH, SoT, and charge/discharge history. The prediction of SoF is quite significant for the reliable operation of the EV battery system.

2.1.5 State of Balance

Due to the difference in the manufacturing process, the concept of the state of balance (SoB) is used to define the battery cell-to-cell consistency. Once with imbalance, overcharging may happen in a cell after all others get fully charged, which may cause distortion, leakage, rise in pressure, and even explosion. In contrast, over-discharging may occur in a cell after all others get fully discharged, which will shorten the cycle life. Accordingly, the detection and management of SoB will benefit the cell balancing in a battery pack.

2.2 *Basic Principles*

Physical damage, performance degradation, and thermal runaway will lead to battery malfunctions or catastrophic failure. A BMS is deployed to prevent the EV battery from experiencing such adverse cases. The BMS is a complex system including data acquisition, modeling, state estimation, charging control, thermal management, fault diagnosis, and communication. The parameters of voltage, current, and temperature of each cell are sensed and processed in a BMS. Wherein, the data handling is massive, and the data communication is usually based on a controller area network transferring data to microprocessors and other units. Accordingly, the main functions of the battery management system (BMS) are summarized as shown in Fig. 8.1, and they can be elaborated as follows.

(1) Battery state estimation, including SoC, SoH, SoT, etc.
(2) Thermal management to avoid thermal runaway in batteries.
(3) Fault diagnosis and assessment, battery protection, and data acquisition.
(4) Cell balancing and equalization.
(5) Coordination with vehicle control unit and other units including the power management and charge–discharge control [9].

As a series of advanced functions are integrated into a BMS, Fig. 8.2 shows a whole block diagram of a BMS to introduce the correlation between management functions and data communication [10]. A block of controlled transceiver serves for transmitting and receiving data. The parameters of voltage V_k, current I_k, and temperature T_k are measured at the cell level by using a sensing block. These parameters are

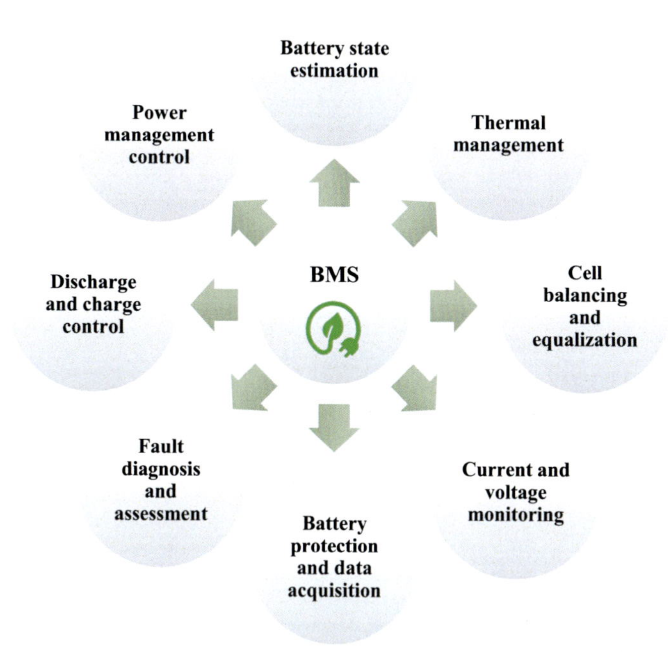

Fig. 8.1 Battery management system in hybrid and electric vehicles

used for battery state estimation of SoC, SoH, SoT, and so on. For battery thermal management, the fan or heater will be controlled to maintain the battery temperature within an optimal range. Meanwhile, the battery capacity estimation can serve to assess the energy capacity, diagnose the health status, and produce the limitations of charge–discharge current. Besides, the cell equalizer is in charge of cell balancing, and multi-dimensional constraints will be generated to prevent the irregularities of over-charge and over-discharge in partial cells. A block of fault diagnosis offers crucial functions of fault prognostics and troubleshooting, thus guaranteeing the operating safety of the battery.

In Fig. 8.3, the battery management technologies mainly include four primary parts: (1) battery modeling, (2) battery state estimation, (3) safety prognostics and health diagnosis, and (4) emerging management technologies. Wherein, the data-driven method is currently recognized as one of the most promising methods for battery management. The emerging management technologies can be further divided into four secondary parts: (1) multi-mode co-estimation, (2) artificial intelligence technology, (3) cloud computing technology, and (4) blockchain technology. In the following content, all relevant technologies will be introduced one by one.

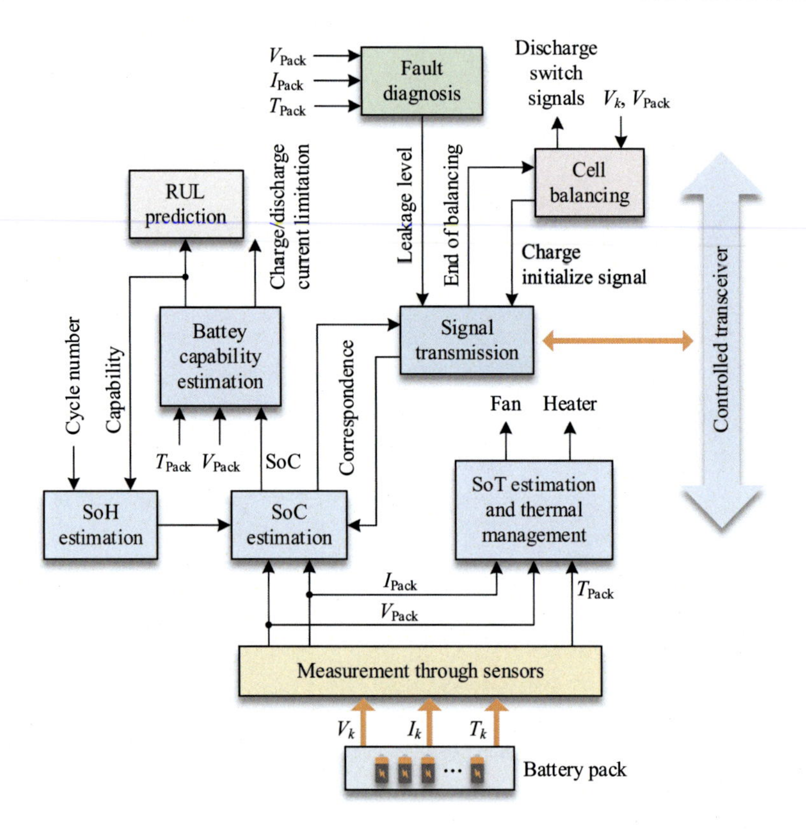

Fig. 8.2 Block diagram of battery management system

3 Battery Modeling Method

To collect the experiment data, Fig. 8.4 shows a battery testing system for the cycling test of battery cells or modules. This battery testing system mainly contains the following five parts: (1) Battery cells or modules, (2) a thermal chamber, (3) a switch board, (4) a battery cycle tester, and (5) a computer with ethernet cables. The computer is in charge of battery state monitoring and data acquisition under controlled ambient temperatures. The dataset from battery cycling is used to build or train the battery model that can extract and redefine the features of battery cells or modules. On top of the battery model, the battery state estimation, fault diagnosis, and health prognostic can be achieved with the help of various advanced algorithms. The electrochemical impedance spectroscopy (EIS) technique is usually used to measure the alternating-current (AC) impedance at different frequencies. Meanwhile, the Nyquist plot of AC impedances may assist to choose a suitable type of battery model that is an important prerequisite for battery management.

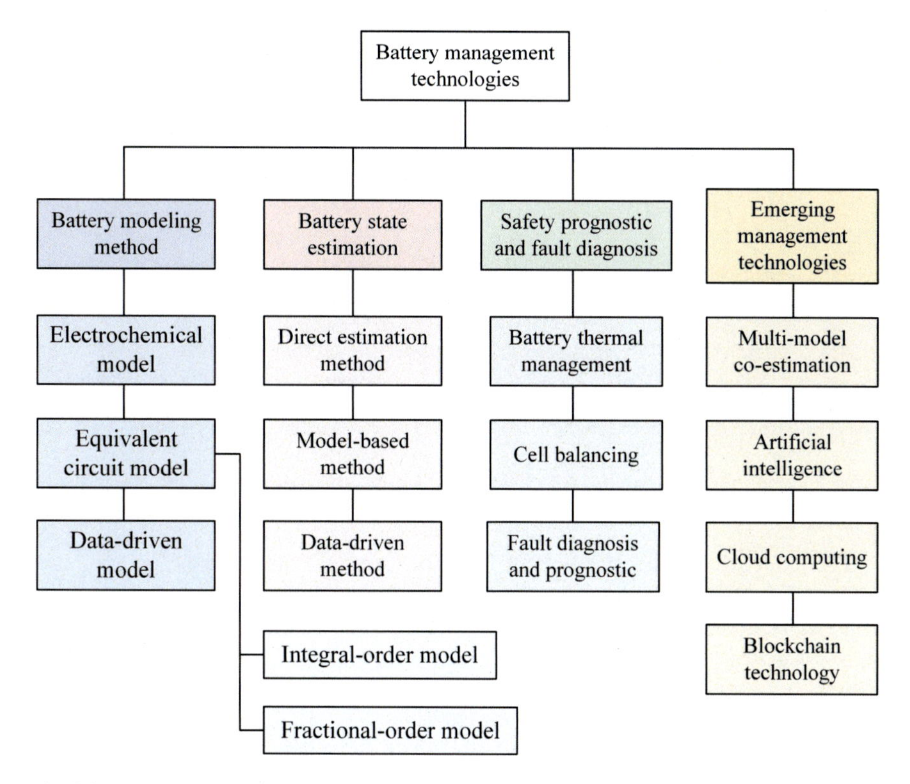

Fig. 8.3 Battery management technologies in hybrid and electric vehicles

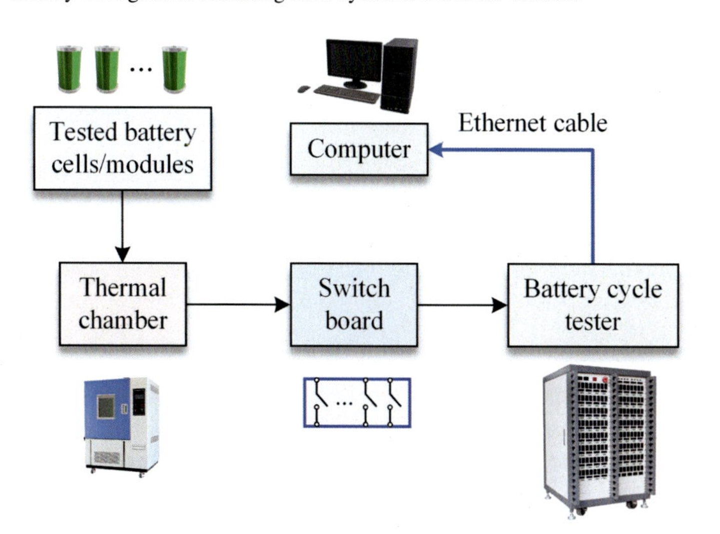

Fig. 8.4 Battery testing system for experiment data collection

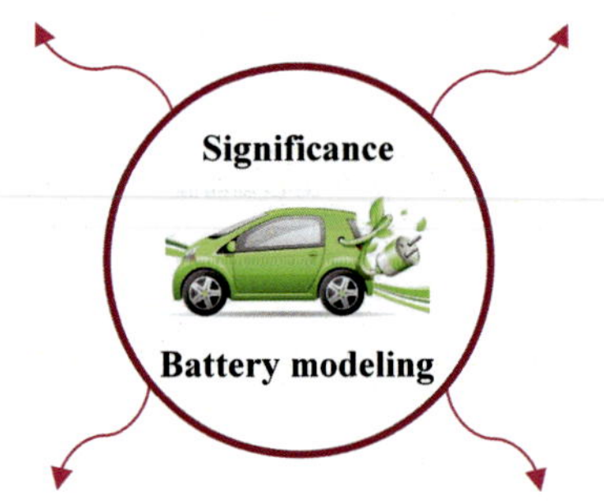

Fig. 8.5 Main significances of battery modeling for battery management

The main significances of battery modeling methods are summarized as shown in Fig. 8.5 for battery management technologies. To offer an accurate representation of a battery, three battery modeling methods are introduced in detail, including (1) electrochemical models, (2) equivalent circuit models (ECMs), and (3) data-driven models. Figure 8.6 shows the current evolution trends of three battery modeling approaches for battery management in EV applications [11]. Thanks to the high model accuracy and acceptable complexity, the data-driven method is recognized as one of the most promising methods at the current stage.

3.1 Electrochemical Model

Battery, exhibited as an electrochemical device, can be represented by physics-based approaches, which can exhibit a consistency of external characteristics between the practical battery and its model [12]. Wherein, by ignoring the influence of concentration distribution and potential on the terminal voltage, the single-particle model in Fig. 8.6 (bottom) owns the features of good simplicity and mature technology but relatively low accuracy [13]. According to the complex non-linear chemical reactions, the electrochemical models are to address the cores of batteries at the microscopic scale. Thus, there exists no doubt that these models represent the most accurate and detailed information about a battery. However, the main barriers blocking the widespread use of electrochemical models are two aspects. On the one hand, global optimization approaches should be used to solve plenty of non-analytical solutions. On the other hand, the correlation between control equations and boundary conditions suffers from a strong coupling [12]. Besides the huge memory requirements and computational burdens, the optimization procedure is inevitably time-consuming due

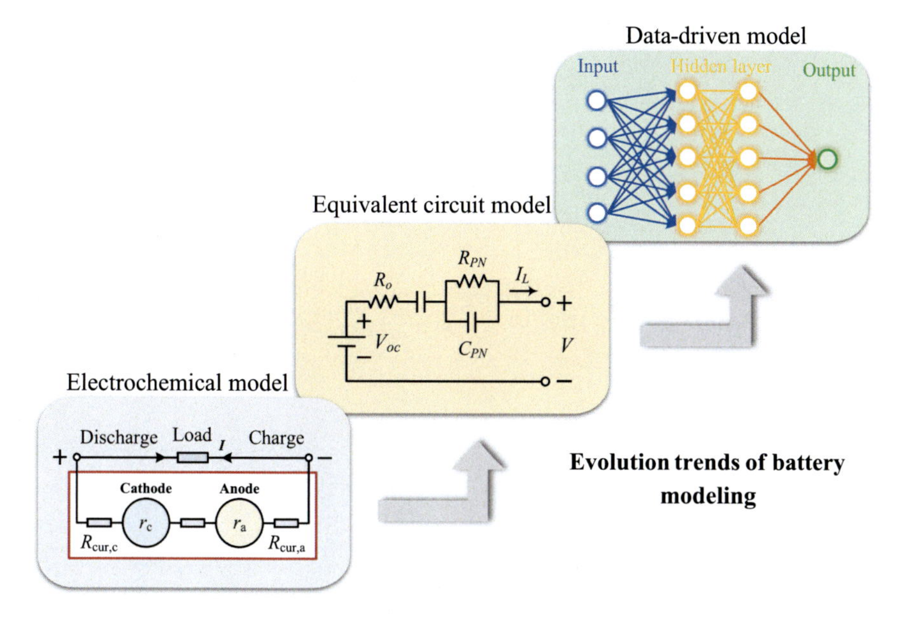

Fig. 8.6 Evolution trends of battery modeling for battery management

to the concern of convergence rate. Fruitful research works are focusing on the development of battery electrochemical models, such as a pseudo-two-dimensional model [14] and a reduced-order electrochemical model [15]. Nevertheless, the variations of battery temperatures and aging effects will significantly increase the difficulty in improving the accuracy of battery modeling. Accordingly, a thermal-electrochemical model develops into a research hotspot recently [16].

3.2 Equivalent Circuit Model

As aforementioned, the complexity of the electrochemical model limits its popularization and practicality, thus leading researchers to investigate another model, namely, the equivalent circuit models (ECMs). Four basic ECMs are shown in Fig. 8.7 [17], which adopt the lump resistor(s), inductor(s), or capacitor(s) for battery representation. Hence, the basic electrical components are used to build such a kind of battery model, which can imitate the battery behaviors and thus increase the model applicability. The ECMs are recognized as a more straightforward method by researchers, which can help the BMS control the power flow of EV batteries. There are two promising types of high-order ECMs: (1) Integral-order models (IOMs), and (2) fractional-order models (FOMs), as shown in Fig. 8.8a and b [3], respectively.

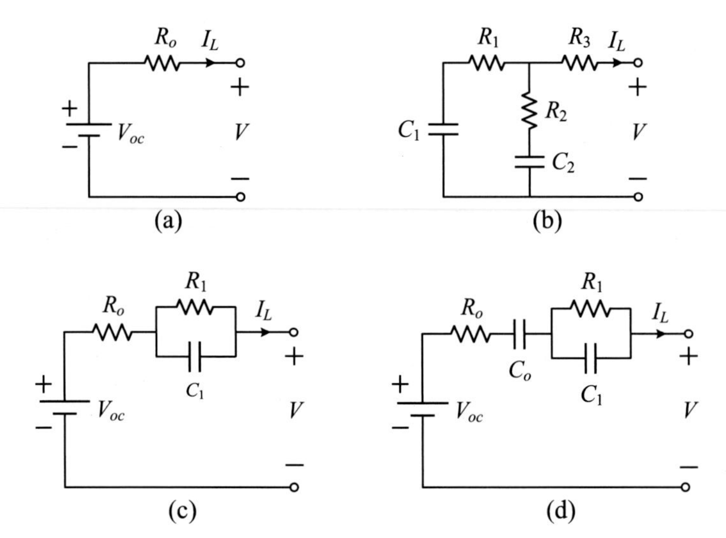

Fig. 8.7 Equivalent circuit models of battery cell

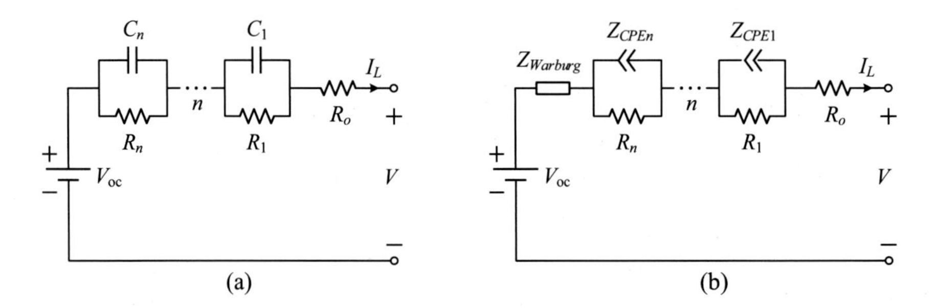

Fig. 8.8 High-order equivalent circuit models of battery cell. **a** Integral-order equivalent circuit model. **b** Fractional-order equivalent circuit model

3.2.1 Integral-Order Model

It should be noted that some basic ECMs in Fig. 8.7 can be seen as the IOMs. The Rint model in Fig. 8.7a has the simplest form using one voltage source and one resistor in series. However, the phenomenon of internal polarization fails to be exhibited in this Rint model [18]. Figure 8.7c shows a Thevenin model with an additional parallel resistor–capacitor (RC) circuit. For presenting the battery dynamic characteristics better, the high-order IOM in Fig. 8.8a adds two or more RC circuits to improve the model accuracy [19]. The advantages of IOMs are the fewer model parameters and faster simulation speed, while the disadvantages are the unreliable extrapolation and the failed prediction of battery's internal electrochemical states. Higher-order IOMs add more RC circuits, which is at the expense of higher difficulty in model parameter identification. All the performances of model accuracy, complexity, and computational burden shall be considered comprehensively for confirming the optimal number of RC circuits.

3.2.2 Fractional-Order Model

Within the whole frequency range, the IOMs using RC circuits only cannot indicate the internal electrochemical characteristics, in particular, for the middle-frequency range. To deal with this issue, the FOMs were suggested to replace the capacitor in the RC circuits by using a constant phase element [20]. As a result, the suggested FOMs can improve the model accuracy for the imitation of physical phenomena in a whole frequency range. However, the improvement of model accuracy sacrifices the simulation speed and adds the model complexity of the ECMs. Once a proper ECM is determined, such as the FOM, the parameters of the battery model can be figured out by using the methods of EIS techniques [21]. Thus, the accurate FOMs can simulate a practical battery and predict battery states precisely, thus conducting battery management effectively.

3.3 Data-Driven Model

As one of the most effective methods, data-driven models are developed according to the battery's external characteristics. This kind of battery model can be treated as a black-box model. The use of experiment data can train out a mathematical model that can offer an advantage of good reflection to the batteries' nonlinear electrochemical reactions. The data-driven modeling method can extract the hidden information effectively from the experiment dataset. Also, the generalization capability of this data-driven modeling method is superior to those of others for battery state estimation [12]. With the rapid development of microprocessors and computer science, the data-driven modeling method has been attracting more and more attention from researchers recently. Consequently, rich outcomes have been achieved in this hot topic. For example, machine learning methods including support vector machines [22] and neural networks [23] have been actively explored for battery management, especially for battery state estimation.

To provide critical comparisons of the aforementioned three battery modeling methods, Table 8.2 lists their key features of model accuracy, interpretability, complexity, and typical applications [20, 24]. As listed in Table 8.2, the electrochemical model can achieve medium accuracy but suffers from high model complexity. The ECM has low complexity and high interpretability, but it fails to increase the model accuracy significantly. The data-driven model is identified with outstanding merits because the model accuracy is highest while the model complexity is acceptable as compared with the others. Importantly, the data-driven modeling method can also exhibit superior performances even taking into account the variations of

Table 8.2 Critical comparisons of battery modeling methods

Battery modeling methods	Electrochemical model	Equivalent circuit model	Data-driven model
Model accuracy	☆☆ (medium)	☆☆	☆☆☆
Model interpretability	☆ (low)	☆☆☆	☆
Model complexity	☆☆ ☆ (high)	☆	☆☆
Typical applications	Battery design	Battery state estimation of SoC and state of power	Battery state estimation of SoH, SoT, and RUL

temperature and aging effects. It should be noted that the use of different training datasets and algorithms will influence the performances of data-driven modeling methods. Furthermore, the implementation time is also another key performance indicator to evaluate the battery modeling methods. Nevertheless, the data-driven modeling method is believed as a potential approach for future battery management, and technological breakthroughs will be made to improve this battery modeling method continuously.

4 Battery State Estimation

Generally, there are three methods for battery state estimation as follows.

(1) The direct estimation method can be further split into the direct measurement method and the look-up table method.
(2) The model-based method includes two subcategories: (1) Filter-based method, and (2) observer-based method.
(3) The data-driven method can integrate with other intelligent technologies, such as machine learning and neural network, for battery state estimation of SoC, SoH, SoT, and so on.

4.1 Direct Estimation Method

It is a straightforward method to conduct the battery state estimation. The open-circuit voltage, internal resistance, electromotive force, or EIS of a battery can be chosen to be used in the direct estimation method. Wherein, a typical estimation method is to measure the open-circuit voltage and create a parameter table. Subsequently, this open-circuit voltage estimation method will look up the established table and find the SoC. It is simple but accurate enough [25]. Because the estimation errors are inevitable due to the hysteresis characteristics of a battery, these estimation errors are not desirable in high-precision applications, such as for the aviation and military areas. In another way, the internal resistance of a battery can be used for predicting the battery states, such as SoC, and capacity, but its accuracy is not acceptable due to the low value of battery internal resistance [26]. Although the SoC can be predicted more straightforwardly by using an ampere-hour integral approach, the sensor errors will be inevitably accumulated to increase the final estimation errors.

Significantly, the influences of temperature and aging may reduce the accuracy of battery state estimation. Also, the surface temperature of a battery can be readily sensed and collected, while the internal temperature cannot. To solve this problem, temperature sensors are implanted between the cell internal layers to estimate the SoT inside a battery. In such a way, the manufacturing difficulty and safety issues should be highly concerned. As a potential direction of technical breakthroughs, the joint estimation method integrating the direct estimation deserves to be explored to improve the state estimation performances, especially for estimation precision and robustness.

4.2 Model-Based Method

For battery state estimation, the filter-based methods mainly include the particle filter approach and the Kalman filter approach, while the observer-based methods mainly involve the H-infinity observer, sliding mode observer, and Luenberger observer. Both kinds of model-based methods are developed to improve the estimation accuracy and reliability. It is preferable to use a highly accurate model while with a relatively low computational burden. To estimate the SoC, SoH, or RUL, an adaptive extended Kalman filter incorporated with a Thevenin ECM was designed and verified against dynamic temperature variations [27]. Also, to estimate both the surface temperature distributions and internal temperature distributions, thermal models of a battery were actively designed by integrating with the Kalman filter. Besides the Kalman filter, the particle filter-based method has been applied to the battery state estimation. It is suitable for dealing with nonlinear and non-Gaussian problems. Therefore, the health prognosis can be achieved readily by using the particle filter-based method. Promisingly, the joint Kalman particle filter method and multi-model particle filter method are actively explored to predict the RUL and battery capacity, which can achieve higher prediction accuracy and stability.

The observer-based method relies on a high-precision observer for predicting battery states. Wherein, the H-infinity observer, sliding mode observer, Luenberger observer, and proportional-integral observer are developed for advancing the model-based methods, respectively. For example, in a fractional-order state estimator, the Luenberger observer and sliding mode observer were in charge of guaranteeing the error convergence and robustness improvement, respectively [28]. Besides, an H-infinity observer was adopted to estimate the SoC by observing the electrochemical impedance [29]. On the other hand, some other types of observers, such as nonlinear observers, have also been studied for battery state estimation. Finally, further improvements of estimation precision and robustness are desired with an acceptable computational cost.

4.3 Data-Driven Method

As a black-box method, the data-driven method avoids the need of prior knowledge of electrochemical mechanisms, which can directly extract the correlations hidden in the measured dataset. The experiment dataset can train the battery model for state estimation, fault diagnosis, and health prognostic. The data-driven methods include several advanced approaches of the artificial neural network [23], machine learning technique, genetic algorithm, support vector machine [22], etc. As technology advances, more scholars and institutes recognized that the data-driven electrothermal model is a potential method for battery state estimation. It will well integrate both the battery thermal model and battery electric model by considering the influences of different temperatures. This data-driven electrothermal model will be used to predict the battery capacity, RUL, and SoT in real time, and it can generate the optimal current reference for battery charging control.

The capacity degradation, SoT, and SoH are highly related to the reliable operation of EV batteries. Possible failures of state prediction may cause battery malfunctions and even more severe problems, such as battery leakage, explosion, or fire. Therefore, early prediction of these key states is very significant for ensuring the battery's safe operation. Until now, many research works are focusing on these data-driven methods. For example, a data-driven method integrated with machine learning technology was used to estimate the cycle life [30], while a joint data-driven could suppress the estimation errors of SoH online [31]. A machine learning framework was implemented to reduce the early-cycle prediction errors of RUL effectively, and it contains three steps including (1) feature extraction, (2) feature selection, and (3) state prediction [32]. It is worth noting that the variations of operating conditions (such as temperature), cell voltage imbalances, and uneven aging effects will significantly increase the difficulty and decrease the accuracy of data-driven state estimation.

Table 8.3 provides in-depth surveys and comparisons of battery state estimation methods including the aforementioned three methods, cloud computing method, and blockchain technology [3, 33]. As listed in Table 8.3, their detailed approaches, advantages, and disadvantages are discussed in detail. The main comments can be summarized as follows:

(1) The direct estimation method is straightforward but mainly suffers from low accuracy and robustness in practice.
(2) The model-based method has better accuracy and robustness that highly rely on the model accuracy, experiment data, and computational ability.

Table 8.3 Surveys and comparisons of state estimation methods

State estimation methods	Detailed approaches	Advantages	Disadvantages
Direct estimation method	• Internal resistance • Open circuit voltage • Impedance spectroscopy • Electromotive force • Embedding sensors	• Low computational burden • Direct and simple for implementation • Joint estimation with model-based methods	• Off-line • Inaccurate in practice • Long resting time • Sensitivity to sensor precision
Model-based method	• Filter-based method - Particle filter - Kalman-filter • Observer-based method - H-infinity observer - Sliding mode observer - Luenberger observer	• Insensitive to initial state • Online and real-time • High accuracy • Fast convergence • Robustness to sensor noise	• High computational burden • Precision depends on model accuracy • Requiring more experiment data and validation
Data-driven method	• Neural network • Machine learning • Genetic algorithm • Support vector machine • Fuzzy logic	• Less pre-tests required • Independent of model • High accuracy • Robustness to conditions and noises • Dynamic data-driven electrothermal model for predicting SoT, SoH, etc	• Relying on training samples • High computational burden • Requirements on efficiency and portability of algorithms
Cloud computing	• Vehicular cloud computing technology	• Ability for running complex algorithms • Collaboration with cloud computing centers • Leveraging resources of participating EVs	• More complicated due to high mobility and wide range of EVs • Leaking information and compromising privacy possibly
Blockchain technology	• Private blockchain • Consortium blockchain	• Public ledger system • Data sharing and tracking • Protecting user privacy • More driving data	• Not mature technology • Some research gaps (latency and throughput) • Expecting to improve usability

(3) The data-driven method can be trained by a huge dataset and owns better accuracy and robustness to the selections of models and the variations of operating conditions.

(4) As the rapid development of communication technology and computer science, cloud computing and blockchain technology are identified as very potential methods for battery state estimation for a large number of EVs and HEVs.

In the next decade, the reduction of computational cost and the acceleration of processing rate are both important directions for future technological breakthroughs. The introduction of cloud computing and blockchain technology can help share the computational capability and vehicular data information among EVs, HEVs, and vehicular internet.

5 Safety Prognostic and Fault Diagnosis

Battery health will determine the safety of EV driving. There are many measures for maintaining the battery health so as to perform its optimal performance. First, to ensure the reliability of electric populations, a battery thermal management system (BTMS) is a very important part to prevent the occurrence of thermal runaway in a BMS. Second, cell balancing may protect the cells from experiencing the imbalance of voltage, capacity, or SoC, which is significant for improving the uneven cell aging effects. Third, fault diagnosis and prognostic are essential for troubleshooting various faults in EV batteries [34]. More significantly, the real-time online diagnosis and prognostic are preferable and deserve to be developed for a smart BMS.

5.1 Battery Thermal Management

A BTMS is deployed to monitor and manage the temperature of batteries within an optimal range. For example, LIBs require a typical temperature of 20~40 °C [35]. Figure 8.9 shows the battery thermal management methods for EVs and HEVs, including the typical preheating control and cooling control. Wherein, the preheating methods can be further divided into five approaches: (1) Air heating, (2) liquid heating, (3) electrical heating, (4) phase change material (PCM) heating, and (5) other heating. On the other hand, the cooling methods can also be classified into five approaches: (1) Air cooling, (2) liquid cooling, (3) PCM cooling, (4) heat pipe cooling, and (5) combination. The main principle, advantages, and disadvantages of each heating or cooling method have been presented in detail in [36, 37].

The BTMS is quite important for the temperature regulation and uniformity of cells [17]. Reliable thermal management enables the battery's safe operation and avoids the malfunctions of leakage, fire, and explosion. In contrast, abnormal temperatures will greatly harm the battery's performance, including its energy capacity and cycle life. Hence, these harms are discussed in terms of different thermal conditions as follows.

(1) If the temperature is over low, the LIBs will generate the lithium dendrites. The fault of short circuit, starting failure, or others occurs possibly [36]. Meanwhile,

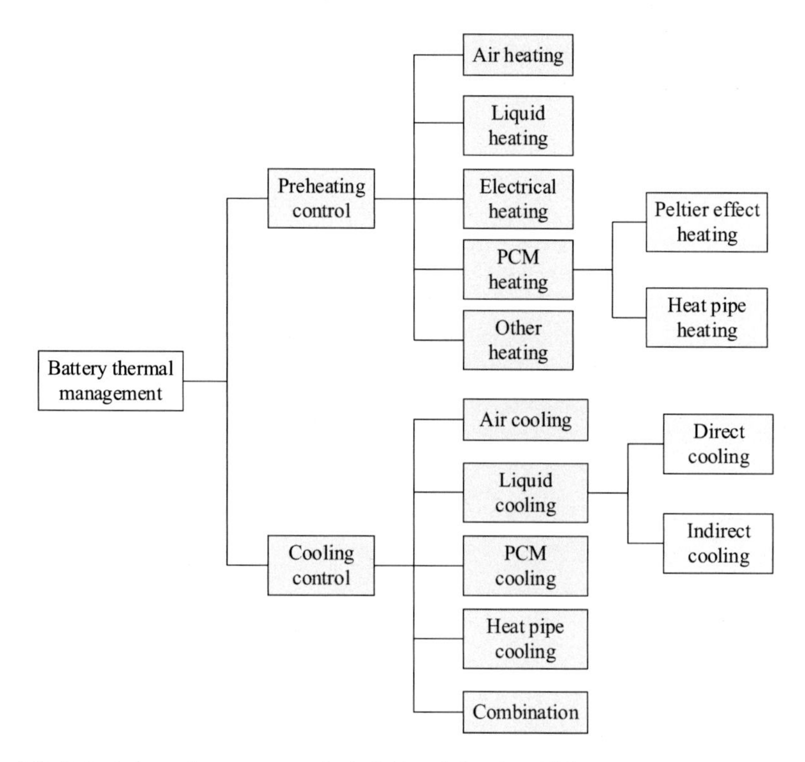

Fig. 8.9 Battery thermal management for hybrid and electric vehicles

the battery internal resistance may increase, and the inactivity of electrochemical reactions will be aggravated, thus downgrading the battery performances inevitably.

(2) If the temperature is over high, a thermal runaway may happen inside thc battery, thus causing the fire and explosion.

(3) If the non-uniformity of battery temperature happens, partial degradation and uneven cell aging will be worsened and accelerated.

5.2 Cell Balancing

Cell balancing can buy the extra run time and battery life. Generally, the cell balancing methods include (1) voltage balancing, (2) capacity balancing, and (3) SoC balancing. Accordingly, their algorithms can be mainly classified into the following three kinds: (1) voltage uniformity approach, (2) capacity uniformity approach, and (3) SOC uniformity approach [38]. The manufacturing differences and thus reaction differences can hardly be avoided among all cells, which may lead to the inconsistency of cells' voltages, SoCs, aging rates, and capacity fade rates [39]. Therefore, cell balancing is the same important as thermal management. Otherwise, the severe inconsistency will cause battery leakage, fire, or explosion, like the thermal runaway. Figure 8.10 shows two typical circuits of (1) passive cell equalizer, and (2) active cell equalizer [39]. Wherein, the passive cell equalizers use the shunting resistors to realize the cell balancing. Differing from the use of switches, they can be further divided into two subcategories: (1) Passive cell equalizer using fixed shunting resistors, and (2) passive cell equalizer using switched shunting resistors. Their circuits are shown in Fig. 8.11a and b, respectively. These kinds of cell equalizers have the advantage of low complexity but suffer from the disadvantage of low efficiency due to the joule loss.

In contrast, the active cell equalizers improve the system efficiency and operation stability effectively but at the expense of slightly adding system complexity. As shown in Fig. 8.10, these active cell equalizers are further classified into four

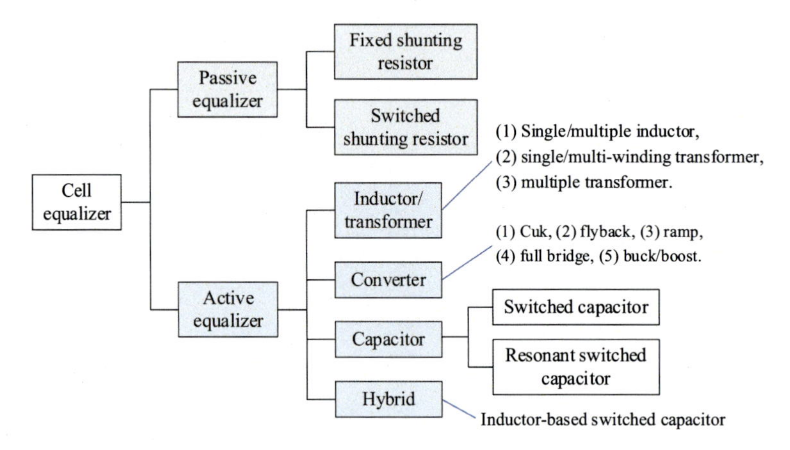

Fig. 8.10 Categories of cell equalizers for electric vehicle batteries

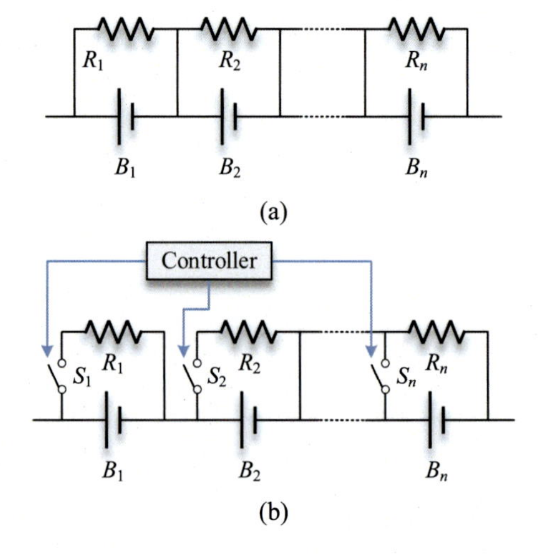

Fig. 8.11 Passive cell equalizers for voltage balancing. **a** Fixed shunting resistor. **b** Switched shunting resistor

subcategories: (1) Inductor/transformer-based types, (2) converter-based types, (3) capacitor-based types, and (4) hybrid types. Typically, their representative circuits are shown in Fig. 8.12a–d, respectively. Further detailed types of cell equalizers are given in Fig. 8.10. As a supplement, the switched capacitor (SC)-based cell equalizers include (1) conventional SC, (2) single SC, (3) double-tiered SC, (4) modularized SC, (5) chain structure, (6) coupling SC, and (7) series–parallel SC. On the other hand, the resonant SC-based cell equalizers include (1) resonant SC, (2) conventional resonant type, (3) quasi-resonant type, (4) modularized type, and (5) chain type. First, the multi-winding transformer-based cell equalizer can be configured as either the flyback topology in Fig. 8.12a or the forward topology. Second, in Fig. 8.12b, the buck/boost converter serves to deliver the surplus energy rapidly from the high-voltage cells to the low-voltage cells through a direct-current (DC) bus. Third, in Fig. 8.12c, to equalize the n cells, n–1 quasi-resonant converters are required, where $2n$ switches are populated. The resonant converters can balance the energy among the cells while maintaining the soft-switching operation. Fourth, in Fig. 8.12d, the inductor-based equalizer using a chain structure can increase the speed of cell balancing. However, its switching loss increases because of adding switches, and the voltage stress and current stress of capacitors may rise with the increasing number of cells. Besides the hardware parts, the software parts including the microcontrollers and algorithms are also very important for achieving the cell balancing reliably.

5.3 *Fault Diagnosis and Prognostic*

There are numerous types of faults related to power batteries, including (1) internal or external short-circuit fault, (2) over-charge or over-discharge fault, (3) BTMS fault, (4) sensor fault, and (5) actuator fault [34]. Accordingly, the functions of fault diagnosis and prognostic are significant to troubleshoot various battery faults in the applications of EVs and HEVs. Otherwise, some catastrophic accidents might happen, which may cause a severe threat to our personal and property safety. To identify the types of various faults, Fig. 8.13 shows the categories of fault diagnosis and prognostic for EV batteries [26, 40], mainly including (1) distributed methods, (2) centralized methods, and (3) joint methods. Wherein, the distributed methods can be further classified into two branches, namely quantitative analysis methods and qualitative analysis methods. The qualitative analysis methods have the advantage of high interpretability but suffer from the failed applicability of complex systems and overreliance on the representativeness and integrity of knowledge [40]. Besides, the model-based methods can diagnose the faults accurately. The data-driven methods can be classified into three main types: (1) signal processing method, (2) machine learning method, and (3) information fusion method [41]. For the fault prognostic, the data-driven method offers an impressive accuracy by using the experiment data of a few early cycles only. Also, the data-driven method can present good reliability and robustness to disturbances.

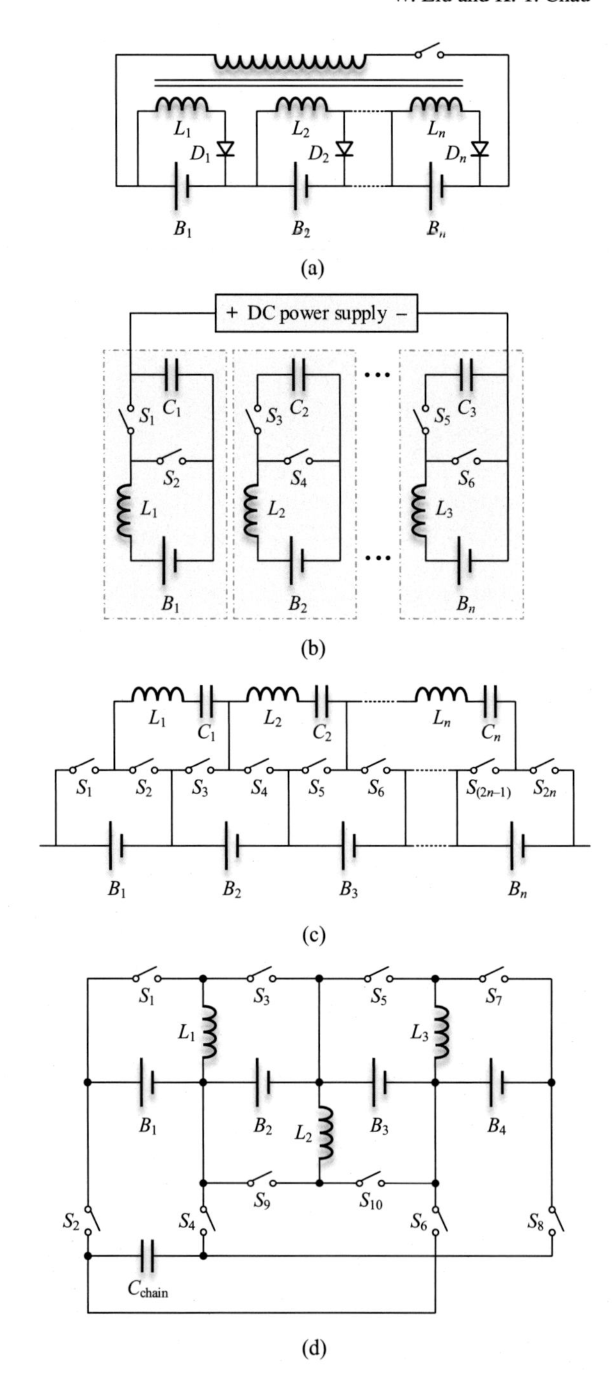

Fig. 8.12 Active cell equalizers for voltage balancing. **a** Multi-winding transformer. **b** Buck/boost. **c** Quasi-resonant converter. **d** Inductor-based switched capacitor

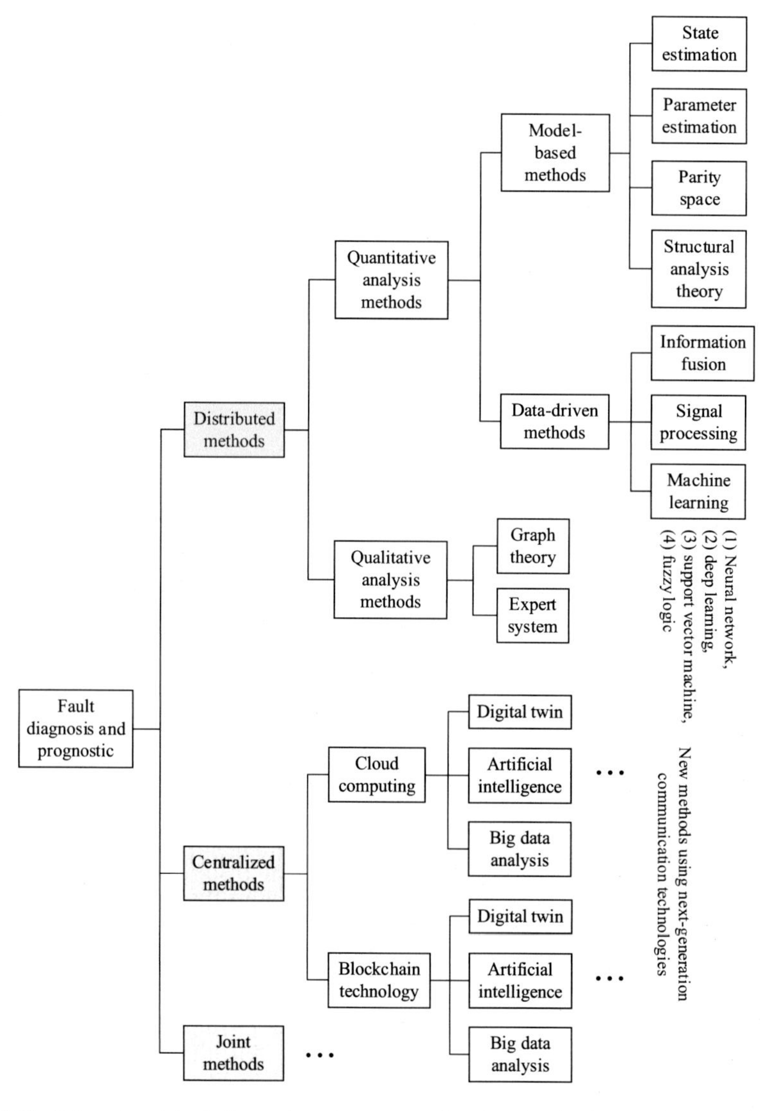

Fig. 8.13 Categories of fault diagnosis and prognostic for electric vehicle batteries

For the centralized methods, cloud computing and blockchain technology can be developed to realize the online fault diagnosis and prognostic for regional EV networks. Accordingly, cloud computing is to satisfy the computational requirements [42] with the help of modern communication technologies. In such a way, three emerging technologies of the digital twin, artificial intelligence, and big data analysis (or data mining) can be applied to build a fault diagnosis and prognostic system for large-scale EVs and HEVs equipped with batteries. Finally, the joint methods will build a strong collaboration with the distributed and centralized methods, thus integrating their advantages to compensate for each other. As a result, they will help establish a local and regional network to diagnose and prognose various faults in real time.

6 Emerging Management Technologies

In recent years, new battery management technologies emerge for achieving the functions of battery state estimation, thermal management, fault diagnosis, and health prognostic. Typically, some representative technologies will be discussed as follows, including (1) multi-model co-estimation, (2) artificial intelligence technology, (3) cloud computing technology, and (4) blockchain technology. Benefiting from the fifth-generation communication and next generations, the rapid development of these emerging technologies will further advance the smart battery management in turn.

6.1 Multi-Model Co-Estimation

For improving the accuracy and robustness of battery state estimation, Fig. 8.14 shows the multi-model co-estimation methods for battery management, which involve multiple models to perform the joint estimation [43, 44]. For example, in Fig. 8.14, the electrochemical model, ECM, and data-driven model can be promisingly used to present a high-fidelity battery model, thus integrating the strengths of each method. Besides, the advanced microcontrollers allow the battery models to be more complex than ever before with higher accuracy and faster estimation speed.

Multi-mode co-estimation methods have drawn many researchers' attention focusing on improving the performance of battery management. Accordingly, two concepts of fusion estimation [44] and joint estimation [31] have been developed and verified for effective battery state prediction, such as for the SoC, SoT, SoH, or SoP. For example, multiple algorithms, such as linear regression, random forest, and support vector regression, can be fused readily for estimating the SoC [45]. Besides, to conduct the real-time multi-state joint estimation for EV batteries, a data-driven method can flexibly collaborate with other methods, such as the model-based unscented particle filter and support vector machine, for advancing the battery management performance.

6.2 Artificial Intelligence

The artificial intelligence technology has been well developed for various applications in recent decades. The battery management technologies have actively

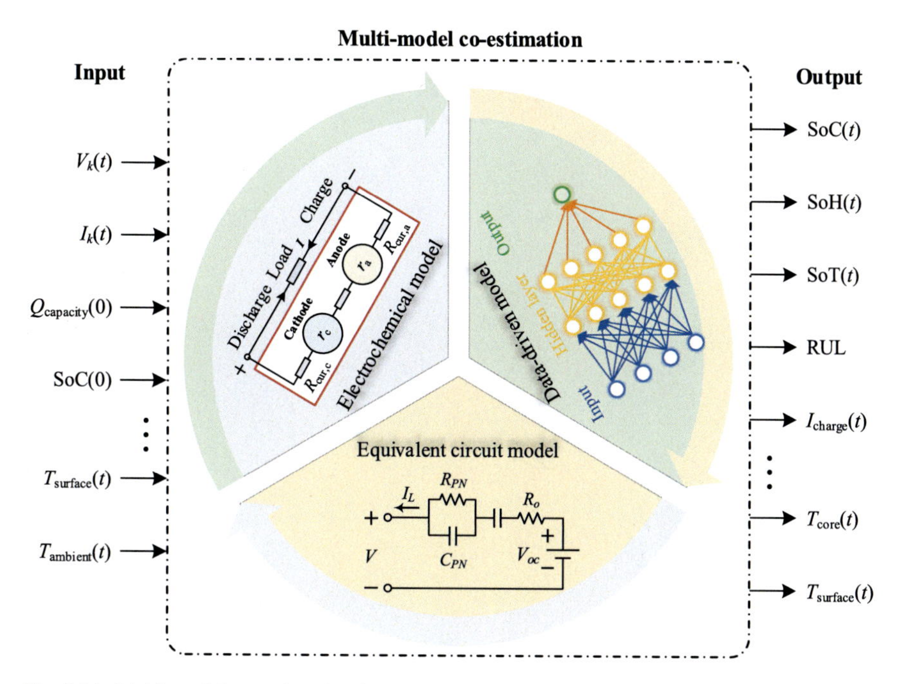

Fig. 8.14 Multi-model co-estimation for battery management.

embraced the artificial intelligence methods. Focusing on this field, some artificial intelligence methods have attracted increasing attention and have been explored widely, including machine learning methods, recurrent neural networks, and support vector machines [46, 47]. To implement the intelligent methods, huge data should be acquired by using various sensors and Internet-of-Thing devices in advance. Subsequently, these data will be further processed to train the artificial intelligence algorithms and extract the key features. Finally, a digital embodiment can be established for modeling the EV batteries accurately. Accordingly, Fig. 8.15 shows a deep neural network algorithm for EV battery management [48], where the battery voltages, currents, surface temperature, and ambient temperature are all required as the inputs, while the targeted battery states (such as the SoC, SoC, and SoH), RUL, and core temperature can be predicted as the outputs.

The application of artificial intelligence enables the upgradation of smart BMS. Typically, the digital twin technology may be used to develop a battery-information twin for intelligent battery management [49]. Relevant studies have been initiated and achieved rich outcomes. For example, by using an artificial neural network model, the SoH of LIBs was predicted precisely in a passive BTMS against different operating conditions [24]. Nevertheless, artificial intelligence also brings new challenges in interdisciplinary fields, such as data sensing, data computing, and data security, and hence continuous breakthroughs are still expected in these relevant fields.

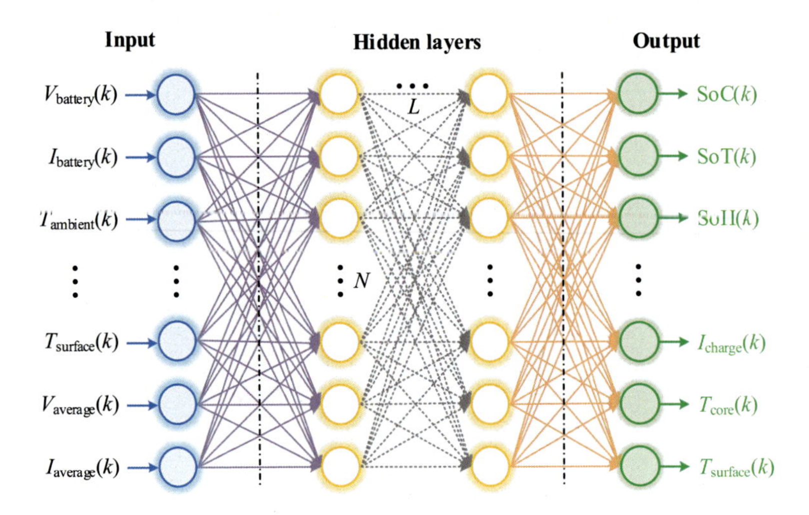

Fig. 8.15 Neural network model for battery management

6.3 Cloud Computing

To manage the batteries well for a large-scale EV network, the cloud computing technology can be appointed as a superior manager who can analyze and process the data and make optimal decisions and predictions. Figure 8.16 shows the schematic of cloud computing technology incorporating artificial intelligence methods for regional battery management. The whole management system mainly consists of three parts, including (1) EVs and HEVs, (2) communication technology, and (3) cloud framework [50]. Moreover, the process of vehicular cloud computing networks can be derived into four steps: (1) Battery data acquisition, (2) data communication, (3) artificial intelligence, and (4) battery model. A local server is in charge of gathering the vehicular battery data and uploading the data to a cloud computing center, where intelligent algorithms can be applied to further data processing [51]. At the cloud computing center rather than an onboard BMS and a local server, both the data mining and big data analysis can be readily performed to handle such a huge dataset.

The rapid development of communication network technologies, such as the fifth-generation technology, sixth-generation technology, and next generations, can be quite competent for the data transmitting and receiving. Consequently, a vehicular Internet-of-Thing can be networking for real-time data sharing and computing power sharing. Moreover, the cloud battery management center is capable of battery state estimation, fault diagnosis, and health prognostic, especially taking into account the aging effects on capacity and power fades [52]. Finally, the cloud BMS will deliver the computing results and recommendations to the local BMSs, namely local servers and onboard BMSs. Regional battery management can be completed successfully by remote monitoring and control.

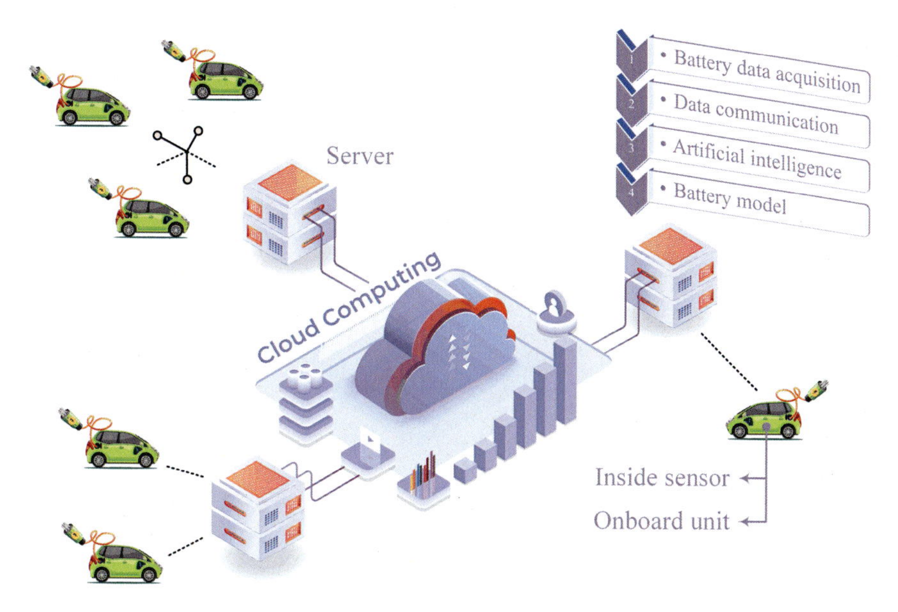

Fig. 8.16 Cloud computing technology using artificial intelligence for regional battery management

6.4 Blockchain Technology

Figure 8.17 shows a scheme of blockchain technology, which serves for the battery management of EVs and HEVs. This blockchain technology has a typical two-layer hierarchy that includes one consortium blockchain and multiple private blockchains. With the help of advanced communication technologies, vehicular battery data can be shared in a league of multiple regions. Significantly, blockchain technology can ensure the data security and integrity, thus outperforming other technologies. On the one hand, each private blockchain is capable of recording both the encrypted data and public data, which are originated from every EV and HEV. On the other hand, the consortium blockchain is in charge of collecting both the public indexes and secure indexes, which are researchable and generated from the public data and encrypted data, respectively [50]. Furthermore, various malicious cyber-physical attacks can be blocked to protect the Internet-connected BMSs. This blockchain technology can offer intelligent monitoring, prognostic, and control [53] for precise battery management in a regional EV league [54]. Having privacy protection for stakeholders, this blockchain technology enables anonymous transactions among EVs and energy routers, which may serve the battery management and power distribution in a virtual EV power network [55].

Finally, a VIEI is envisioned for energy and data sharing in Fig. 8.18, which can help relieve the overdependence on batteries and onboard BMSs in a network of EVs and HEVs. Accordingly, the blueprint of VIEI contains six parts: (1) Vehicular power network [56, 57], (2) energy internet and router [55], (3) artificial intelligence,

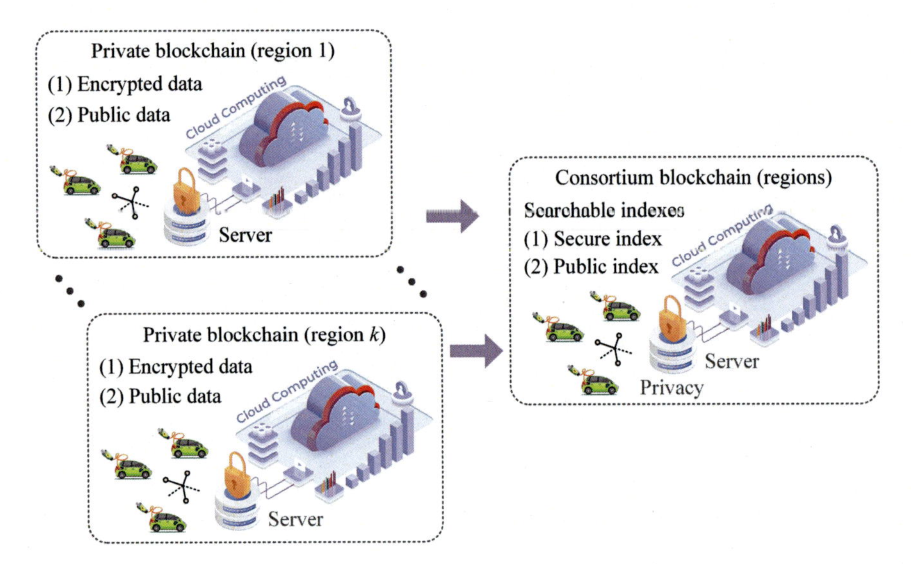

Fig. 8.17 Blockchain technology of regional battery management for hybrid and electric vehicles

(4) vehicle-to-grid, (5) vehicle-to-home, and (6) vehicle-to-vehicle. In the regime of VIEI, effective sharing of vehicular data and computing power may collaborate with the autonomous driving, hence advancing the electrified transportation [58]. Moreover, the multi-internet merging will embrace the vehicular Internet of Things [59, 60], thus enabling new functions and widening the scales of information, energy, and humanity internets. Thanks to the assistance of artificial intelligence and cloud computing technologies, both the EVs and HEVs will provide more types of service, more than pure transportation tools. To block malicious attackers, the emerging technologies, such as data-driven artificial intelligence and blockchain technology, will be applied to guarantee both the security and privacy of data and energy [61], therefore benefiting the achievement of a smarter VIEI.

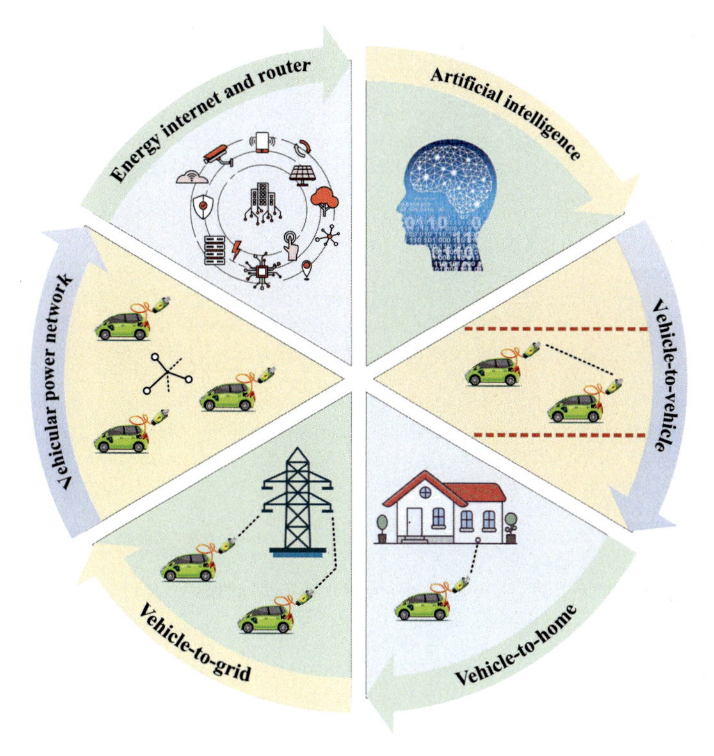

Fig. 8.18 Blueprint of vehicular information and energy internet

7 Conclusion

This chapter analyzed and discussed the state-of-the-art technologies of battery management in HEVs and pure EVs thoroughly. Various advanced and emerging methods were introduced for battery modeling, battery state estimation, safety prognostic, and fault diagnosis. Meanwhile, their main features, advantages, and disadvantages were also presented. Significantly, the data-driven method can be identified as a promising technology for developing a dynamic data-driven electrothermal model, which may offer accurate state estimation and reliable health prognostic by using a dataset of fewer early cycles only. Finally, with the help of communication technologies and intelligent management methods, a blueprint of VIEI was envisioned to support the vehicular data and energy sharing and smarter battery management for a network of HEVs and EVs.

Acknowledgements This work was partially supported by a grant from the Hong Kong Research Grants Council, Hong Kong Special Administrative Region, China, under Project No. T23-701/20-R, and partially supported by a grant from The Hong Kong Polytechnic University, under Project No. P0048560.

References

1. Chau KT (2016) Energy systems for electric and hybrid vehicles. The IET
2. China's New Energy Automobile Industry Development Plan for 2021 to 2035. International council on clean transportation. Available online: https://theicct.org/publication/chinas-new-energy-vehicle-industrial-development-plan-for-2021-to-2035
3. Liu W, Placke T, Chau KT (2022) Overview of batteries and battery management for electric vehicles. Energy Rep 8:4058–4084
4. Liu C, Chau KT, Wu D, Gao S (2013) Opportunities and challenges of vehicle-to-home, vehicle-to-vehicle, and vehicle-to-grid technologies. Proc IEEE 101(11):2409–2427
5. Liu W, Chau KT, Hua Z (2021) Overview of batteries for electric vehicle propulsion. In: Proceeding of 34th international electric vehicle symposium & exhibition, Nanjing, China, pp 1–12
6. Schmuch R, Wagner R, Hörpel G, Placke T, Winter M (2018) Performance and cost of materials for lithium-based rechargeable automotive batteries. Nat Energy 3(4):267–278
7. Duffner F, Kronemeyer N, Tübke J, Leker J, Winter M, Schmuch R (2021) Post-lithium-ion battery cell production and its compatibility with lithium-ion cell production infrastructure. Nat Energy 6(2):123–134
8. Chau KT, Wong YS, Chan CC (1999) An overview of energy sources for electric vehicles. Energy Convers Manage 40(10):1021–1039
9. Chau KT, Wong YS (2002) Overview of power management in hybrid electric vehicles. Energy Convers Manage 43(15):1953–1968
10. Hannan MA, Lipu MSH, Hussain A, Mohamed A (2017) A review of lithium-ion battery state of charge estimation and management system in electric vehicle applications: Challenges and recommendations. Renew Sustain Energy Rev 78:834–854
11. Tamilselvi S, Gunasundari S, Karuppiah N et al (2021) A review on battery modelling techniques. Sustainability 13(18):10042
12. Zhou W, Zheng Y, Pan Z, Lu Q (2021) Review on the battery model and SOC estimation method. Processes 9(9):1685
13. Fotouhi A, Auger DJ, Propp K, Longo S, Wild M (2016) A review on electric vehicle battery modelling: from lithium-ion toward lithium–sulphur. Renew Sustain Energy Rev 56:1008–1021
14. Doyle M, Fuller TF, Newman J (1993) Modeling of galvanostatic charge and discharge of the lithium/polymer/insertion cell. J Electrochem Soc 140(6):1526
15. Li C, Cui N, Wang C, Zhang C (2021) Reduced-order electrochemical model for lithium-ion battery with domain decomposition and polynomial approximation methods. Energy 221
16. Li D, Yang L, Li C (2021) Control-oriented thermal-electrochemical modeling and validation of large size prismatic lithium battery for commercial applications. Energy 214:119057
17. Wang Q, Jiang B, Li B, Yan Y (2016) A critical review of thermal management models and solutions of lithium-ion batteries for the development of pure electric vehicles. Renew Sustain Energy Rev 64:106–128
18. Johnson V (2002) Battery performance models in ADVISOR. J Power Sources 110(2):321–329
19. Xia B, Sun Z, Zhang R, Lao Z (2017) A cubature particle filter algorithm to estimate the state of the charge of lithium-ion batteries based on a second-order equivalent circuit model. Energies 10(4):457
20. Liu C, Hu M, Jin G, Xu Y, Zhai J (2021) State of power estimation of lithium-ion battery based on fractional-order equivalent circuit model. J Energy Storage 41:102954
21. Ruan H, Sun B, Jiang J et al (2021) A modified-electrochemical impedance spectroscopy-based multi-time-scale fractional-order model for lithium-ion batteries. Electrochim Acta 394:139066
22. Yao L, Fang Z, Xiao Y, Hou J, Fu Z (2021) An intelligent fault diagnosis method for lithium battery systems based on grid search support vector machine. Energy 214:118866
23. Lindgren J, Asghar I, Lund PD (2016) A hybrid lithium-ion battery model for system-level analyses. Int J Energy Res 40(11):1576–1592

24. Jaliliantabar F, Mamat R, Kumarasamy S (2022) Prediction of lithium-ion battery temperature in different operating conditions equipped with passive battery thermal management system by artificial neural networks. Mater Today: Proc 48:1796–1804
25. Dong G, Wei J, Zhang C, Chen Z (2016) Online state of charge estimation and open circuit voltage hysteresis modeling of $LiFePO_4$ battery using invariant imbedding method. Appl Energy 162:163–171
26. Lu L, Han X, Li J, Hua J, Ouyang M (2013) A review on the key issues for lithium-ion battery management in electric vehicles. J Power Sources 226:272–288
27. Jiang C, Wang S, Wu B, Fernandez C, Xiong X, Coffie-Ken J (2021) A state-of-charge estimation method of the power lithium-ion battery in complex conditions based on adaptive square root extended Kalman filter. Energy 219:119603
28. Zou C, Hu X, Dey S, Zhang L, Tang X (2018) Nonlinear fractional-order estimator with guaranteed robustness and stability for lithium-ion batteries. IEEE Trans Indus Electron 65(7):5951–5961
29. Chen N, Zhang P, Dai J, Gui W (2020) Estimating the state-of-charge of lithium-ion battery using an H-infinity observer based on electrochemical impedance model. IEEE Access 8:26872–26884
30. Severson KA, Attia PM, Jin N et al (2019) Data-driven prediction of battery cycle life before capacity degradation. Nat Energy 4(5):383–391
31. Song Y, Liu D, Liao H, Peng Y (2020) A hybrid statistical data-driven method for on-line joint state estimation of lithium-ion batteries. Appl Energy 261:114408
32. Fei Z, Yang F, Tsui K-L, Li L, Zhang Z (2021) Early prediction of battery lifetime via a machine learning based framework. Energy 225:120205
33. Wang Y, Tian J, Sun Z et al (2020) A comprehensive review of battery modeling and state estimation approaches for advanced battery management systems. Renew Sustain Energy Rev 131:110015
34. Xiong R, Sun W, Yu Q, Sun F (2020) Research progress, challenges and prospects of fault diagnosis on battery system of electric vehicles. Appl Energy 279:115855
35. Jilte R, Afzal A, Panchal S (2021) A novel battery thermal management system using nano-enhanced phase change materials. Energy 219:119564
36. Zhang X, Li Z, Luo L, Fan Y, Du Z (2022) A review on thermal management of lithium-ion batteries for electric vehicles. Energy 238:121652
37. Zichen W, Changqing D (2021) A comprehensive review on thermal management systems for power lithium-ion batteries. Renew Sustain Energy Rev 139:110685
38. Ouyang Q, Chen J, Zheng J, Fang H (2018) Optimal cell-to-cell balancing topology design for serially connected lithium-ion battery packs. IEEE Trans Sustain Energy 9(1):350–360
39. Das UK, Shrivastava P, Tey KS et al (2020) Advancement of lithium-ion battery cells voltage equalization techniques: a review. Renew Sustain Energy Rev 134:110227
40. Hu X, Zhang K, Liu K, Lin X, Dey S, Onori S (2020) Advanced fault diagnosis for lithium-ion battery systems: a review of fault mechanisms, fault features, and diagnosis procedures. IEEE Ind Electron Mag 14(3):65–91
41. Dai H, Jiang B, Hu X, Lin X, Wei X, Pecht M (2021) Advanced battery management strategies for a sustainable energy future: Multilayer design concepts and research trends. Renew Sustain Energy Rev 138:110480
42. Kim T, Makwana D, Adhikaree A, Vagdoda JS, Lee Y (2018) Cloud-based battery condition monitoring and fault diagnosis platform for large-scale lithium-ion battery energy storage systems. Energies 11(1):125
43. Lin C, Mu H, Xiong R, Cao J (2017) Multi-model probabilities based state fusion estimation method of lithium-ion battery for electric vehicles: state-of-energy. Appl Energy 194:560–568
44. Li Y, Wang C, Gong J (2017) A multi-model probability SOC fusion estimation approach using an improved adaptive unscented Kalman filter technique. Energy 141:1402–1415
45. Wang Q (2019) Battery state of charge estimation based on multi-model fusion. Chinese Automation Congress (CAC), Hangzhou, China, pp 2036−2041

46. Vidal C, Malysz P, Kollmeyer P, Emadi A (2020) Machine learning applied to electrified vehicle battery state of charge and state of health estimation: state-of-the-art. IEEE Access 8:52796–52814
47. Xi Z, Wang R, Fu Y, Mi C (2022) Accurate and reliable state of charge estimation of lithium ion batteries using time-delayed recurrent neural networks through the identification of overexcited neurons. Appl Energy 305:117962
48. Chemali E, Kollmeyer PJ, Preindl M, Emadi A (2018) State-of-charge estimation of Li-ion batteries using deep neural networks: a machine learning approach. J Power Sources 400:242–255
49. Wu B, Widanage WD, Yang S, Liu X (2020) Battery digital twins: perspectives on the fusion of models, data and artificial intelligence for smart battery management systems. Energy AI 1:100016
50. Hu X, Xu L, Lin X, Pecht M (2020) Battery lifetime prognostics. Joule 4(2):310–346
51. Li S, Zhao P (2021) Big data driven vehicle battery management method: a novel cyber-physical system perspective. J Energy Storage 33:102064
52. Li W, Rentemeister M, Badeda J, Jöst D, Schulte D, Sauer DU (2020) Digital twin for battery systems: cloud battery management system with online state-of-charge and state-of-health estimation. J Energy Storage 30:101557
53. Kim T, Ochoa J, Faika T et al (2022) An overview of cyber-physical security of battery management systems and adoption of blockchain technology. IEEE J Emerg Select Top Power Electron 10(1):1270–1281
54. Florea BC, Taralunga DD (2020) Blockchain IoT for smart electric vehicles battery management. Sustainability 12(10):3984
55. Liu W, Chau KT, Chow CCT, Lee CHT (2022) Wireless energy trading in traffic internet. IEEE Trans Power Electron 37(4):4831–4841
56. Yi P, Tang Y, Hong Y et al (2014) Renewable energy transmission through multiple routes in a mobile electrical grid. In: Proceedings of the IEEE PES Innovative Smart Grid Technologies Conference, Washington, DC, USA, pp 1−5
57. Lam AYS, Leung K, Li VOK (2017) Vehicular energy network. IEEE Trans Transp Electrification 3(2):392–404
58. Peng C, Wu C, Gao L, Zhang J, Alvin Yau KL, Ji Y (2020) Blockchain for vehicular internet of things: recent advances and open issues. Sensors 20(18):5079
59. Farman H, Jan B, Khan Z, Koubaa A (2020) A smart energy-based source location privacy preservation model for Internet of Things-based vehicular ad hoc networks. Trans Emerg TelecommunTechnol 1−14
60. Du Z, Wu C, Yoshinaga T, Yau KLA, Ji Y, Li J (2020) Federated learning for vehicular internet of things: recent advances and open issues. IEEE Open J Comput Soc 1:45–61
61. Zhang J, Zhong H, Cui J, Xu Y, Liu L (2020) An extensible and effective anonymous batch authentication scheme for smart vehicular networks. IEEE Internet Things J 7(4):3462–3473

Fast Chargers for Plug-In Electric and Hybrid Vehicles

Chris Mi, Siqi Li, and Sizhao Lu

Abstract This chapter introduces the technical background of fast chargers for plug-in electric and hybrid vehicles, including technique development for on-board chargers and fast charging stations, standard, regulations, and future trends. The development of wide-bandgap (WBG) devices is the basic technology and the biggest driving force for innovation of fast chargers. The fundamentals of WBG devices and the unipolar and bipolar WBG power semiconductors commonly used in fast chargers are introduced. To take advantages of WBG devices, state-of-the-art topologies for on-board chargers and fast-charging stations are presented. The medium frequency transformers and the control strategies for modular SST-based fast charging stations are also introduced.

1 Introduction

Electric vehicles (EVs) have enjoyed a very fast pace of growth in the past 10 years, with many countries promoting zero-emission vehicles by 2035–2050, and major automotive manufacturers to abolish internal combustion engine powered vehicles altogether in the next 15–30 years. Traditionally, EVs have three bottlenecks, including limited range, long charging time, and high cost. With the cost of lithium-ion batteries dropping dramatically in the past ten years, as well as the increase in energy density of lithium-ion batteries, the cost and range anxiety have been mitigated to some degree, with automotive manufacturers installing more and more batteries on various types of EVs. For example, the 2022 Rivian R1T is equipped with a 135

C. Mi (✉)
San Diego State University, San Diego, USA
e-mail: cmi@sdsu.edu

S. Li · S. Lu
Kunming University of Science and Technology, Kunming, China
e-mail: lisiqi@kust.edu.cn

S. Lu
e-mail: lusz10@kust.edu.cn

C. H. T. Lee (ed.), *Emerging Technologies for Electric and Hybrid Vehicles*,
Green Energy and Technology, https://doi.org/10.1007/978-981-99-3060-9_9

kWh battery back which provides a driving range of 314 miles, and the 2022 T Model S Long Range is equipped with a fairly large 100 kWh battery pack and can be driven for 405 miles in a single charge as shown in Figs. 9.1 and 9.2, respectively. However, charging these large battery packs can be a challenge. For example, most EVs are equipped with an onboard charger (OBC) and most homes can only depend on a maximum of 1.9 kW without a dedicated electric vehicle supply equipment (EVSE) or 6.5 kW power with a dedicated EVSE, which will take nearly 20 h to charge a 100 kWh pack. Therefore, fast, efficient, and convenient charging is crucial for the further penetration of EVs.

The Society of Automotive Engineers (SAE) has developed an EV Charging Standard, J1772, entitled "SAE Surface Vehicle Recommended Practice J1772, SAE Electric Vehicle Conductive Charge Coupler." The SAE J1772-2017 standard defines four levels of charging: AC Level 1, AC Level 2, DC Level 1, and DC Level 2. The maximum charging power is 1.92 kW according to Level 1 profile. For Level 2 AC charging, the output power of the OBC can be built up to 19.2 kW with a dedicated EVSE or using a 208-V/240-V outlet. The power of an OBC is limited by the cost, size, and weight constraints of the vehicle. Most original preinstalled OBC for light-duty vehicles (LDVs) are far below 19.2 kW. At this rate or below, it takes many hours to restore the EV's full range. Hence, to alleviate range anxiety, tens to hundreds of kilowatts of charging power are urgently needed to charge the battery pack in a short period of time. Suitable fast or ultra-fast DC charging devices that provide DC current directly to the battery have now been developed. The SAE J1772 defines 80 kW and 400 kW as the maximum value for dc Level 1 and dc

Fig. 9.1 The 2022 Rivian R1T

Fig. 9.2 The 2022 T Model S long range

Level 2 respectively, which are two dc fast-charging power levels. Other DC fast charging standards such as CHAdeMO, Tesla Supercharger, and GB/T 20,234.3 are also competitive. In 2019, Tesla's supercharger v2 can output 145 kW. The latest Supercharger v3 has a maximum charging power of 250 kW. Figure 9.3 shows the plug of J1772.

Wireless charging of electrical vehicles has also gained considerable attention in the last ten years. Wireless charging is supposed to be more convenient and safer.

Fig. 9.3 SAE J1772 plugs

However, foreign object detection, electromagnetic field emissions, cost, and interoperability are preventing automotive manufacturers to roll out wireless charging systems yet. Dynamic online wireless charging can reduce battery capacity, thereby remarkably reducing the cost of electric vehicles. However, the disadvantage of this charging technology is that it requires significant investment in road infrastructure [1]. The SAE developed SAE J2954, its first recommended practice for wireless charging of electric vehicles, in 2016, and revised it in two rounds in 2017 and 2019.

To realize fast charging, whether wired or wireless, the power from the AC grid must be rectified, isolated, and regulated to charge the EV batteries. The cost, size, and weight of fast charging equipment makes it impractical to install the fast DC charger onboard an EV. In addition, thermal management is important to the charging equipment during charging. For example, for a DC fast charger rated at 300 kW, a 97% efficiency will result in 9 kW of losses in the charging equipment. This heat needs to be actively dissipated into the ambient through some sort of cooling medium such as a liquid cooling system. Therefore, the efficiency of the fast charging system is critical. Additionally, when large numbers of electric vehicles are deployed, small gains in efficiency can help save a lot of energy as a whole.

As a result, the focuses of EV fast charging are focused on improving efficiency, power density, and power capacity while reducing costs. To that end, wide-band-gap (WBG) power devices have become the most popular choice for EV fast charging due to their superior properties [2].

2 Wide Bandgap Power Semiconductor Devices in EV Fast Chargers

Efficiency is critical for high-power EV fast chargers, as large power losses can lead to heat dissipation difficulties and a significant increase in energy consumption. Improving the efficiency of charging equipment has been the pursuit of academia and industry for many years. Power semiconductor devices can be seen as the basis for all other key technologies in EV chargers. The performance and characteristics of the power device determine how we choose the topology. Every time, the breakthrough in power devices will bring technological and performance revolutions to charging equipment. Power devices made of silicon have been developed for over fifty years, and the last major performance boost was the super junction technology invented 30 years ago. At present, power devices made of silicon are very close to the theoretical limit of the material [3]. The wide bandgap devices are considered as the next generation of power semiconductor materials [4]. The theoretical limit of WBG materials is tens to hundreds of times that of silicon materials. WBG devices make the EV charger smaller, more efficient, and more reliable than using the traditional silicon counterpart [5].

2.1 WBG Material Characteristics

Before we introduce the WBG material characteristics, we need to know what is the wide bandgap. Figure 9.4 visualizes the difference between conductors, semiconductors, and insulators using band theory. The conduction band is the band of electron orbitals, in which the electron's energy is high enough that they can move freely through the material, making the material conductive. The valence band is another band of electron orbitals, in which the electron's energy is not high enough to move freely. However, the electrons in the valance band can bounce up into the conduction band when energized. The minimum energy required for an electron to transition from the valence band to the conduction band, that is, the energy gap between the two bands, is called the bandgap energy.

For conductors, the energy gap is small or even overlaps, and there are a large number of free electrons at the ground state. For semiconductor and insulator materials, there is a distinct energy gap between the conduction and valence bands. This gap is also known as the forbidden band. Semiconductors and insulators have very few electrons in the conduction band and are insulating at their ground state. We expect that the bandgap energy of a good insulating material should be much greater than 5 eV. Semiconductor materials have a bandgap energy between conductors and insulators, which means that the electrons in semiconductor materials can be easily energized to the conduction band. When the bandgap energy of a semiconductor material is higher than 2.2 eV, we call it a wide bandgap material [6].

The most widely studied WBG materials for power semiconductors and their properties are shown in Table 9.1. The intrinsic carrier density of wide-bandgap materials is much lower than that of silicon, giving WBG devices distinct advantages in high-temperature operation, low leakage current, and fast reverse recovery. In addition to the wide bandgap, the characteristics of the WBG materials also depend on other properties that are accompanied by the wide bandgap. The critical electric field refers to the electric field strength at which breakdown of the material occurs. The high critical electric field makes WBG devices more advantageous in high-voltage applications. Additionally, the high critical electric field means that thinner layers can be used to block high voltages to ensure low on-resistance. High electron mobility is also beneficial for reducing the on-resistance of the device. Thermal conductivity

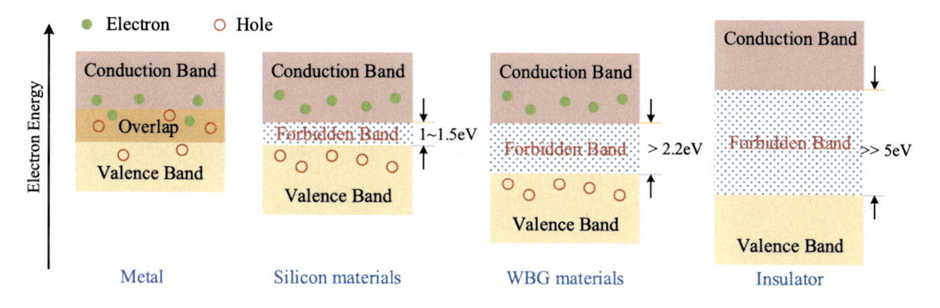

Fig. 9.4 Illustration of material bandgap energy

determines the ability to dissipate heat, which is crucial to the power density of the device. When the WBG layer is grown on a substrate of different materials, special attention should be paid to the coefficient of thermal expansion (CTE). A large CTE mismatch will cause greater stress during thermal cycling, which may lead to device failure due to physical cracks.

In order to comprehensively evaluate semiconductor materials or devices, a few definitions of Figure of Merit (FOM) have been proposed. In 1965, the Johnson figure of merit (JFOM) was defined by the power frequency product [7]. In 1982, the Baliga figure of merit (BFOM) was introduced, which is related to the power handling capability per unit area of a power device wafer [8]. Alex Huang proposed three new FOM definitions in 2004 [9], which are Huang material FOM (HMFOM), Huang chip area figure-of merit (HCAFOM), Huang thermal figure-of merit (HTFOM). Different FOM definitions characterize the performance of the device from different aspects. The JFOM, BFOM and HMFOM of different materials are also given in Table 9.1. It is clear that diamond has the best theoretical performance among the known semiconductor materials. With the development of homoepitaxial growth and doping control technique, various diamond power devices have been fabricated in laboratories. Some of them have shown certain advantages, especially at high operating temperature. However, due to the high cost of the material itself and the immaturity of the manufacturing process, such as edge termination, ion implantation control, etc., there is still a long way to go before the diamond power device can be practically used in high power applications.

Table 9.1 Comparison of material properties and FOM [10–12]

Materials properties and FOM[a]	Si	3C-SiC	4H-SiC	6H-SiC	Wurtzite 2H-GaN	Diamond
Bandgap energy E_g (eV)	1.1	2.2	3.26	3.0	3.39	5.45
Critical electric field E_c (MV/cm)	0.3	1.2	2.0	2.4	3.3	5.6
Electron mobility μ (cm^2/V·s)	1350	900	720[b]/650[c]	370[b]/50[c]	900	1900
Saturation drift velocity v_s (10^7 cm/s)	1.0	2.0	2.0	2.0	2.5	2.7
Relative dielectric constant ε	11.8	9.6	10	9.7	9.0	5.5
Thermal conductivity λ (W/cm·K)	1.5	4.5	4.5	4.5	1.3	20
CTE (ppm/K)	2.6	2.77	5.12	5.12	5.4–7.2	0.8
JFOM[a] $= E_c \cdot v_s/2\pi$	1.0	8.0	13.3	16.0	27.5	50.4
BFOM[a] $= \varepsilon \cdot \mu \cdot E_c^3$	1.0	34.7	133.9	115.4	676.8	4267
HMFOM[a] $= E_c \cdot \sqrt{\mu}$	1.0	3.27	4.87	4.19	8.98	22.15

Note [a]All FOM data are normalized against silicon, [b]mobility along *a*-axis, [c]mobility along *c*-axis

SiC and GaN based WBG devices are the most practical options now and in the near future. Power devices based on silicon carbide (SiC) and gallium nitride (GaN) have been studied for many years and are already in mass production [13–19]. Silicon carbide has been studied since the 1940s or even earlier. To date, more than 200 polytypes of SiC have been discovered. Among the various polytypes, 3C-SiC, 6H-SiC, and 4H-SiC are the most widely studied where the letter C or H represents whether the crystal system is cubic or hexagonal. The number in front of the letter indicates how many Si–C bilayers a repeating unit cell consists of. 6H-SiC and 3C-SiC are more popular in the early studies, a cubic 3C-SiC MOSFET was reported in 1986. In terms of critical electric field strength, 4H-SiC exhibits a greater advantage than 3C-SiC. Compared with 6H-SiC, 4H-SiC has obvious advantages in terms of higher electron mobility and much lower mobility anisotropy. The comprehensive properties of 4H-SiC are the best among the three materials. As the wafer growth method matures, it has become the exclusive material for SiC power devices. For GaN, the most popular crystal polytype is wurtzite 2H polytype. In the rest of this chapter, we will refer to SiC for 4H-SiC and GaN for 2H-GaN unless otherwise specific.

The manufacturing process of SiC wafers matured rapidly. 6-inch and 8-inch SiC wafers are already in mass production. In contrast, the manufacturing process of GaN wafer development is relatively slow. In 2018, the cost of free-standing GaN wafers per unit area was about 20 times more than that of SiC wafers [20]. To reduce costs, a process has been developed to grow GaN material on silicon substrates, making GaN-on-Si wafers rather than free-standing wafers. GaN-on-Si wafers can leverage existing semiconductor production lines, resulting in significant cost reductions. In 2018, the cost per unit area of GaN-on-Si wafers is as low as 1/5 of that of SiC wafers [21].

If we compare the properties of 2H-GaN and 4H-SiC in Table 9.1, we can see that GaN material performs significantly better than SiC material. However, the performance of WBG devices is not only related to the material properties but also to the structure of the device, which will be introduced in the next section.

2.2 *Structure and Performance of WBG Power Devices*

The structure of a semiconductor device can be divided into a lateral structure and a vertical structure, wherein the vertical structure is more suitable for high-voltage and high-power application scenarios. Vertical structure devices typically require a free-standing wafer substrate of WBG material. Due to the lack of bulk GaN wafer fabrication processes, GaN-on-GaN power devices, which are made from free-standing GaN wafers are still in the laboratory stage. GaN-on-Si power devices have become a research hotspot due to the maturity and cost advantages. Therefore, GaN power devices are typically in lateral forms, while SiC devices are more in vertical forms.

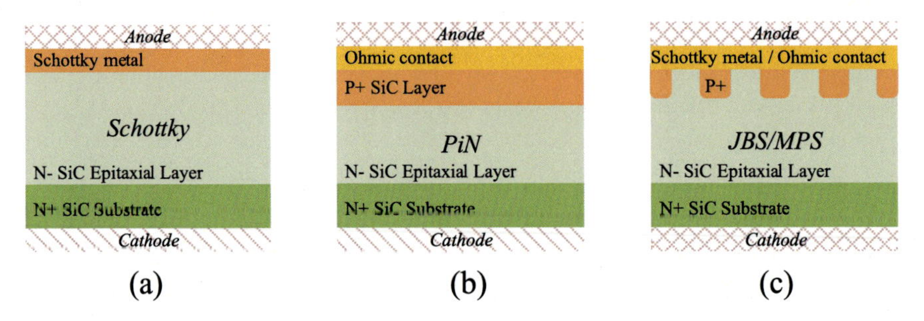

Fig. 9.5 SiC diodes structures. **a** Schottky. **b** PiN. **c** JBS/MPS

2.2.1 WBG Power Diodes

The structure of commonly seen SiC diodes is given in Fig. 9.5. Figure 9.5a shows the structure of SiC Schottky diodes. The Schottky barrier is formed by the N-epitaxial layer and the anode metal. SiC Schottky diodes can reach a withstand voltage of 3300 V, while the highest voltage rating of commercial silicon Schottky diodes is about 300 V. The PiN diode structure shown in Fig. 9.5b adds an additional N-intrinsic layer, which helps the diode withstand multi-kilovolts voltage. The advantage of PiN diodes is the low dynamic on-resistance brought by the conductivity modulation effect, and the disadvantage is the high on-voltage drop caused by the large energy bandgap. By combing the structures of Schottky and PiN diodes, a junction-barrier Schottky (JBS) shown in Fig. 9.5c can be obtained [22], which is also referred to as a merged PiN Schottky (MPS) diode. JBS/MPS diodes combine the advantages of the low on-voltage drop of Schottky diodes and the low dynamic on-resistance of PiN diodes.

Two typical GaN-on-Si diode structures are shown in Fig. 9.6. Figure 9.6a is a vertical structure, which is theoretically suitable for high-power applications, but the existence of the silicon substrate brings great difficulty to fully exploiting the performance [23]. Figure 9.6b shows a lateral form of GaN-on-Si diodes. The AlN layer and the GaN layer form a heterojunction, in which a two-dimensional electron gas (2DEG) with high electron density and mobility is generated, enabling high conductivity between the anode and cathode.

2.2.2 Unipolar WBG Power Transistors

A unipolar transistor is a field effect transistor that uses either electrons (N-type) or holes (P-type) for conduction. Unipolar devices typically have faster switching speeds, but their voltage and current ratings cannot be too high, making them suitable for low to medium power applications. Metal oxide semiconductor field effect transistors (MOSFETs) and junction field effect transistors (JFET) are two common

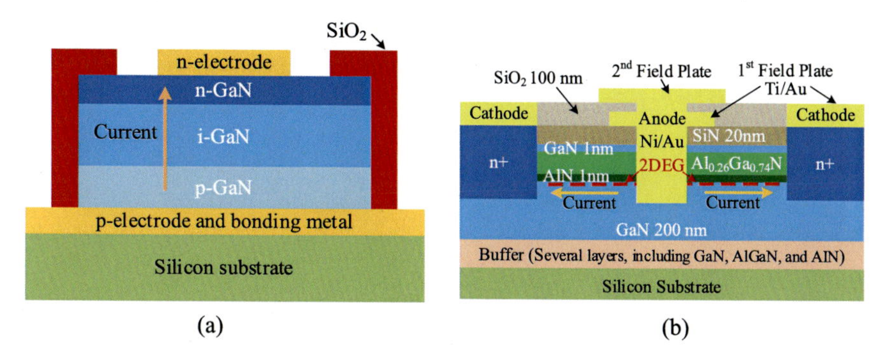

Fig. 9.6 GaN-on-Si diodes structures. **a** Vertical PiN [24]. **b** Lateral Schottky [25]

types of unipolar transistors. Due to the maturity of SiC wafers and SiO_2 manufacture processes, the study and commercialization of SiC MOSFETs have developed rapidly.

The lateral structure shown in Fig. 9.7a is suitable for applications where the device voltage rating is not too high. With gate, drain, and source configured on the top layer of the lateral SiC MOSFET, it is particularly suitable for integration with driving, sampling, and protection circuits. The vertical structure shown in Fig. 9.7b is usually adopted for higher power and voltage. The structure of SiC MOSFET is almost the same as that of silicon MOSFET, which has prompted its rapid development. There are many off-the-shelf SiC MOSFETs on the market from manufacturers such as Infineon, BASiC, wolfspeed, and ROHM, with voltage ratings ranging from 600 to 1700 V.

JFETs have lower on-resistance and higher reliability than MOSFETs in the same wafer area. However, the main disadvantage of JFETs is that they are normally-on devices [26], which means JFETs are in one state before having the control power and are undesirable for power electronics applications. A normally-off SiC FET can be obtained by cascading a low-voltage Si-MOSFET with a JFET. UnitedSiC is one of the major companies offering cascode SiC FETs with voltage ratings ranging from 650 ~ 1700 V and on-resistance as low as 6 mΩ.

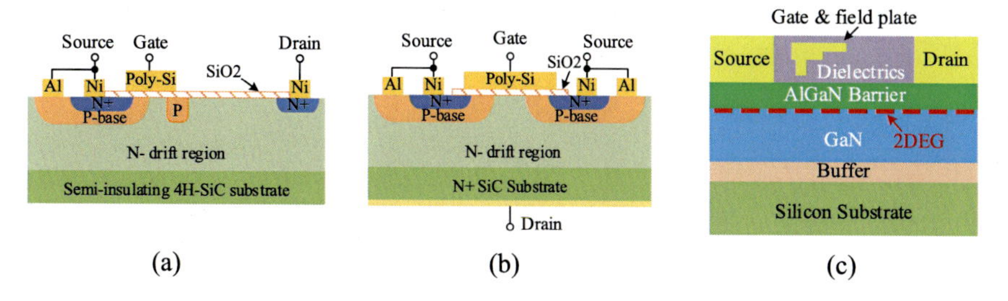

Fig. 9.7 WBG transistor structures. **a** Lateral MOSFETs [28]. **b** vertical MOSFETs [29]. **c** Lateral GaN HEMT [27]

GaN-on-Si devices have received a lot of attention for their compatibility with depreciated silicon wafer production lines. A typical GaN-on-Si FET structure is shown in Fig. 9.7c. The GaN-on-Si FET adopts a lateral structure. It utilizes the high electron mobility in the 2DEG to conduct current and is called a high electron mobility transistor (HEMT).

In Fig. 9.7c, there is a 2DEG even with zero gate-source bias voltage, which means the device is a normally-on depletion mode HEMT. By configuring a cascode low-voltage Si-MOSFET together with the GaN HEMT, normally-off GaN devices can be realized. Alternatively, a normally-off GaN enhancement mode HEMT (E-HEMT) can be achieved by implanting a P-GaN area between the gate and the AlGaN barrier [27].

2.2.3 Bipolar WBG Power Transistors

A bipolar transistor is a field effect transistor that uses both electrons and holes for conduction. The conductivity modulation effect in bipolar devices enables injection of minority carriers in the drift region, thereby significantly reducing the on-resistance at high current densities. As there is stored charge in the drift region, silicon bipolar devices can only operate at low switching frequency. For WBG devices, the high electric field strength of the material makes the drift region much thinner, which means that the drift region stores much less charge. The lifetime of minority carriers in WBG devices is two orders of magnitude shorter than that of silicon devices. The above two factors make the injection and extraction process of minority carriers much faster than the silicon counterpart, enabling higher switching frequencies.

SiC bipolar devices have similar structures to the corresponding silicon devices. Common bipolar WBG devices are SiC bipolar junction transistors (BJTs) [30], SiC insulated gate bipolar transistors (IGBTs) [31], and SiC gate turn-off thyristors (GTOs) [32]. The lower collector–base current gain of a BJT requires a larger drive current, which complicates the drive circuit and introduces additional losses. Figure 9.8 shows a comparison of applicable voltage and current ratings for SiC MOSFETs, SiC IGBTs, and SiC GTOs/thyristors.

2.3 *Performance of WBG Power Devices*

The conduction performance of power device can be represented using the relationship between the specific on-resistance R_{onsp}, and the breakdown voltage V_B. Using the unipolar one-dimensional (1-D) device model [34], the theoretical R_{onsp} limit can be calculated from the material properties, as shown in Eq. (9.1).

$$R_{\mathrm{onsp}} = \frac{4V_B^2}{\mu \varepsilon E_C^3} \tag{9.1}$$

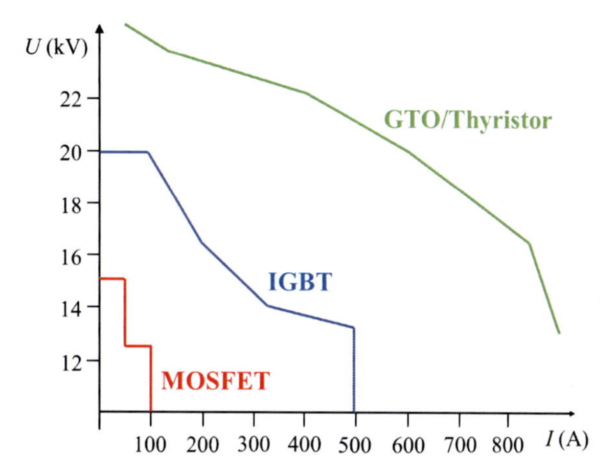

Fig. 9.8 Comparison of applicable ratings of typical biopolar SiC devices [33]

Figure 9.9 shows the R_{onsp} and V_B of some recently developed WBG diodes. Clearly, all WBG diodes exceed the Si material limit. In 1999, the performance of SiC diodes has reached the material limit [35]. The performance of GaN diodes has exceeded the limit of SiC materials [35, 36], which is promising, but the mass production process of GaN wafers and devices has affected its promotion. As bipolar devices, PiN diodes can further push the theoretical limit of unipolar devices and are extremely attractive in high-voltage and high-power applications [38].

The super-junction (SJ) structure proposed in the 1990s, which forms a two-dimensional (2-D) structure through interleaved multiple thin P-type and N-type layers, has pushed the device performance beyond the theoretical limit of Eq. (9.1) [39]. The material theoretical limit of the two-dimensional structure can be obtained from Eq. (9.2).

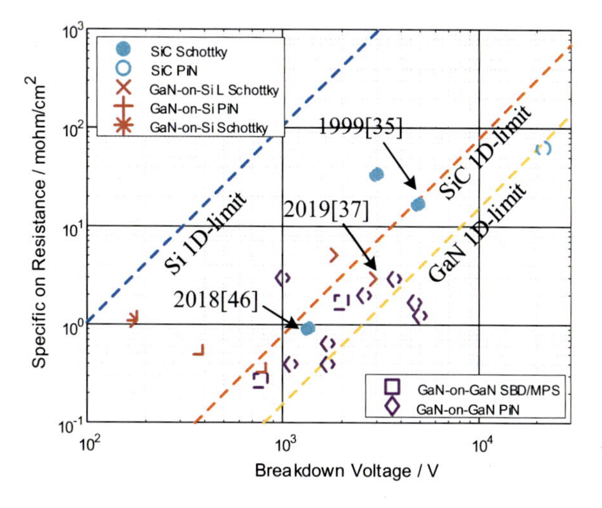

Fig. 9.9 R_{onsp} versus breakdown voltage of WBG diodes

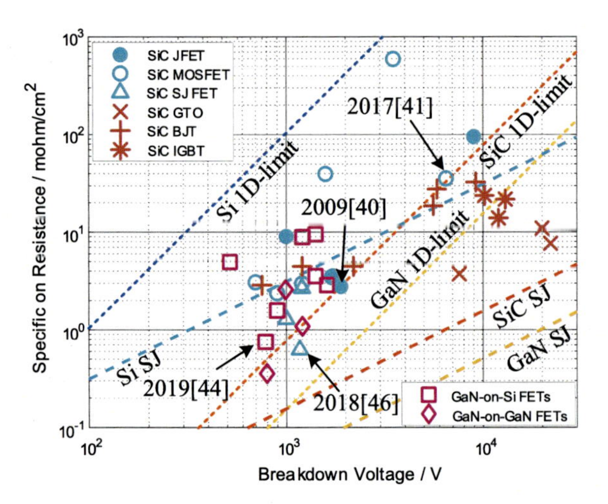

Fig. 9.10 R_{onsp} versus breakdown voltage of WBG transistors and thyristors

$$R_{onsp-2D} = \frac{4dV_B}{\mu\varepsilon E_c^2} \tag{9.2}$$

where d represents the thickness of the P-type and N-type layers. In Fig. 9.10, the R_{onsp} limit for different materials adopting the SJ structure is given when d is 1 μm. It is clear that the R_{onsp} limit of the SJ structure is much lower than that of conventional 1-D devices. Figure 9.10 shows the R_{onsp} and V_B of the WBG power devices. The maturity of SiC wafers has promoted the rapid development of SiC power devices. In 2009, SiC JFETs had reached the one-dimensional limit of SiC materials [40] and exceeded the limits of silicon-based super-junction structures [41]. However, the device performance of GaN devices has not yet reached the theoretical limit of the material [41–43]. Overall, the conduction performance of WBG power devices is improved by 1–2 orders of magnitude compared to silicon devices [44, 45]. SiC bipolar power devices can achieve low R_{onsp} under high voltage, and have obvious advantages in high-power application scenarios.

In addition to the conduction characteristics, the switching performance of WBG power devices has also made a qualitative leap compared with Si diodes. There is almost no reverse recovery process of SiC Schottky diodes. Due to the high electrical strength of WBG materials, the drift region of high voltage WBG PiN diodes can be much thinner than that of Si PiN diodes. This means that in the WBG PiN diode, the drift region stores much less charge and leads to much faster switching speed and reverse recovery process. Similar to PiN diodes, bipolar power devices also have conductivity modulation effects. It is easy to understand that WBG bipolar power devices can switch at higher frequencies than their Si counterparts. For a SiC JFET or MOSFET, there is an inherent anti-parallel body diode inside, which has bipolar diode reverse recovery characteristics. Due to the excellent properties of SiC material, the reverse recovery time of this diode is much shorter than that in Si devices. For the structure of the GaN HEMT, there are no parasitic diodes, thus eliminating the reverse recovery process. GaN HEMTs are expected to operate at higher switching frequencies, especially under hard switching conditions.

3 On-Board Fast Chargers

The on-board charging can be implemented as a non-integrated OBC or an integrated OBC [47]. The no-integrated OBC is a separate charging circuit that is only used as a propulsion battery charger. It can be realized as the single- or three-phase topologies. However, the single-phase-based OBC can only provide slow charging because the charging power is less than 10 kW [47–49]. For the fast charging, the three-phase-based OBC is required. the three-phase-based OBC is mainly divided into single-stage and two-stage according to the number of power conversion stages [47]. The integrated OBC includes two categories. One is the propulsion system integrated on-board chargers, which can reduce the cost and weight of the OBC by reusing the existing inverter and motor components for the propulsion battery charging circuit [50]. The other one is the auxiliary power module integrated on-board chargers, which can provide significant improvements in size, weight, efficiency, cost, and reliability [47].

3.1 Single-Stage On-Board Fast Chargers

In the single-stage on-board fast charger, the DC-link capacitor for the rectifier is eliminated and an AC-link capacitor is employed. Because the AC-link capacitor is not used as an energy storage capacitor, so the single-stage on-board fast charger can achieve a high-power density.

In [5], a single-stage, bidirectional, and isolated dual active bridge (DAB) ac–dc converter is adopted as the single-phase unit, which is shown in Fig. 9.11. The power rating of each single-phase unit is 7.2 kW and three phase units form a three-phase OBC which can deliver a 22 kW charging power to the propulsion battery. An H-bridge is employed as the active front end (AFE) converter which works in low-frequency synchronous rectification mode. The dc/dc stage controls the ac side power factor and the dc output power and directly delivers the ac power to the battery, so only a small AC-link capacitance is needed and a high-power density can be achieved. With the back-to-back switches, a three-phase single-stage OBC topology is proposed in [51] based on the matrix converter, which is shown in Fig. 9.12. This topology can achieve bidirectional power flow and reactive power compensation to the ac grid as well as high-power density. In [52], an isolated three-phase bidirectional OBC is proposed based on a three-port high frequency transformer and back-to-back switches, which is shown in Fig. 9.13. It can achieve a compact design and small output capacitor as well as bidirectional power flow. However, the disadvantages of these two topologies include a high semiconductor count because of the back-to-back switches and complex control schemes of the matrix converter.

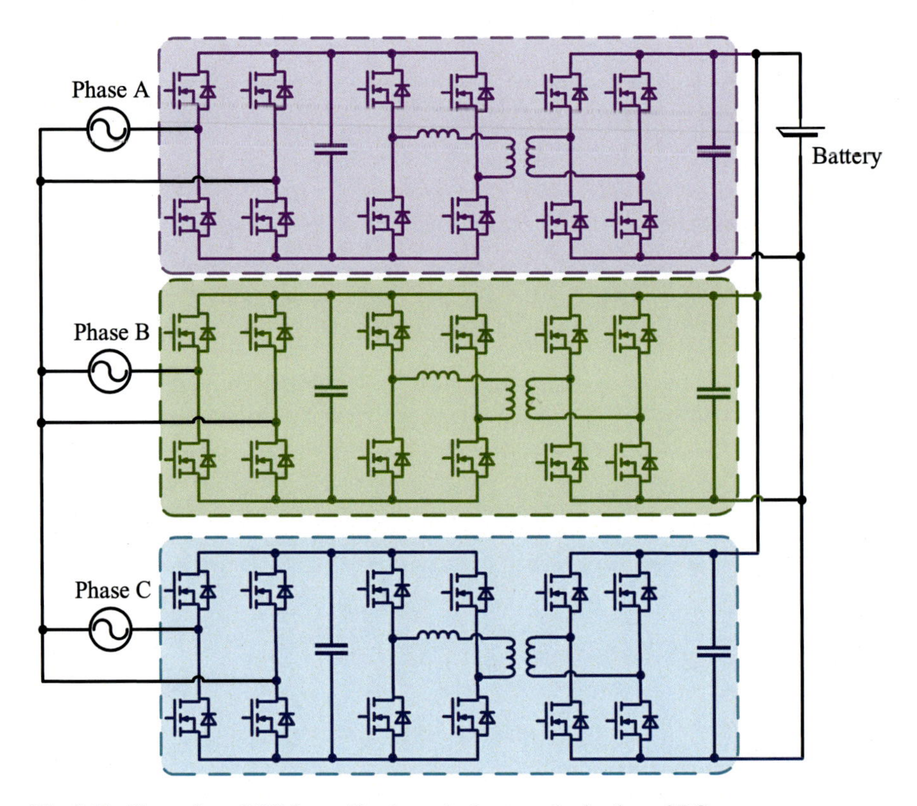

Fig. 9.11 Three-phase OBC formed by three single-stage single-phase OBCs

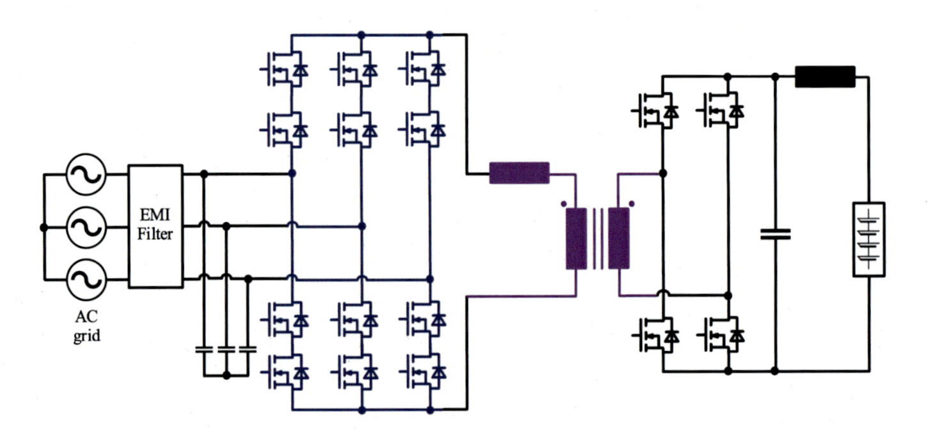

Fig. 9.12 Three-phase single-stage OBC based on the matrix converter

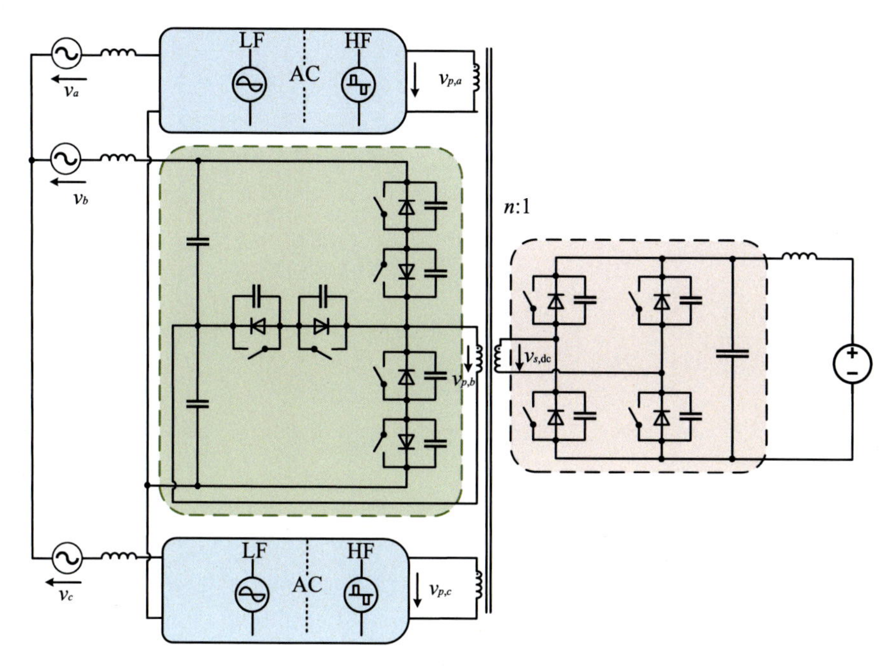

Fig. 9.13 Three-phase single-stage OBC based on a three-port high frequency transformer

3.2 *Two-Stage On-Board Fast Chargers*

In the two-stage on-board fast charger, a PFC is usually employed to reduce the input current distortion. Meanwhile, an energy storage DC-link capacitor is implemented at the PFC output, which can stabilize the output voltage. The DC-link capacitor is followed by an isolated DC/DC converter which provides the charging power to the propulsion battery.

In [53], a 10.5 kW OBC is proposed which can be plugged into a single-phase grid or a three-phase grid. It is formed by three single-phase OBC modules which can deliver full power from the ac grid to the propulsion battery, as shown in Fig. 9.14. Each single-phase OBC is composed of a diode rectifier, a boost converter PFC and a half bridge LLC resonant converter. Similar architecture can also be found in [54], where the power rating is 22 kW. An interleaved boost PFC and two H-bridge LLC converters with the phase-shift control are adopted in each phase unit, which is shown in Fig. 9.15. A bidirectional power follow can also be achieved by using this topology. In [55], a 10 kW bidirectional OBC is realized by using a three-phase boost-type PFC as the AFE converter, as shown in Fig. 9.16. The AFE converter is followed by a bidirectional isolated LLC soft switching dc-dc converter, which is connected to the propulsion battery. A similar AFE converter is adopted in [56], and an input-series-output-parallel (ISOP) half-bridge LLC converter is used to deliver a 20 kW charging power to the propulsion battery, which is shown in Fig. 9.17.

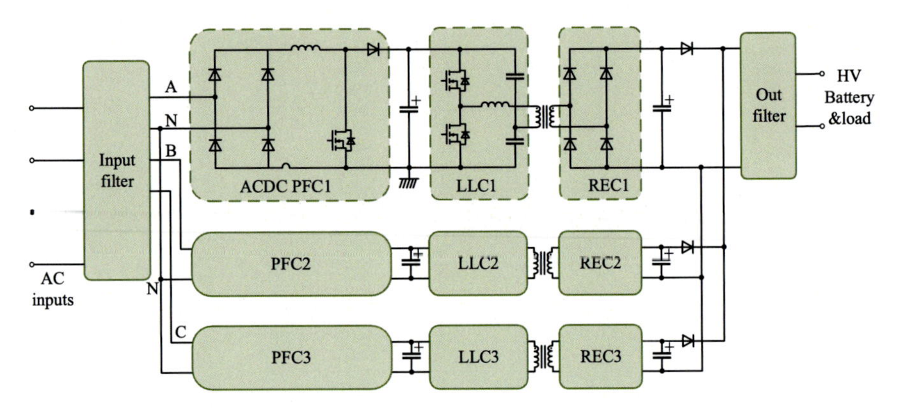

Fig. 9.14 10.5 kW three-phase OBC based on three unidirectional single-phase OBCs

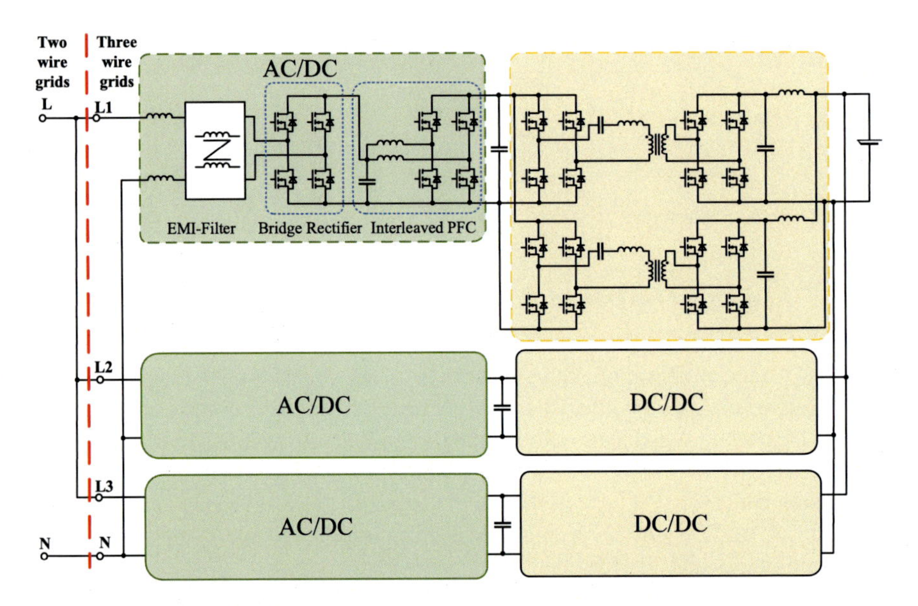

Fig. 9.15 22 kW three-phase OBC based on three bidirectional single-phase OBCs

3.3 Propulsion System Integrated On-Board Fast Chargers

The concept of propulsion system integrated OBC was firstly presented in 1985 [57], which is based on a thyristor inverter. Integrating the OBCs with the propulsion system is the most popular implementation to build integrated OBC because the propulsion system also includes the switches, diodes, inductors, and capacitors, which is the same as a typical OBC system. The main challenges of the integrated OBC are the average torque production during the charging mode and the necessary hardware reconfiguration switches to switch between the propulsion mode and the

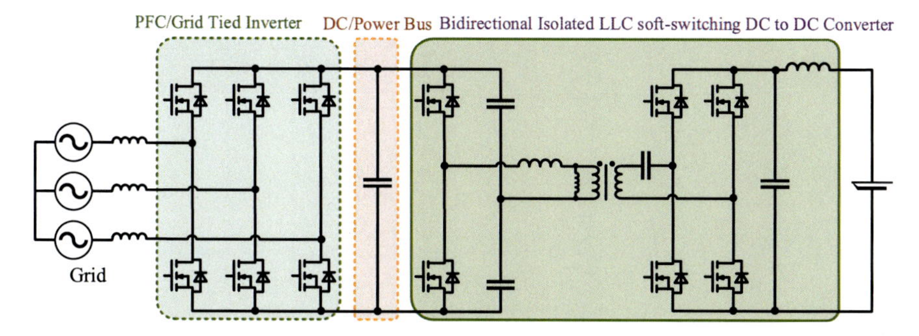

Fig. 9.16 Three-phase OBC based on a three-phase boost-type PFC and a bidirectional isolated LLC soft switching dc-dc converter

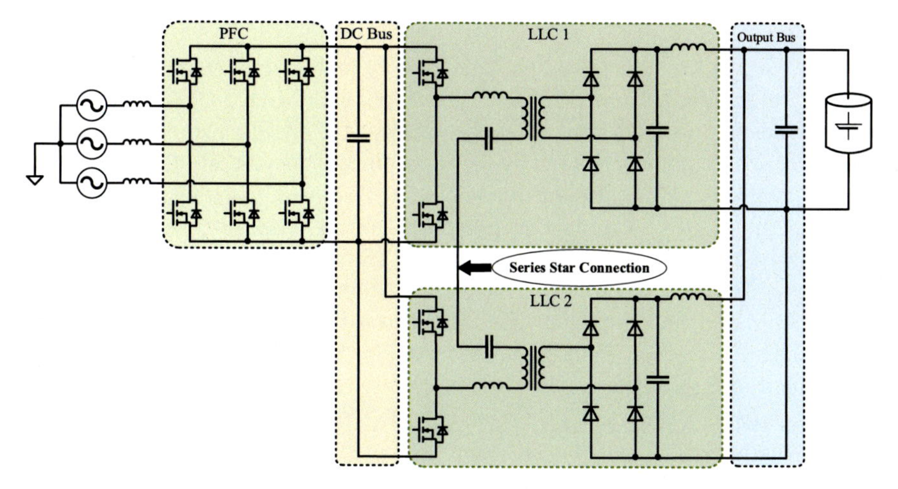

Fig. 9.17 Three-phase OBC based on a three-phase boost-type PFC and an ISOP half-bridge LLC converter

charging mode [50]. For the propulsion system integrated on-board fast chargers, it is usually realized by the three-phase topologies because the three-phase topologies can deliver higher power compared with the single-phase topologies. The three-phase propulsion system can be directly connected to the three-phase ac grid with an additional buck-type three-phase full-bridge converter, which forms a three-phase integrated OBC, as shown in Fig. 9.18 [58]. This three-phase integrated OBC reuses the motor inductors as a coupled dc inductor for the charger and its charging power could be as high as that of the propulsion system. Meanwhile, the three-phase ac side PFC and the battery voltage/current regulations could also be achieved. Moreover, no modification for the motor and no special design requirements for the propulsion system are needed [47].

For the multi-phase motors, they can realize zero average torque production during the charging mode [50]. Meanwhile, a multi-phase propulsion system can provide

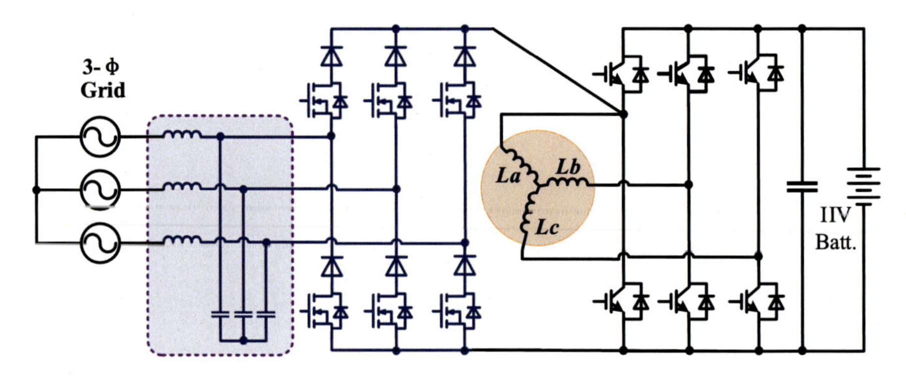

Fig. 9.18 Three-phase integrated OBC with an additional buck-type three-phase full-bridge converter

all the necessary components of the OBC system and the bidirectional power flow can also be achieved. Therefore, various three-phase integrated OBCs based on the multi-phase propulsion system are proposed [58–60]. These three-phase integrated OBCs reuse the motor inductors as the input filter inductors and the multi-phase propulsion inverter is utilized as a three-phase AFE converter. In [59] and [60], three-phase integrated OBCs based on the five-phase and six-phase propulsion system are presented, which are shown in Figs. 9.19 and 9.20. The three-phase AFE converters made by the multi-phase propulsion inverter are directly connected to the propulsion battery. In [61], a three-phase integrated OBC based on an asymmetrical nine-phase motor and an inverter is proposed, as shown in Fig. 9.21. In this integrated OBC, the three-phase AFE converter is followed by a dc-dc converter which is connected to the propulsion battery. The main disadvantage of these multi-phase propulsion system based integrated OBCs is that the implementation capabilities of these charging techniques are limited because the current EV propulsion system is dominated by the three-phase propulsion motors.

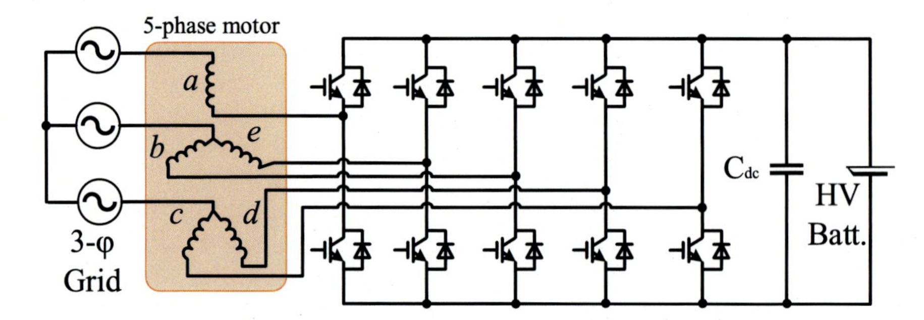

Fig. 9.19 Three-phase integrated OBC based on the five-phase propulsion system

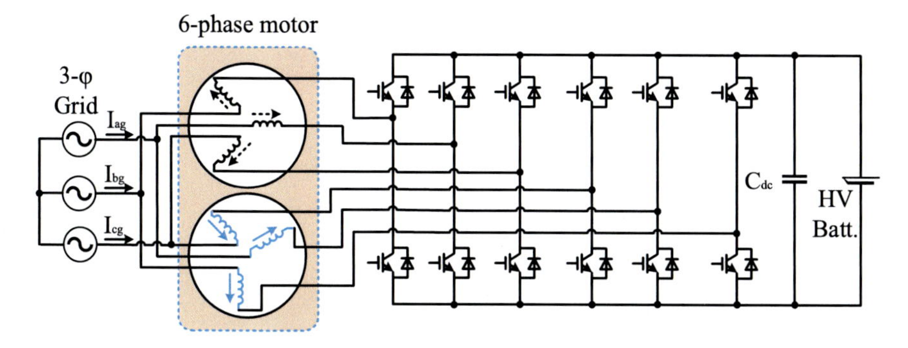

Fig. 9.20 Three-phase integrated OBC based on the six-phase propulsion system

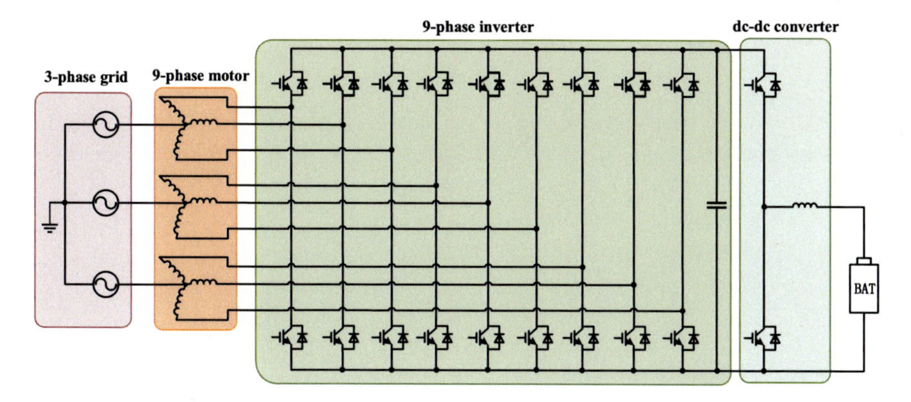

Fig. 9.21 Three-phase integrated OBC based on the nine-phase propulsion system

3.4 Auxiliary Power Module Integrated On-Board Fast Chargers

Because the low-voltage auxiliary battery in every EV needs to be charged by an auxiliary power module, the concept of the auxiliary power module integrated on-board charger is introduced in [62], in which a selective switch is used to realize multi-function and it can charge the auxiliary battery via the propulsion battery. In [63], a 6.6 kW OBC and 1.9 kW APM are integrated by reusing the semiconductor power devices and mechanical elements. The volume of the APM is significantly reduced to 1.87 L, which is 85% of the conventional nonintegrated APM. The peak efficiencies of the OBC and the APM are 97.3% and 93.13%, respectively. To realize the bidirectional power flow, a 3.3 kW dual-output dc-dc resonant converter is proposed based on a three-winding integrated transformer [64]. Overall efficiency of 94% and a power density of 11.7 W/in^3 are realized. However, the integrated OBCs proposed in [63] and [64] are based on single-phase AC-grid, the power level can only push

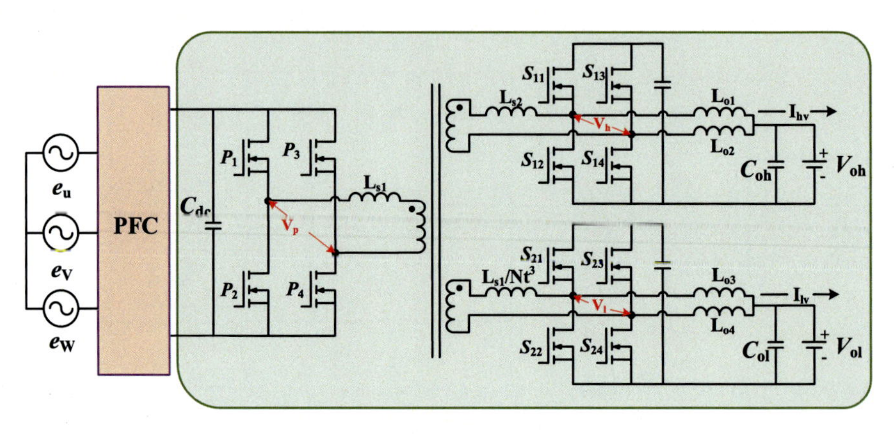

Fig. 9.22 Auxiliary power module integrated on-board fast chargers based on a current-fed three-port DC/DC converter

up to 6.6 kW. To realize fast charging, an auxiliary power module integrated on-board fast charger based on a three-port current-fed triple-active bridge converter is proposed in [65], which realizes an 11 kW OBC plus 3.5 kW APM integration as shown in Fig. 9.22. The oversize of the converter is 7 L and the calculated power density is 2.05 kW/L, which is higher than that of the state-of-art APM.

4 Fast Charging Stations

Because of volume, weight, and cost constraints in an EV, the power level of the OBC is limited. With a level 2 OBC rated at 7.2 kW, it would require approximately 10 h to add 200 mi range to the EV. With the charging power increasing, the required charging time can be reduced. The charging time for adding 200mi can be reduced to 1 h by using a 50 kW fast charger and to 27 min by using a 135 kW T supercharger [66]. The range anxiety can be alleviated if a considerable cruising range can be replenished in a few minutes, which can be achieved by charging the battery with tens to hundreds of kilowatts [2]. The fast charging station is introduced to deliver such high power to the EV [66].

4.1 Traditional Fast Charging Stations

Traditional fast charging stations can be divided into the ac bus architecture and the dc bus architecture as shown in Fig. 9.23 [67]. Both of these two architectures are based on low frequency transformers. In the ac bus architecture, each charger is formed by an ac-dc converter and a dc-dc converter. The ac-dc converter is directly connected to the ac bus and it converts the ac power to the dc power. The ac-dc converter is followed by a dc-dc converter which interfaces with the EVs. In the dc bus architecture, a central front-end ac-dc converter is implemented to create a common dc bus. A dc-dc converter is employed as the charger, it is directly connected to the common dc bus and delivers the power to the EVs. Compared with the common ac bus architecture, the common dc bus architecture requires fewer power conversion stages between the common dc bus and EVs or renewable energy sources, which can reduce system complexity, cost, and power losses [64]. Therefore, the common dc bus architecture is preferred in the traditional fast charging stations compared with the common ac bus architecture. The common dc bus can be created as a unipolar dc bus or a bipolar dc bus [68]. In [69], a unipolar dc bus is built by a two-level AFE converter as shown in Fig. 9.24. The unipolar dc bus is connected by a three-phase interleaved buck converter which is used as a charger to control the battery charging current. This unipolar dc bus structure is widely employed by the commercialized dc fast chargers due to its simple structure, low cost, and well-established control schemes [64]. The bipolar dc bus can be created by using a three-level diode neutral-point clamped converter (DNPC) as shown in Fig. 9.25, which is followed by a two-phase interleaved three-level buck converter as a charger [67]. The bipolar dc bus structure cannot only provide more power capacity but also offer more flexible ways to connect loads to the dc bus.

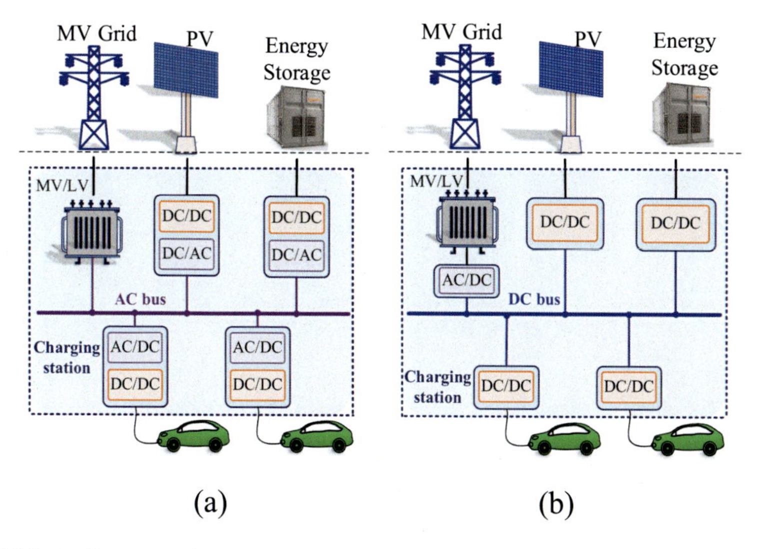

Fig. 9.23 Architectures of traditional fast charging stations. **a** Common ac bus architecture. **b** Common dc bus architecture

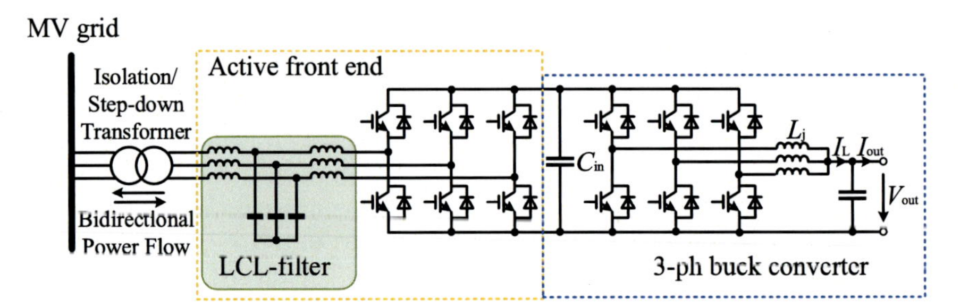

Fig. 9.24 Unipolar dc-bus using two-level AFE converter and three-phase interleaved buck converter as the charger

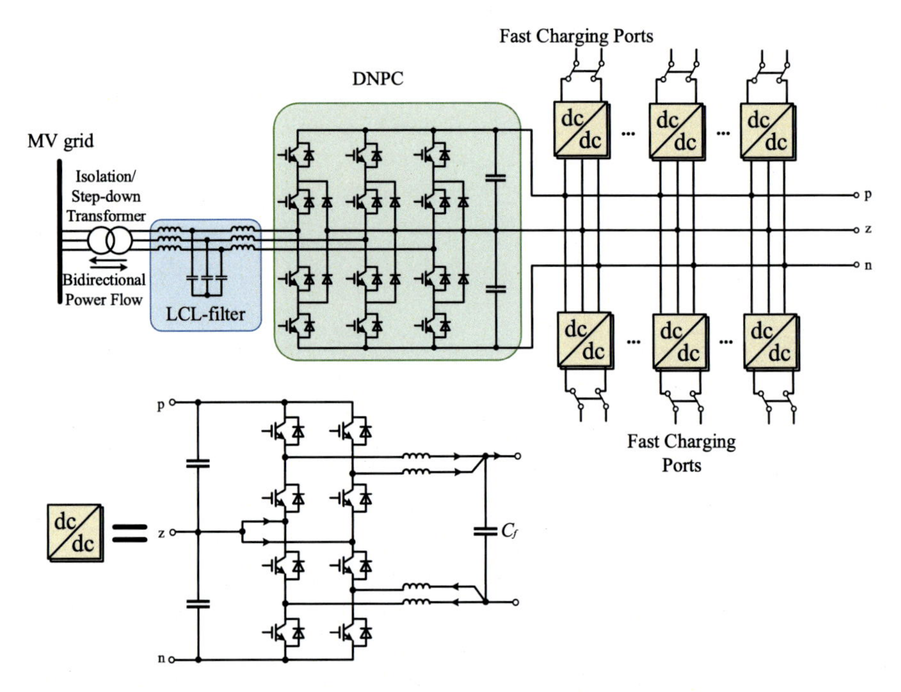

Fig. 9.25 Bipolar dc-bus using three-level DNPC and a two-phase three-level interleaved buck converter as the charger

4.2 *Modular SST-Based Fast Charging Stations*

The solid-state-transformers (SSTs) can be employed to build the fast charging station, in which the AC/Dc converter is directly connected to the medium voltage (MV) grid, as shown in Fig. 9.26. In the SST-based fast charging station, the low-frequency transformer is eliminated and the AC/DC converter is followed by isolated DC/DC converters which also provide the electrical isolation for the system. Then a low voltage dc bus is created by the outputs of the isolated DC/DC converters. The

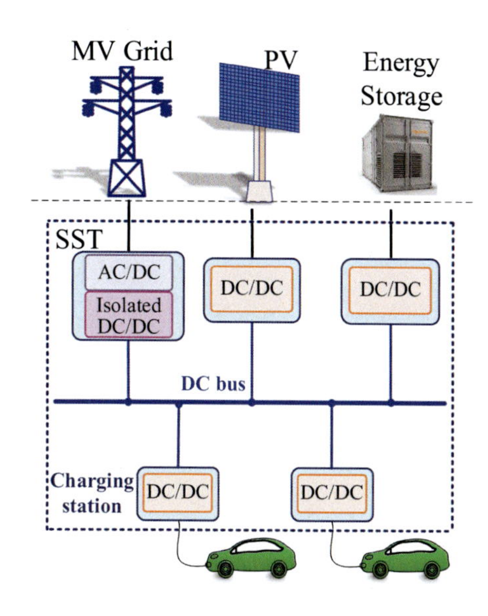

Fig. 9.26 Architecture of SST-based fast charging station

EVs and the renewable energy sources are connected to this dc bus by using non-isolated dc-dc converters [70]. According to the structure of the SST, the SST-based fast charging station can be divided into the modular SST-based architecture or the single-module SST-based architecture.

4.2.1 Architectures of Modular SST-Based Fast Charging Stations

For the modular SST-based architecture, the modular AFE converter, such as a cascaded H-bridge converter, is usually used to directly connect to the MV grid, as shown in Fig. 9.27 [70, 71]. Due to the modular structure of the cascaded H-bridge converter, low voltage rating commercialized power devices can be employed in the system. The output of the cascaded H-bridge converter is followed by an ISOP-isolated DC-DC converter, which offers electrical isolation for the system. In [70], a three-level DNPC-bridge-based AFE converter is employed for the modular SST-based fast charging station, which is shown in Fig. 9.28. Fewer isolation components, simple structure, better THD, smaller filter, high efficiency and lower cost can be achieved by using the three-level DNPC-bridge. In Fig. 9.28, the output of the three-level DNPC-bridge based AFE converter is followed by an ISOP three-level LLC resonant DC-DC converter, which creates a 1 kV dc bus. The EVs are connected to this dc bus by using a three-level interleaved buck converter as the charger. Meanwhile, the system efficiency is improved by adopting 1200 V/1700 V SiC MOSFETs in the three-level DNPC-bridges and low voltage side H-bridges.

In [73], a 25 kW all-SiC MVAC-LVDC SST based on the isolated AFE converter is proposed as shown in Fig. 9.29. It serves as the interface between 6.6 kV MV grid and 400 V low dc bus. The output of the isolated AFE converter is a low voltage

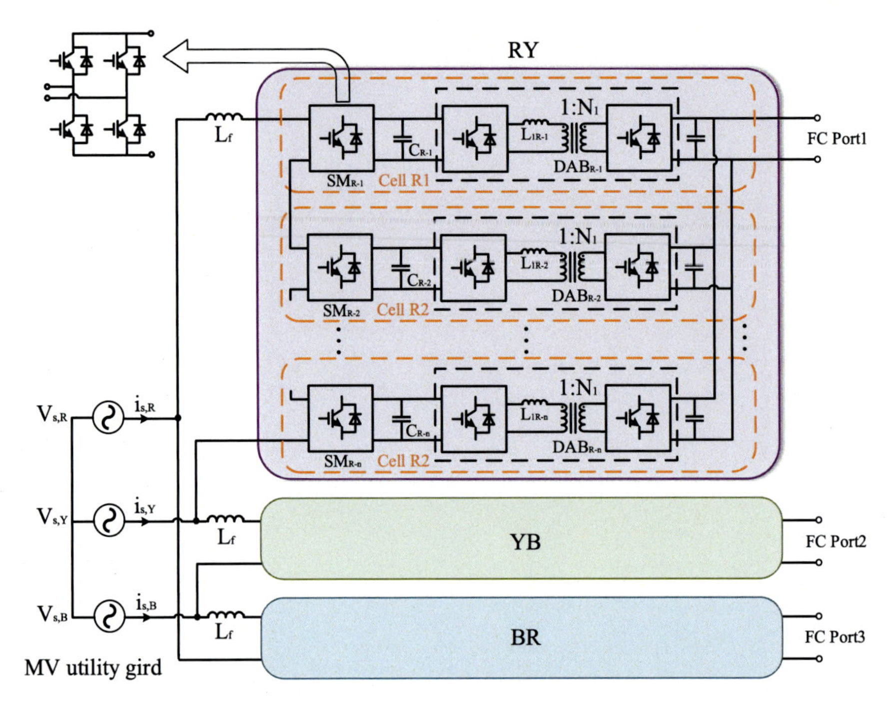

Fig. 9.27 Modular SST-based fast charger using cascaded H-bridge converters and DABs

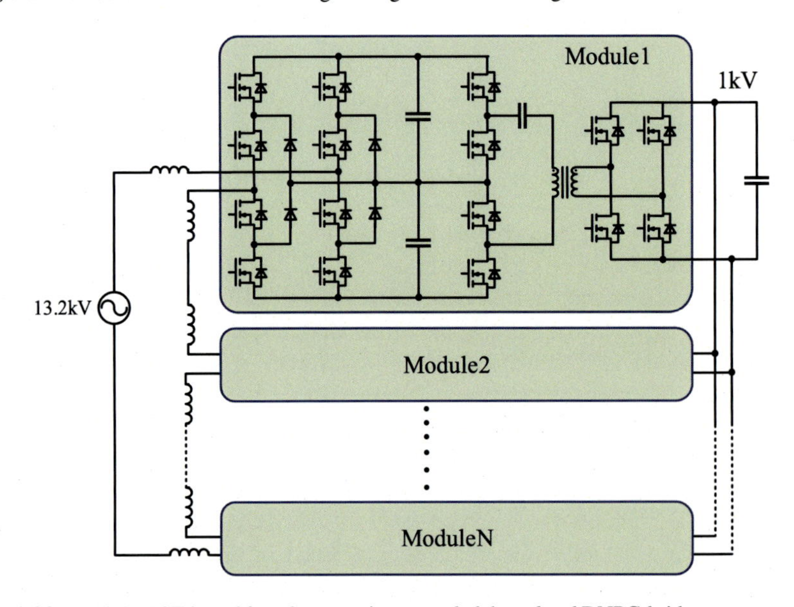

Fig. 9.28 Modular SST-based fast charger using cascaded three-level DNPC-bridge converters and LLC converters

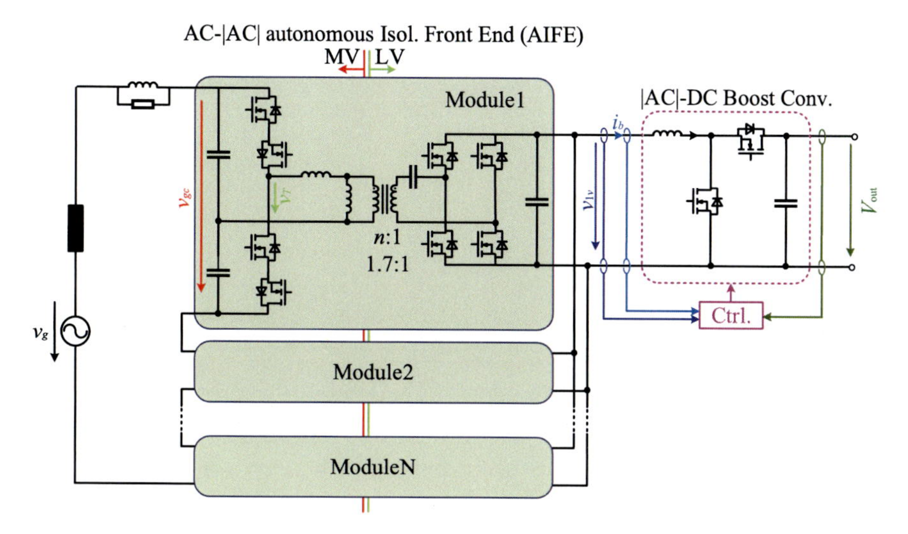

Fig. 9.29 Modular SST-based fast charger using isolated AFE converter and a boost converter

half-wave ac voltage, which is converted to a constant dc voltage by using a boost PFC. By adopting uncontrolled diode bridge rectifiers, the number of active power devices can be reduced as shown in Figs. 9.30 and 9.31 [73, 74]. These topologies can not only reduce the system cost but also improve efficiency and compactness. However, only unidirectional power flow can be realized by using these topologies and the reactive power control is limited. In the modular SST-based fast charger, the voltage rating of each module is low, so commercialized low voltage rating power devices can be used. However, many modules are needed because the ac-dc converter is directly connected to MV grid, which induces the complicated topology structure and multiple control loops.

4.2.2 Medium Frequency Transformer for Modular SST-Based Fast Charging Stations

The medium frequency transformer (MFT) is the core component of the modular SST-based fast charging station, and its performance directly determines the efficiency, power density, and isolation voltage of the modular SST-based fast charging station. The existing MFTs mainly adopt the solenoidal structure [75–78] and the coaxial structure [80], as shown in Figs. 9.32 and 9.33, respectively. The solenoidal structure winds the windings on the magnetic core in the form of a solenoid, which mainly includes the core-type structure, the shell-type structure, and the matrix structure. The coaxial structure uses a conducting copper tube as the low-voltage side winding and multiple turns of wire at the center of the tube as the high-voltage side windings. Meanwhile, high permeability toroidal magnetic cores are employed along with the conducting copper tube. The leakage inductance of the solenoidal MFT is relatively

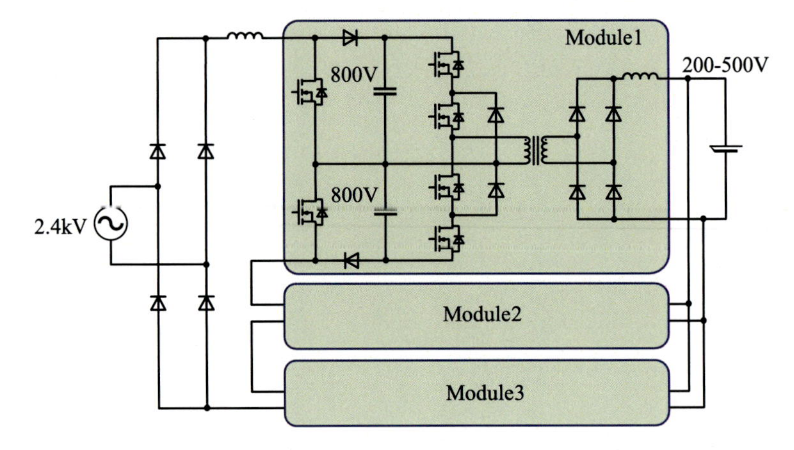

Fig. 9.30 Modular SST-based fast charger using unidirectional multi-cell boost topology with three ISOP dc/dc converter modules

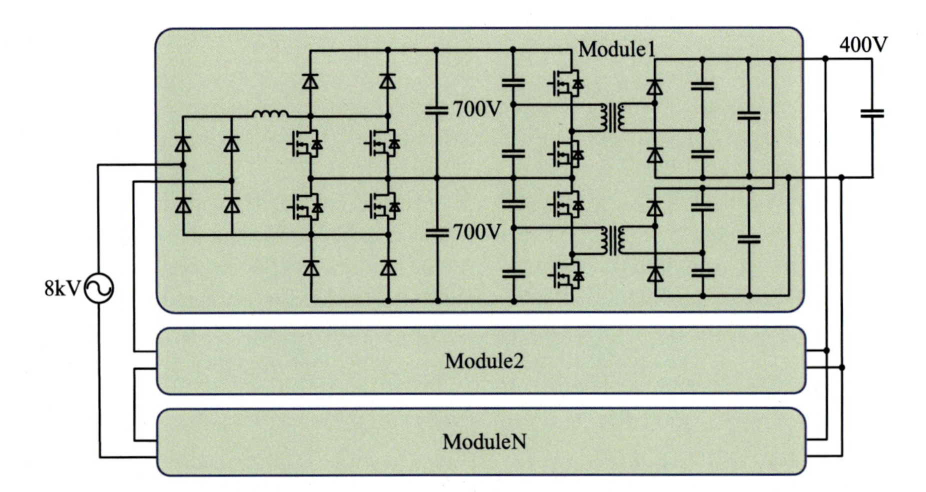

Fig. 9.31 Modular SST-based fast charger using three-level AFE boost converters and half-bridge LLC-based dc-dc converters

large and it is not easy to control accurately. Consequently, the loss of the solenoidal MFT is also relatively large. Meanwhile, the use of a non-circular magnetic core would cause edge effects and bring difficulties to heat dissipation. However, the solenoidal MFTs have the advantages of simple manufacturing process and low cost. Moreover, a flexible voltage transformation ratio can be achieved by controlling the number of turns of the primary and secondary winding [81]. When the MFT adopts the coaxial structure, an extremely low leakage inductance can be achieved and the magnetic field distribution and the thermal field distribution are more uniform [81]. Meanwhile, the magnitude of the leakage inductance can also be controlled by adding toroidal magnetic cores between the coils [82]. However, it is difficult to deal with the insulation issue between the primary and secondary sides. Moreover, it is difficult to design the low-voltage winding of the coaxial MFT as a multi-turn winding, which limits the voltage transformation ratio of the coaxial MFT [81]. In [83], an 80 kW MFT as shown in Fig. 9.34 is proposed based on the split planar structure, a power density of 21.1 kW/L and a peak efficiency of 99.33% are achieved. Due to the advantages of flexible voltage transformation ratio, simple manufacturing process and low cost, most of the existing MFTs adopt the solenoidal structure [75–77, 83].

4.2.3 Control Architectures of Modular SST-Based Fast Charging Stations

In the modular SST, many modules are needed and multiple control loops are employed to control the variables of the converter. For a specific modular SST topology, a dedicated control strategy should be developed to make the system achieve a good performance. Therefore, different modular SST topologies need different control strategies. In the existing modular SST, the cascaded H-bridge converter is usually employed as the AFE converter, which is followed by multiple isolated DC-DC converters. These isolated DC-DC converters are generally connected as the input series and output parallel. For the cascaded H-bridge converter-based modular SST, the power balance of the isolated DC-DC converter should be achieved by using proper control strategies. It means that the input capacitor voltage of the isolated DC-DC converter should be balanced and a good output current sharing of the isolated DC-DC converter should be guaranteed. There are two control methods to realize the power balance control of the isolated DC-DC converter. One is the AFE converter to control the input voltage balance of the isolated DC-DC converter, and the isolated DC-DC converter realizes the output current sharing. The other one is the input capacitor voltage balance and output current sharing are all realized by the isolated DC-DC converter [85]. The second control method has the advantages of high quality of AC side input current [86], good capacitor voltage balance, less required computing resources, and good soft start performance. Meanwhile, the control structure is simpler, the robustness of control parameters is stronger and the operation reliability is higher compared with the first control method [84, 85]. However, when the power balance control is realized by the isolated DC-DC converter, closed-loop control methods should be applied to the isolated DC-DC converters. When the DAB converter or dual full-bridge LC resonant converter is employed in the modular SST, the phase-shift angle between the primary side and the secondary side of the isolated DC-DC converter should be adjusted to realize the power balance control [86]. When the bidirectional LLC resonant converter or the bidirectional CLLC resonant converter is employed in the modular SST, the switching frequency of the isolated DC-DC converter should be adjusted to realize the power balance control [87].

In [72], a control strategy for the modular SST based on the cascaded H-bridge converter is presented. The control scheme of the AFE cascaded H-bridge converter is shown in Fig. 9.35, in which a grid synchronization control and a power balancing

Fig. 9.32 Solenoidal MFT structure [79]

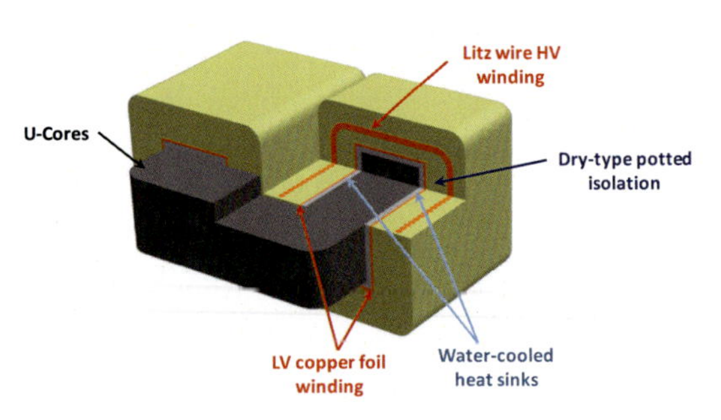

(a) Core-type transformer.

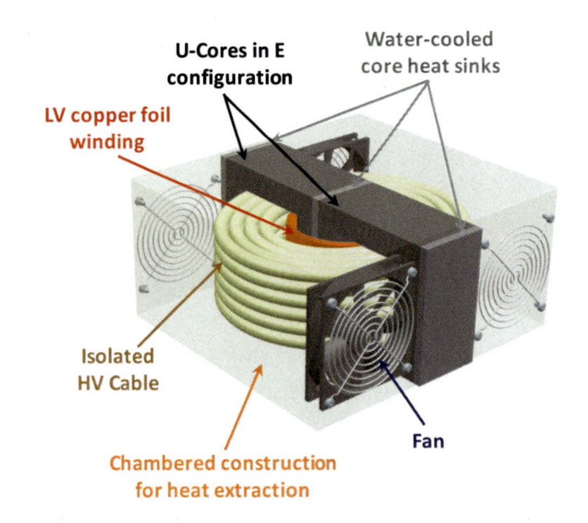

(b) Shell-type transformer.

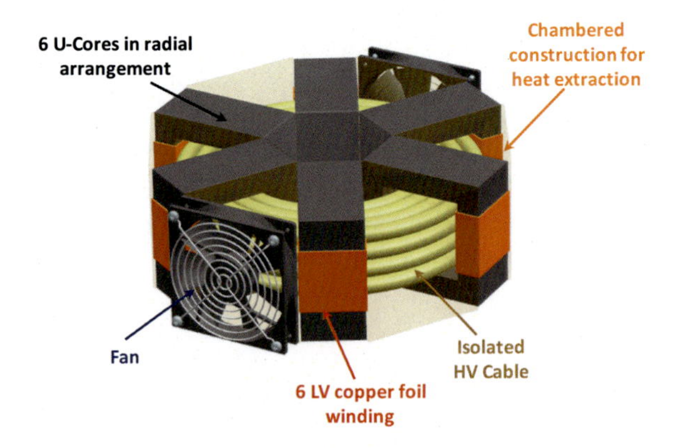

(c) Maxtrix transformer.

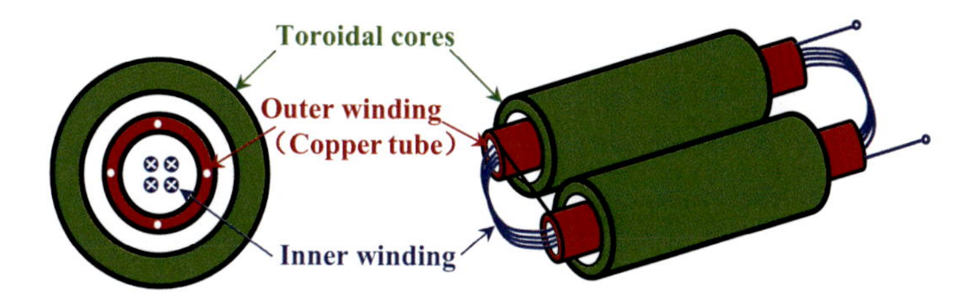

Fig. 9.33 Coaxial MFT structure [80]

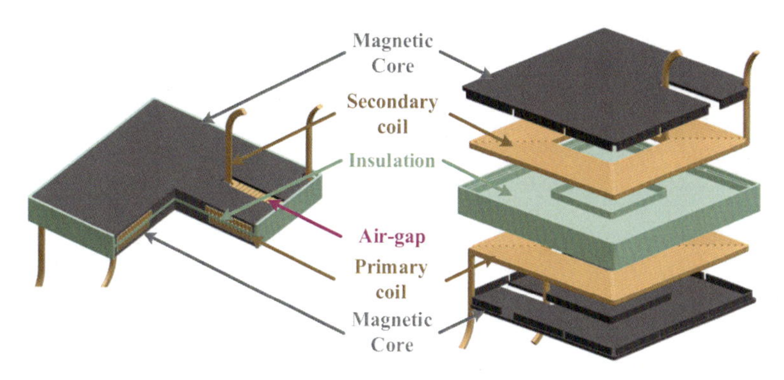

Fig. 9.34 Split planar MFT structure [83]

control are employed. The grid synchronization control is based on the synchronous rotation reference frame, which is used to control the input current and the output DC voltage of the AFE cascaded H-bridge converter. Meanwhile, a power balancing control is used to balance the power among the three phases. The control architecture of the output DAB is shown in Fig. 9.36, which is made up of an inner current loop and an outer voltage loop. In [88], the control architecture of a fast charging station is presented. Its control architecture includes a system controller, a central controller, and the module controllers, which is shown in Fig. 9.37. The module controller is made up of the primary controller and the secondary controller. The system controller plays the function of measuring environmental parameters in each cabinet of the entire SST and communicating with the control center. The central controller is responsible for the input AC voltage equalization control and the power module internal capacitor voltage balancing control. The primary and secondary controllers in the power module are in charge of AC side current tracking control and output voltage regulation. The control block diagram of the AFE converter is shown in Fig. 9.38. The central controller and the primary controller cooperate to control the capacitor voltage equalization among the power modules to maintain the same DC link voltage for the modules. An outer voltage control loop and an inner current control loop (using d-q axis current control) are employed to regulate the system input current and meet the power factor requirements. The control block diagram of the DC/DC converter in the power module is shown in Fig. 9.39. A

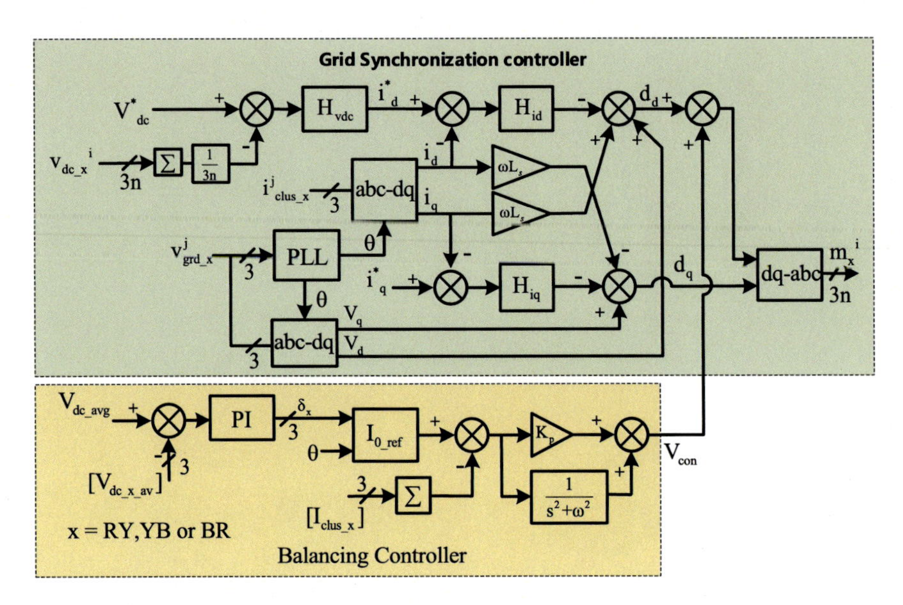

Fig. 9.35 Control scheme of the front-end cascaded H-bridge converter

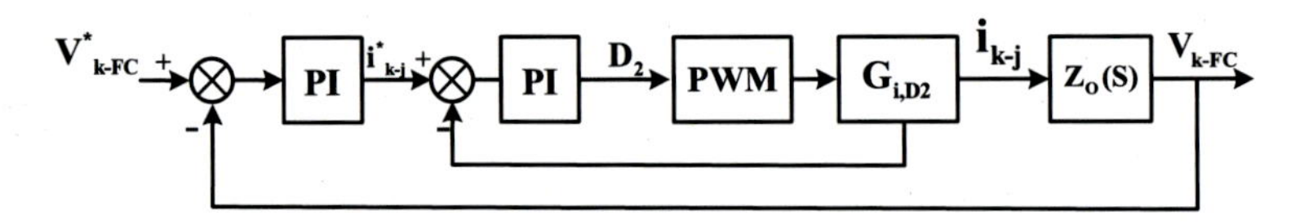

Fig. 9.36 Control scheme of the output DAB

coordination control based on the central controller and the secondary controller is used in order to achieve the output voltage regulation and the current sharing among the power modules.

In [74], a control scheme for multi-cell boost topology with three ISOP modules is presented. The control diagram for each ISOP module is shown in Fig. 9.40, where three PI control loops are implemented. The total dc bus voltage is regulated by the input stage main loop. The voltage balancing loop is used to balance the capacitor voltages and the output voltage is regulated by the NPC loop. In [75], the control architecture for a 8 kV, 25 kW modular SST is proposed. The basic control block diagram is shown in Fig. 9.41, in which a DSP controller is employed. The DSP controls the input current and power factor correction for the AFE converter and regulates the output voltage. The LLC converter is operated in the open-loop with 100% duty cycle to maximize efficiency.

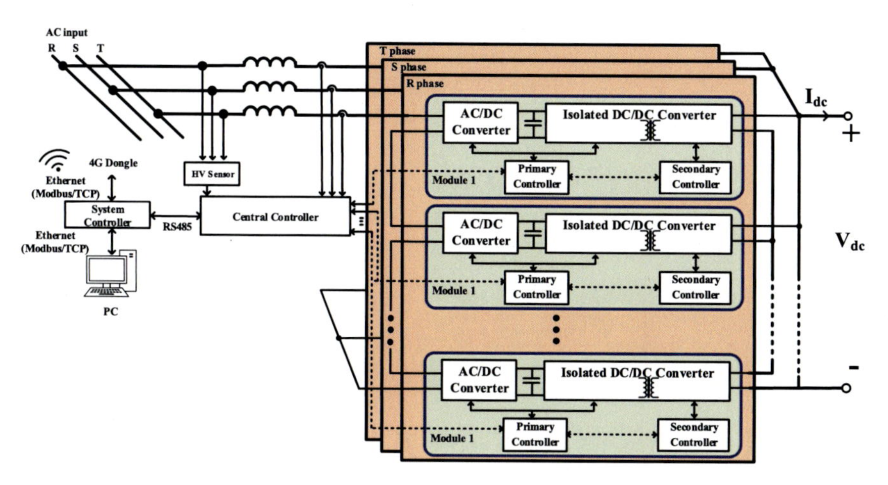

Fig. 9.37 Control architecture of modular SST-based fast charging station

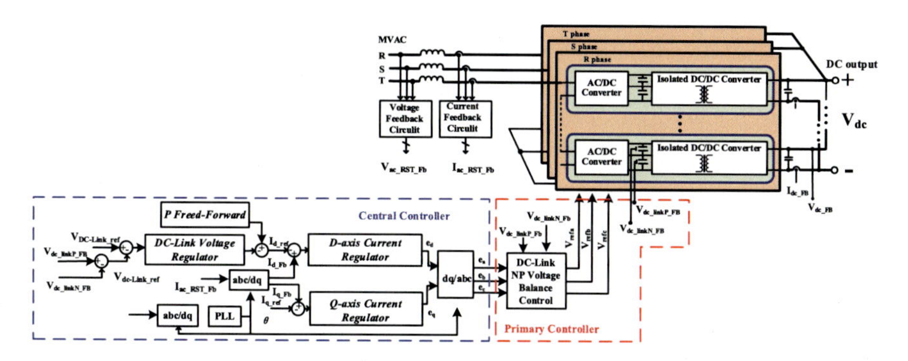

Fig. 9.38 Control block diagram of AC-DC converter

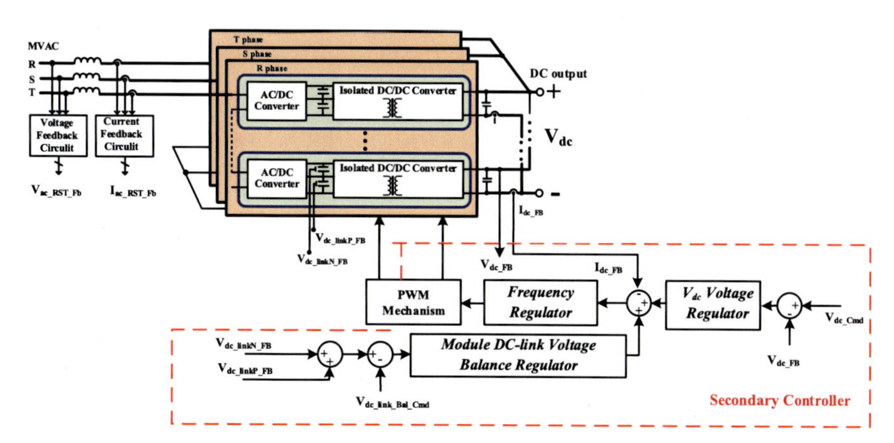

Fig. 9.39 Control block diagram of the DC/DC converter

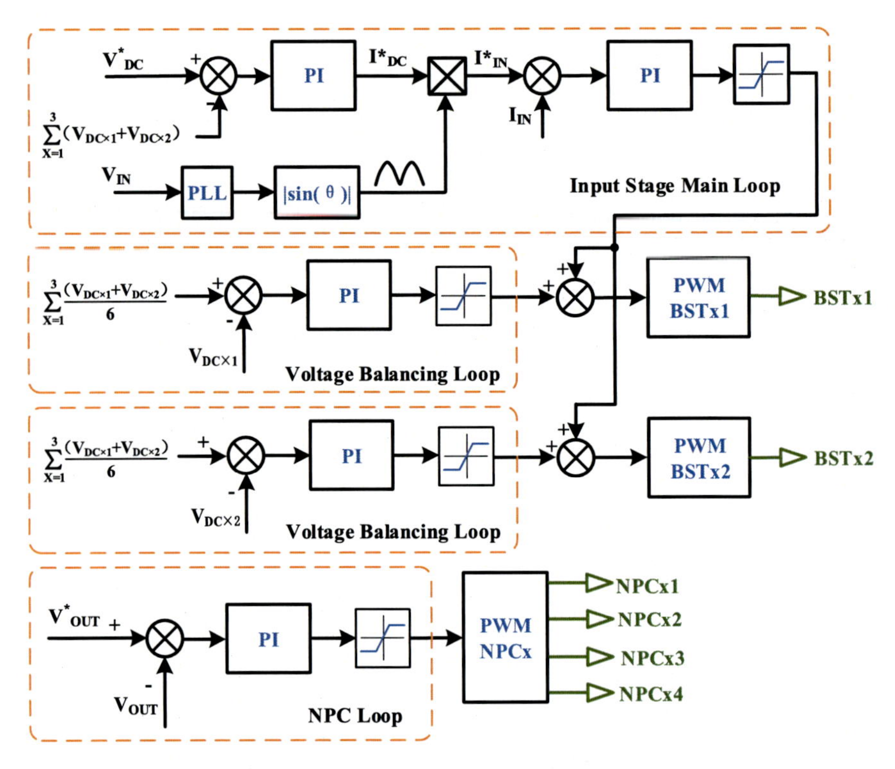

Fig. 9.40 Control block diagram of modular SST-based fast charger using unidirectional multi-cell boost topology with three ISOP dc/dc converter modules

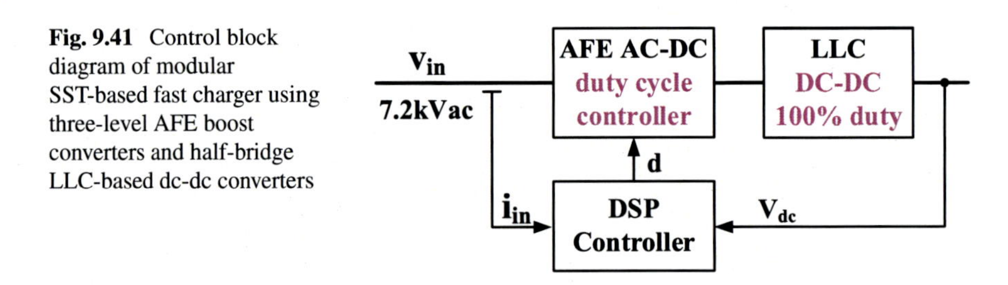

Fig. 9.41 Control block diagram of modular SST-based fast charger using three-level AFE boost converters and half-bridge LLC-based dc-dc converters

4.3 Single-Module SST-Based Fast Charging Stations

With the development of high voltage rating SiC MOSFETs and SiC IGBTs (e.g. 10 kV SiC MOSFETs, 13 kV SiC MOSFETs, and 15 kV SiC IGBTs), the fast charging station can be realized by the single-module SST [66]. The single-module SST-based fast charging station can adopt simple two-level or three-level converters to directly interface MV grid. Consequently, the system complexity can be significantly reduced. In [89], a single-phase 10 kW SST based on 13 kV SiC MOSFETs and junction barrier Schottky diodes is directly connected to 3.6 kV MV grid, which is shown in Fig. 9.42a. A two-level H-bridge is used as the AFE converter and its output voltage is 6.1 kV dc. The isolation stage adopts a dual half bridge (DHB) converter and its output voltage is 400 V dc. In [90] and [91], another single-phase 25 kW SST based on 10 kV SiC MOSFETs is realized as shown in Fig. 9.42b. The AFE converter is connected to 3.8 kV ac grid and the output voltage of the AFE converter is 7 kV. An LC-branch is inserted between the two terminals of the AFE converter H-bridge legs to realize soft-switching over the entire line period. The output of the AFE converter is followed by an LLC resonant converter. In [92], the three-phase 4.16 kV grid voltage is rectified by a three-phase two-level converter with 10 kV SiC MOSFETs, followed by a three-phase DAB, which is shown in Fig. 9.43. The internal dc bus voltage is 7.2 kV and its output dc voltage is 800 V. In [93], The AFE converter adopts a three-phase three-level DNPC based on 15 kV SiC IGBTs and 10 kV SiC MOSFETs, which directly interfaces 13.8 kV MV grid, as shown in Fig. 9.44. The output voltage of the AFE converter is 22 kV. A three-phase DAB is selected as the isolation stage, in which a three-phase three-level DNPC is employed on the primary side and two two-level three-phase converters are used on the secondary side. The output dc voltage of the three-phase DAB is 800 V. Thanks to the high voltage rating of SiC MOSFETs and SiC IGBTs, the single-module SST-based fast charger can adopt simple two-level or three-level converters and a higher efficiency and power density can be achieved. However, these high voltage rating power devices are still developing and have not been commercialized.

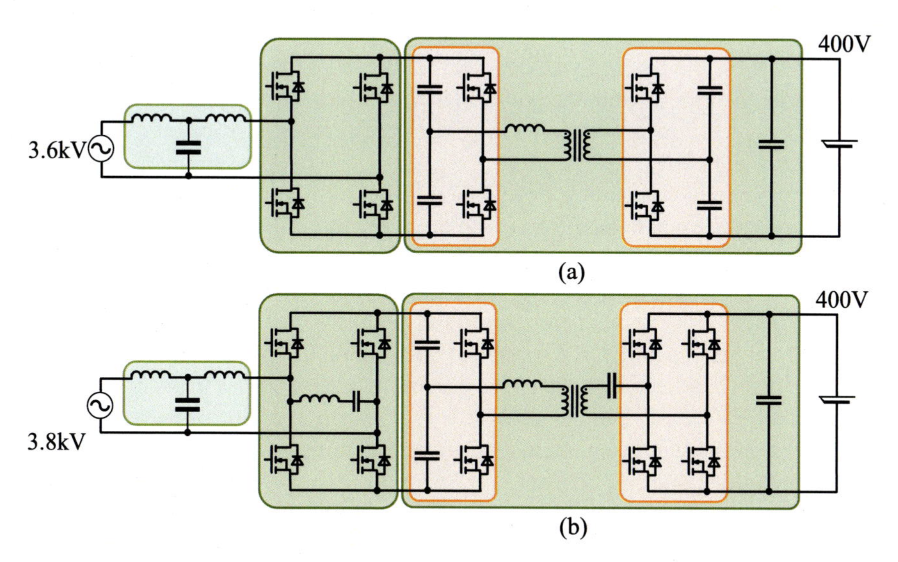

Fig. 9.42 Single-phase single-module SST-based fast charger based on two-level converters

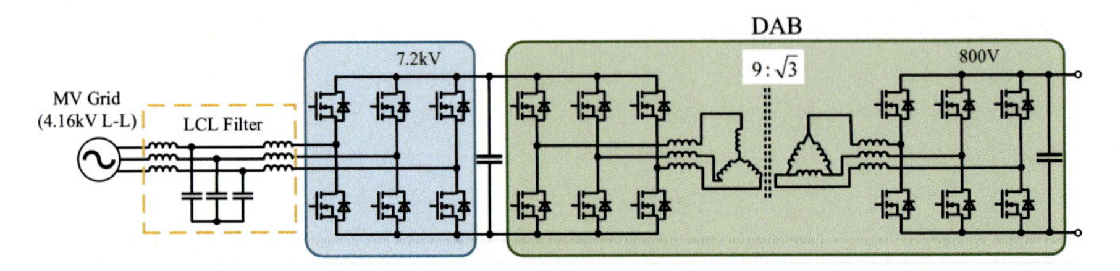

Fig. 9.43 Three-phase single-module SST-based fast charger based on two-level converters

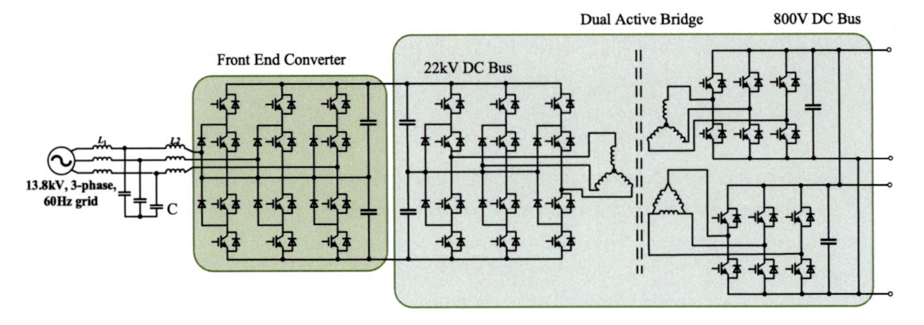

Fig. 9.44 Three-phase single-module SST-based fast charger based on three-level converters

5 Future Trends of EV Fast Chargers

In summary, WBG devices will continue to expand its market share due to their superior performances from the power electronics perspective, and fast charging of EVs will continue to evolve as power electronics devices and battery technology continues to advance.

Due to potential irreversible degradation that can be caused by fast charging, EV owners will be reluctant to use fast charging very often. It can be foreseeable that fast charging only happens as needed, such as during the middle of a trip.

Homeowners with a garage will likely charge their vehicles at home most of the time at a much lower rate. This is due to the fact that most homes are not equipped with high power capability. EV owners living in multi-unit dwellings, such as high-rise apartment buildings without a fixed parking space are limited in the way they can charge their EVs. Fast charging at a public charge station can be a feasible solution.

Battery swamping, as an alternative to fast charging could also charge the way EVs and their batteries are charged.

Solid state batteries, on the other hand, may help fast charging become more popular. Solid stat batteries, though currently is still under development, seems to have the potential to offer lower price and high energy density so more batteries can be installed on an EV. Charging a large battery pack at home may not be an option. In addition, solid state batteries seem to be able to handle high charging power without inducing fast degradation.

References

1. Ko YD, Jang YJ (2013) The optimal system design of the online electric vehicle utilizing wireless power transmission technology. IEEE Trans Intell Transp Syst 14:1255–1265. https://doi.org/10.1109/TITS.2013.2259159
2. Li S, Lu S, Mi CC (2021) Revolution of electric vehicle charging technologies accelerated by wide bandgap devices. Proc IEEE 109:985–1003. https://doi.org/10.1109/JPROC.2021.3071977
3. Huang AQ (2017) Power semiconductor devices for smart grid and renewable energy systems. Proc IEEE 105:2019–2047. https://doi.org/10.1109/JPROC.2017.2687701
4. Agarwal AK, Casady JB, Rowland LB, Valek WF, White MH, Brandt CD (1997) 1.1 kV 4H-SiC power UMOSFETs. IEEE Electron Device Lett 18:586–588. https://doi.org/10.1109/55.644079
5. Lu J, Bai K, Taylor AR, Liu G, Brown A, Johnson PM, McAmmond M (2018) A modular-designed three-phase high-efficiency high-power-density EV battery charger using dual/triple-phase-shift control. IEEE Trans Power Electron 33:8091–8100. https://doi.org/10.1109/TPEL.2017.2769661
6. Baliga BJ (2018) Wide bandgap semiconductor power devices: materials, physics, design, and applications. Woodhead Publishing
7. Johnson E (1966) Physical limitations on frequency and power parameters of transistors. In: 1958 IRE international convention record, pp 27–34
8. Baliga BJ (1982) Semiconductors for high-voltage, vertical channel field-effect transistors. J Appl Phys 53:1759–1764. https://doi.org/10.1063/1.331646
9. Huang AQ (2004) New unipolar switching power device figures of merit. IEEE Electron Device Lett 25:298–301. https://doi.org/10.1109/LED.2004.826533
10. Bhatnagar M, Baliga BJ (1993) Comparison of 6H-SiC, 3C-SiC, and Si for power devices. IEEE Trans Electron Devices 40:645–655. https://doi.org/10.1109/16.199372
11. Hudgins JL, Simin GS, Santi E, Khan MA (2003) An assessment of wide bandgap semiconductors for power devices. IEEE Trans Power Electron 18:907–914. https://doi.org/10.1109/TPEL.2003.810840
12. Elasser A, Chow TP (2002) Silicon carbide benefits and advantages for power electronics circuits and systems. Proc IEEE 90:969–986. https://doi.org/10.1109/JPROC.2002.1021562
13. StG M, Glass RC, Hobgood HM, Tsvetkov VF, Brady M, Henshall D, Jenny JR, Malta D, Carter CH (2000) The status of SiC bulk growth from an industrial point of view. J Cryst Growth 211:325–332. https://doi.org/10.1016/S0022-0248(99)00835-0
14. Treu M, Rupp R, Blaschitz P, Hilsenbeck J (2006) Commercial SiC device processing: Status and requirements with respect to SiC based power devices. Superlattices Microstruct 40:380–387. https://doi.org/10.1016/j.spmi.2006.09.005
15. Veliadis V (2018) Accelerating commercialization of wide-bandgap power electronics [expert view]. IEEE Power Electron Mag 5:63–65. https://doi.org/10.1109/MPEL.2018.2875169
16. Kim ST, Lee YJ, Moon DC, Hong CH, Yoo TK (1998) Preparation and properties of free-standing HVPE grown GaN substrates. J Cryst Growth 194:37–42. https://doi.org/10.1016/S0022-0248(98)00551-X
17. Wakahara A, Yamamoto T, Ishio K, Yoshida A, Seki Y, Kainosho K, Oda O (2000) Hydride vapor phase epitaxy of GaN on $NdGaO_3$ substrate and realization of freestanding GaN wafers with 2-inch scale. Jpn J Appl Phys 39:2399. https://doi.org/10.1143/JJAP.39.2399
18. Sato T, Okano S, Goto T, Yao T, Seto R, Sato A, Goto H (2013) Nearly 4-inch-diameter free-standing GaN wafer fabricated by hydride vapor phase epitaxy with pit-inducing buffer layer. Jpn J Appl Phys 52:08JA08. https://doi.org/10.7567/JJAP.52.08JA08
19. Fujikura H, Yoshida T, Shibata M, Otoki Y (2017) Recent progress of high-quality GaN substrates by HVPE method. In: Gallium nitride materials and devices XII. SPIE, pp 13–20
20. Reese SB, Remo T, Green J, Zakutayev A (2019) How much will gallium oxide power electronics cost? Joule 3:903–907. https://doi.org/10.1016/j.joule.2019.01.011

21. Zhang Y, Yuan M, Chowdhury N, Cheng K, Palacios T (2018) 720-V/0.35-mΩ cm^2 fully vertical GaN-on-Si power diodes by selective removal of Si substrates and buffer layers. IEEE Electron Device Lett 39:715–718. https://doi.org/10.1109/LED.2018.2819642
22. Kimoto T, Cooper JA (2014) Fundamentals of silicon carbide technology: growth, characterization, devices and applications. John Wiley & Sons Singapore Pte. Ltd, Singapore
23. Zou X, Zhang X, Lu X, Tang CW, Lau KM (2016) Fully vertical GaN p-i-n diodes using GaN-on-Si epilayers. IEEE Electron Device Lett 37:636–639. https://doi.org/10.1109/LED.2016.2548488
24. Zhang X, Zou X, Lu X, Tang CW, Lau KM (2017) Fully- and quasi-vertical GaN-on-Si p-i-n diodes: high performance and comprehensive comparison. IEEE Trans Electron Devices 64:809–815. https://doi.org/10.1109/TED.2017.2647990
25. Zhu M, Song B, Qi M, Hu Z, Nomoto K, Yan X, Cao Y, Johnson W, Kohn E, Jena D, Xing HG (2015) 1.9-kV AlGaN/GaN lateral Schottky barrier diodes on silicon. IEEE Electron Device Lett 36:375–377. https://doi.org/10.1109/LED.2015.2404309
26. Treu M, Rupp R, Blaschitz P, Ruschenschmidt K, Sekinger T, Friedrichs P, Elpelt R, Peters D (2007) Strategic considerations for unipolar SiC switch options: JFET vs. MOSFET. In: 2007 IEEE industry applications annual meeting, pp 324–330
27. Chen KJ, Häberlen O, Lidow A, Tsai CL, Ueda T, Uemoto Y, Wu Y (2017) GaN-on-Si power technology: devices and applications. IEEE Trans Electron Dev 64:779–795. https://doi.org/10.1109/TED.2017.2657579
28. Spitz J, Melloch MR, Cooper JA, Capano MA (1998) 2.6 kV 4H-SiC lateral DMOSFETs. IEEE Electron Dev Lett 19:100–102. https://doi.org/10.1109/55.663527
29. Cooper JA, Melloch MR, Singh R, Agarwal A, Palmour JW (2002) Status and prospects for SiC power MOSFETs. IEEE Trans Electron Dev 49:658–664. https://doi.org/10.1109/16.992876
30. Zhang J, Alexandrov P, Burke T, Zhao JH (2006) 4H-SiC power bipolar junction transistor with a very low specific ON-resistance of 2.9 mΩ cm^2. IEEE Electron Dev Lett 27:368–370. https://doi.org/10.1109/LED.2006.873370
31. Ryu S-H, Cheng L, Dhar S, Capell C, Jonas C, Clayton J, Donofrio M, O'Loughlin M, Al Burk, Agarwal A, Palmour J (2012) Development of 15 kV 4H-SiC IGBTs. In: Devaty R, Dudley M, Chow T, Neudeck P (eds). pp 1135–1138
32. Cheng L, Agarwal AK, Capell C, O'Loughlin M, Lam K, Richmond J, Van Brunt E, Burk A, Palmour JW, O'Brien H, Ogunniyi A, Scozzie C (2013) 20 kV, 2 cm^2, 4H-SiC gate turn-off thyristors for advanced pulsed power applications. In: 2013 19th IEEE Pulsed Power Conference (PPC), pp 1–4
33. Palmour JW, Zhang JQ, Das MK, Callanan R, Agarwal AK, Grider DE (2010) SiC power devices for Smart Grid systems. In: The 2010 international power electronics conference—ECCE ASIA, pp 1006–1013
34. Baliga BJ (1989) Power semiconductor device figure of merit for high-frequency applications. IEEE Electron Dev Lett 10:455–457. https://doi.org/10.1109/55.43098
35. McGlothlin HM, Morisette DT, Cooper JA, Melloch MR (1999) 4 kV silicon carbide Schottky diodes for high-frequency switching applications. In: 1999 57th annual device research conference digest (Cat. No.99TH8393), pp 42–43
36. Liu Z, Wang J, Gu H, Zhang Y, Wang W, Xiong R, Xu K (2019) High-voltage vertical GaN-on-GaN Schottky barrier diode using fluorine ion implantation treatment. AIP Adv 9:055016. https://doi.org/10.1063/1.5100251
37. Zhang T, Zhang J, Zhou H, Wang Y, Chen T, Zhang K, Zhang Y, Dang K, Bian Z, Zhang J, Xu S, Duan X, Ning J, Hao Y (2019) A $>$ 3 kV/2.94 mΩ cm^2 and low leakage current with low turn-on voltage lateral GaN Schottky barrier diode on silicon substrate with anode engineering technique. IEEE Electron Device Lett 40:1583–1586. https://doi.org/10.1109/LED.2019.2933314
38. Ohta H, Kaneda N, Horikiri F, Narita Y, Yoshida T, Mishima T, Nakamura T (2015) Vertical GaN p-n junction diodes with high breakdown voltages over 4 kV. IEEE Electron Device Lett 36:1180–1182. https://doi.org/10.1109/LED.2015.2478907

39. Fujihira T (1997) Theory of semiconductor superjunction devices. Jpn J Appl Phys 36:6254. https://doi.org/10.1143/JJAP.36.6254
40. Sheridan DC, Ritenour A, Bondarenko V, Burks P, Casady JB (2009) Record 2.8 mΩ-cm^2 1.9kV enhancement-mode SiC VJFETs. In: 2009 21st international symposium on power semiconductor devices & IC's, pp 335–338
41. Sabri S, Van Brunt E, Barkley A, Hull B, O'Loughlin M, Burk A, Allen S, Palmour J (2017) New generation 6.5 kV SiC power MOSFET. In: 2017 IEEE 5th workshop on wide bandgap power devices and applications (WiPDA), pp 246–250
42. Sun M, Zhang Y, Gao X, Palacios T (2017) High-performance GaN vertical fin power transistors on bulk GaN substrates. IEEE Electron Device Lett 38:509–512. https://doi.org/10.1109/LED.2017.2670925
43. Hwang I, Choi H, Lee J, Choi HS, Kim J, Ha J, Um C-Y, Hwang S-K, Oh J, Kim J-Y, Shin JK, Park Y, Chung U, Yoo I-K, Kim K (2012) 1.6kV, 2.9 mΩ cm^2 normally-off p-GaN HEMT device. In: 2012 24th international symposium on power semiconductor devices and ICs, pp 41–44
44. Wu C-H, Chen J-Y, Han P-C, Lee M-W, Yang K-S, Wang H-C, Chang P-C, Luc QH, Lin Y-C, Dee C-F, Hamzah AA, Chang EY (2019) Normally-off tri-gate GaN MIS-HEMTs with 0.76 mΩ·cm^2 specific on-resistance for power device applications. IEEE Trans Electron Devices 66:3441–3446. https://doi.org/10.1109/TED.2019.2922301
45. Kosugi R, Sakuma Y, Kojima K, Itoh S, Nagata A, Yatsuo T, Tanaka Y, Okumura H (2014) First experimental demonstration of SiC super-junction (SJ) structure by multi-epitaxial growth method. In: 2014 IEEE 26th international symposium on power semiconductor devices & IC's (ISPSD), pp 346–349
46. Masuda T, Saito Y, Kumazawa T, Hatayama T, Harada S (2018) 0.63 mmΩ cm^2/1170 V 4H-SiC super junction V-groove trench MOSFET. In: 2018 IEEE international electron devices meeting (IEDM), pp 8.1.1–8.1.4
47. Khaligh A, D'Antonio M (2019) Global trends in high-power on-board chargers for electric vehicles. IEEE Trans Veh Technol 68:3306–3324. https://doi.org/10.1109/TVT.2019.2897050
48. Yilmaz M, Krein PT (2013) Review of battery charger topologies, charging power levels, and infrastructure for plug-in electric and hybrid vehicles. IEEE Trans Power Electron 28:2151–2169. https://doi.org/10.1109/TPEL.2012.2212917
49. Liu Z, Li B, Lee FC, Li Q (2017) High-efficiency high-density critical mode rectifier/inverter for WBG-device-based on-board charger. IEEE Trans Indus Electron 64:9114–9123. https://doi.org/10.1109/TIE.2017.2716873
50. Metwly MY, Abdel-Majeed MS, Abdel-Khalik AS, Hamdy RA, Hamad MS, Ahmed S (2020) A review of integrated on-board EV battery chargers: advanced topologies, recent developments and optimal selection of FSCW slot/pole combination. IEEE Access 8:85216–85242. https://doi.org/10.1109/ACCESS.2020.2992741
51. Varajao DACP, Miranda LMF, Araujio RME (2018) AC/DC converter with three to single phase matrix converter, full-bridge AC/DC converter and HF transformer
52. Jauch F, Biela J (2013) Modelling and ZVS control of an isolated three-phase bidirectional AC-DC converter. In: 2013 15th European conference on power electronics and applications (EPE), pp 1–11
53. Yang G, Draugedalen E, Sorsdahl T, Liu H, Lindseth R (2016) Design of high efficiency high power density 10.5 kW three phase on-board-charger for electric/hybrid vehicles. In: PCIM Europe 2016; international exhibition and conference for power electronics, intelligent motion, renewable energy and energy management, pp 1–7
54. Schmenger J, Endres S, Zeltner S, März M (2014) A 22 kW on-board charger for automotive applications based on a modular design. In: 2014 IEEE conference on energy conversion (CENCON), pp 1–6
55. Wang X, Jiang C, Lei B, Teng H, Bai HK, Kirtley JL (2016) Power-loss analysis and efficiency maximization of a silicon-carbide MOSFET-based three-phase 10-kW bidirectional EV charger using variable-DC-bus control. IEEE J Emerg Select Top Power Electron 4:880–892. https://doi.org/10.1109/JESTPE.2016.2575921

56. Johnson PM, Bai KH (2017) A dual-DSP controlled SiC MOSFET based 96%-efficiency 20kW EV on-board battery charger using LLC resonance technology. In: 2017 IEEE symposium series on computational intelligence (SSCI), pp 1–5
57. Thimmesch D (1985) An SCR inverter with an integral battery charger for electric vehicles. IEEE Trans Indus Appl IA-21:1023–1029. https://doi.org/10.1109/TIA.1985.349573
58. Shi C, Tang Y, Khaligh A (2018) A three-phase integrated onboard charger for plug-in electric vehicles. IEEE Trans Power Electron 33:4716–4725. https://doi.org/10.1109/TPEL.2017.2727398
59. Subotic I, Bodo N, Levi E (2016) An EV drive-train with integrated fast charging capability. IEEE Trans Power Electron 31:1461–1471. https://doi.org/10.1109/TPEL.2015.2424592
60. Subotic I, Bodo N, Levi E, Jones M, Levi V (2016) Isolated chargers for EVs incorporating six-phase machines. IEEE Trans Indus Electron 63:653–664. https://doi.org/10.1109/TIE.2015.2412516
61. Subotic I, Bodo N, Levi E, Jones M (2015) Onboard integrated battery charger for EVs using an asymmetrical nine-phase machine. IEEE Trans Indus Electron (1982) 62:3285–3295. https://doi.org/10.1109/tie.2014.2345341
62. Kim S, Kang F-S (2015) Multifunctional onboard battery charger for plug-in electric vehicles. IEEE Trans Indus Electron 62:3460–3472. https://doi.org/10.1109/TIE.2014.2376878
63. Kim D-H, Kim M-J, Lee B-K (2017) An integrated battery charger with high power density and efficiency for electric vehicles. IEEE Trans Power Electron 32:4553–4565. https://doi.org/10.1109/TPEL.2016.2604404
64. Tang Y, Lu J, Wu B, Zou S, Ding W, Khaligh A (2018) An integrated dual-output isolated converter for plug-in electric vehicles. IEEE Trans Veh Technol 67:966–976. https://doi.org/10.1109/TVT.2017.2750076
65. Zhu L, Bai H, Brown A, Keuck L (2021) A current-fed three-port DC/DC converter for integration of on-board charger and auxiliary power module in electric vehicles. IEEE, pp 577–582
66. Tu H, Feng H, Srdic S, Lukic S (2019) Extreme fast charging of electric vehicles: a technology overview. IEEE Trans Transp Electrification 5:861–878. https://doi.org/10.1109/TTE.2019.2958709
67. Srdic S, Lukic S (2019) Toward extreme fast charging: challenges and opportunities in directly connecting to medium-voltage line. IEEE Electrification Mag 7:22–31. https://doi.org/10.1109/MELE.2018.2889547
68. Tan L, Wu B, Rivera S, Yaramasu V (2016) Comprehensive DC power balance management in high-power three-level DC–DC converter for electric vehicle fast charging. IEEE Trans Power Electron 31:89–100. https://doi.org/10.1109/TPEL.2015.2397453
69. Aggeler D, Canales F, Zelaya-De La Parra H, Coccia A, Butcher N, Apeldoorn O (2010) Ultra-fast DC-charge infrastructures for EV-mobility and future smart grids. In: 2010 IEEE PES innovative smart grid technologies conference Europe. ISGT Europe, pp 1–8
70. Zhu C, Electroncis D high-efficiency, medium-voltage input, solid-state, transformer-based 400-kW/1000-V/400-a extreme fast charger for electric vehicles
71. Moeini A, Wang S (2018) Design of fast charging technique for electrical vehicle charging stations with grid-tied cascaded H-bridge multilevel converters. In: 2018 IEEE applied power electronics conference and exposition (APEC), pp 3583–3590
72. Nair AC, Fernandes BG (2018) A solid state transformer based fast charging station for all categories of electric vehicles. In: IECON 2018—44th annual conference of the IEEE industrial electronics society, pp 1989–1994
73. Huber JE, Böhler J, Rothmund D, Kolar JW (2017) Analysis and cell-level experimental verification of a 25 kW all-SiC isolated front end 6.6 kV/400 V AC-DC solid-state transformer. CPSS Trans Power Electron Appl 2:140–148. https://doi.org/10.24295/CPSSTPEA.2017.00014
74. Srdic S, Liang X, Zhang C, Yu W, Lukic S (2016) A SiC-based high-performance medium-voltage fast charger for plug-in electric vehicles. In: 2016 IEEE energy conversion congress and exposition (ECCE), pp 1–6

75. Lai J-S, Lai W-H, Moon S-R, Zhang L, Maitra A (2016) A 15-kV class intelligent universal transformer for utility applications. In: 2016 IEEE applied power electronics conference and exposition (APEC), pp 1974–1981
76. Jaritz M, Blume S, Biela J (2017) Design procedure of a 14.4 kV, 100 kHz transformer with a high isolation voltage (115 kV). IEEE Trans Dielectr Electr Insul 24:2094–2104. https://doi.org/10.1109/TDEI.2017.006279
77. Zhao S, Li Q, Lee FC, Li B (2018) High-frequency transformer design for modular power conversion from medium-voltage AC to 400 VDC. IEEE Trans Power Electron 33:7545–7557. https://doi.org/10.1109/TPEL.2017.2774440
78. Mogorovic M, Dujic D (2019) 100 kW, 10 kHz medium-frequency transformer design optimization and experimental verification. IEEE Trans Power Electron 34:1696–1708. https://doi.org/10.1109/TPEL.2018.2835564
79. Ortiz G, Biela J, Bortis D, Kolar JW (2010) 1 Megawatt, 20 kHz, isolated, bidirectional 12kV to 1.2kV DC-DC converter for renewable energy applications. In: The 2010 international power electronics conference—ECCE ASIA, pp 3212–3219
80. Baek S (2014) High-frequency AC-link transformers for medium-voltage DC-DC converters and solid state transformer applications. North Carolina State University, N1—2020-03-20 16:04:00
81. Zixin L, Fanqiang G, Cong Z, Zhe W, Hang Z, Ping W, Yaohua L (2018) Research review of power electronic transformer technologies. Proceedings of the CSEE 38:1593. https://doi.org/10.13334/j.0258-8013.pcsee.172575
82. Baek S, Bhattacharya S (2018) Analytical modeling and implementation of a coaxially wound transformer with integrated filter inductance for isolated soft-switching DC–DC converters. IEEE Trans Industr Electron 65:2245–2255. https://doi.org/10.1109/TIE.2017.2740855
83. Lu S, Kong D, Xv S, Luo L, Li S (2022) A high-efficiency 80kW split planar transformer for medium-voltage modular power conversion. IEEE Trans Power Electron. https://doi.org/10.1109/TPEL.2022.3151796
84. Hannan MA, Ker PJ, Lipu MSH, Choi ZH, Rahman MSA, Muttaqi KM, Blaabjerg F (2020) State of the art of solid-state transformers: advanced topologies, implementation issues, recent progress and improvements. IEEE Access 8:19113–19132. https://doi.org/10.1109/ACCESS.2020.2967345
85. Yang J, Liu J, Zhang J, Zhao N, Zheng TQ (2018) Research on different balance control strategies for a power electronic traction transformer. https://doi.org/10.1109/APEC.2018.8341264
86. Gorla NBY, Kolluri S, Chai M, Panda SK (2019) A comprehensive harmonic analysis and control strategy for improved input power quality in a cascaded modular solid state transformer. IEEE Trans Power Electron 34:6219–6232. https://doi.org/10.1109/TPEL.2018.2873201
87. Lu C, Hu W, Lee FC (2021) Neutral-point voltage balancing methods of series-half-bridge LLC converter for solid state transformer. IEEE Trans Power Electron 36:7060–7073. https://doi.org/10.1109/TPEL.2020.3035150
88. Lu C, Hu W, Zhang W (2021) High-frequency high-efficiency modular solid state transformer for quick charging stations. J Power Supply 19:74–82. https://doi.org/10.13234/j.issn.2095-2805.2021.6.74
89. Wang F, Wang G, Huang A, Yu W, Ni X (2014) Design and operation of A 3.6kV high performance solid state transformer based on 13kV SiC MOSFET and JBS diode. Institute of Electrical and Electronics Engineers Inc., pp 4553–4560
90. Rothmund D, Guillod T, Bortis D, Kolar JW (2019) 99.1% efficient 10 kV sic-based medium-voltage ZVS bidirectional single-phase PFC AC/DC stage. IEEE J Emerg Sel Topics Power Electron 7:779–797. https://doi.org/10.1109/JESTPE.2018.2886140
91. Rothmund D, Guillod T, Bortis D, Kolar JW (2019) 99% efficient 10 kV SiC-based 7 kV/400 V DC transformer for future data centers. IEEE J Emerg Sel Topics Power Electron 7:753–767. https://doi.org/10.1109/JESTPE.2018.2886139
92. Anurag A, Acharya S, Prabowo Y, Jakka V, Bhattacharya S (2018) Design of a medium voltage mobile utilities support equipment based solid state transformer (MUSE-SST) with 10 kV SiC

MOSFETs for grid interconnection. Institute of Electrical and Electronics Engineers Inc., Charlotte, NC, United states
93. Madhusoodhanan S, Tripathi A, Patel D, Mainali K, Kadavelugu A, Hazra S, Bhattacharya S, Hatua K (2015) Solid-state transformer and MV Grid Tie applications enabled by 15 kV SiC IGBTs and 10 kV SiC MOSFETs based multilevel converters. IEEE Trans Ind Appl 51:3343–3360. https://doi.org/10.1109/TIA.2015.2412096

Wireless Power Transfer for Electric and Hybrid Electric Vehicles

C. Q. Jiang, Teng Long, and Daniel E. Gaona

European countries have agreed to accelerate the development of low-emission transport technologies to meet targets set during the Paris climate summit. This includes reducing carbon dioxide emission from the traditional vehicle powered by internal combustion engine (ICE), and the transition to zero-emission by deploying the promoted electric vehicles (EVs), or less-emission by deploying hybrid electric vehicle (HEV) technologies. Recently, with the problem of short driving range per charge of the EV, many researchers proposed different topologies or methodologies to elaborate the next generation chargers with higher convenience and safety. One emerging charging technique, wireless power transfer (WPT), is currently very popular to greatly facilitate the charging process, instead of setting up more supercharging stations with faster EV battery chargers. Apart from the wireless EV charging application, the WPT technologies can be adopted in many other charging required applications, as shown in Fig. 1. Most importantly, since there is no manual plug-in and metal contacts, electric shock accidents that may occur during charging can be effectively avoided, which can make electric vehicles better than ICEVs regarding the user safety for self-charging or refueling. Moreover, wireless EV charging can be bidirectional to facilitate the energy interactions between the power grid and movable energy router (EVs) by offering vehicle-to-home (V2H) or vehicle-to-grid (V2G) technologies, as shown in Fig. 2.

In this chapter, various advanced WPT technologies, including far-field applications and near-field applications, are introduced. Then, different compensation networks are analyzed. Besides, the magnetic pad designs for EVs and HEVs are

C. Q. Jiang (✉)
Department of Electrical Engineering, City University of Hong Kong, Hong Kong, China
e-mail: chjiang@cityu.edu.hk

T. Long
Department of Engineering, University of Cambridge, Cambridge CB3 0FA, UK

D. E. Gaona
Huawei Nuremberg Research Center, 90449 Nuremberg, Germany

C. H. T. Lee (ed.), *Emerging Technologies for Electric and Hybrid Vehicles*,
Green Energy and Technology, https://doi.org/10.1007/978-981-99-3060-9_10

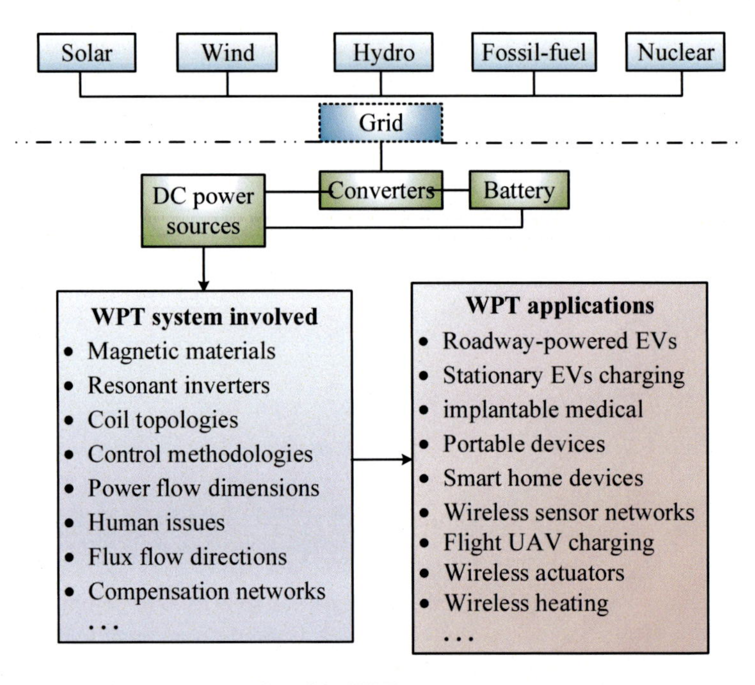

Fig. 1 Several typical usage scenarios of the WPT system

investigated. After revealing the important magnetic core component of the WPT system, the park-and-charge (PAC) system and move-and-charge (MAC) system are introduced. Lastly, magnetic shielding is discussed in detail, as well as the magnetic flux leakage regulations in EVs.

1 Wireless Power Transfer

Due to the prominent advantages of power cable-less, metallic contactless, almost no electrocutions, high flexibility, and better convenience, the WPT technique has drawn extensive attention in many interdisciplinary fields and industrial applications. As one of the most potential and emerging techniques, the WPT is transforming the traditional use of energy in human daily life [1–3]. At the same time, many other applications, such as non-accessible electronics, medical instruments, portable electronic devices, induction heating, and different EVs including park-and-charge system and move-and-charge system [4–6], may have very demanding utilization of the WPT techniques. It is shown in Fig. 3 that the regular configuration of a wireless EV charging system consists of power converters, compensation networks, coupling cores, and transmitter and receiver pads.

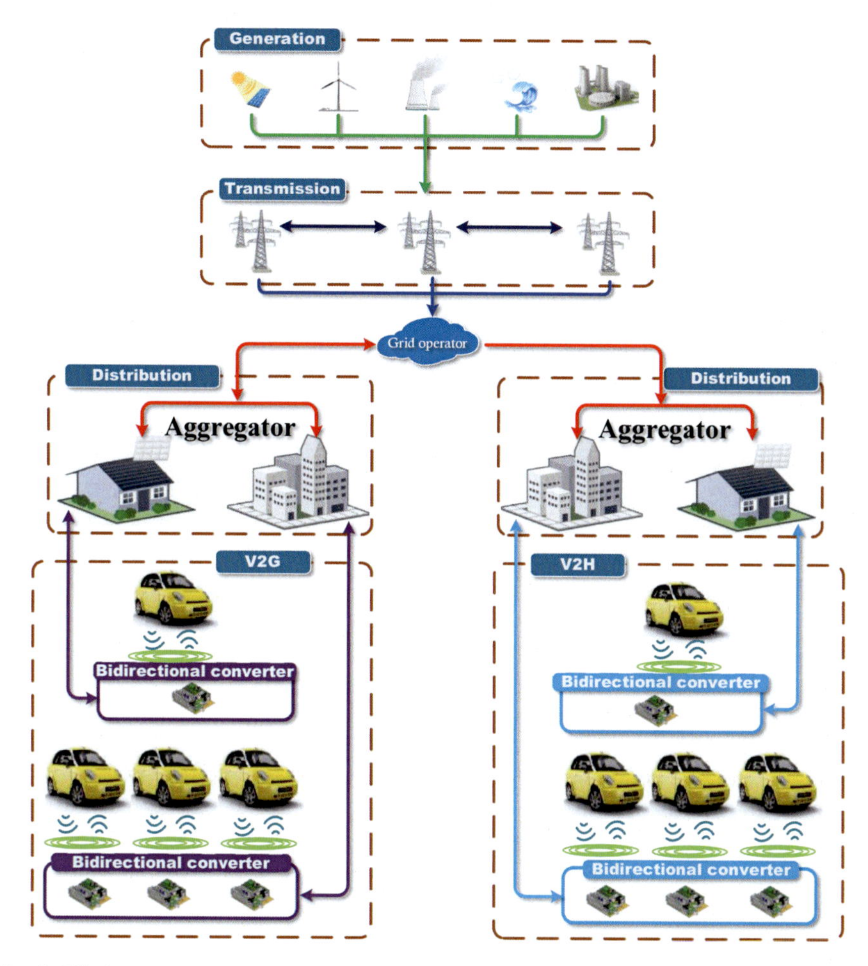

Fig. 2 Wireless EV charging scheme combined with V2G and V2H

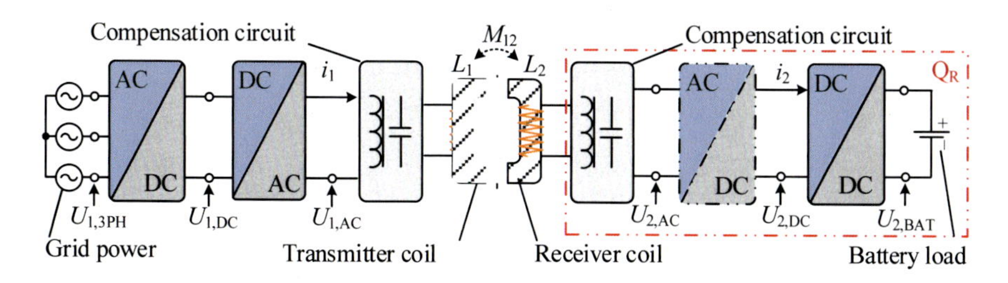

Fig. 3 General WPT configuration for wireless EV charging system

1.1 Classification of WPT Technologies

From the aspect of distance between the transmitter and receiver, the WPT can be classified into two main groups, specifically, the far-field transmission and the near-field transmission [7]. The far-field transmission generally involves the designs for special applications with low power requirement and low importance of system efficiency. In general, the far-field transmission is achieved via microwave methods or laser beams over straight transmission routes, which are line-of-sight. However, when considering the human exposure safety and system efficiency, the far-field transmission is not a good choice for wireless energy delivery applied in our daily life. In terms of the radio wave power transmission, it is often used for signal broadcasting, where the power involved is only microwatts. For wireless EV charging, these far-field transmission methods are not suitable because of the low power level and efficiency.

Because of the lower radiofrequency exposure safety limit and the higher transmission efficiency, near-field transmission is a superior choice when comparing to the laser transmission or the microwave method. Especially, among the various WPT technologies for near-field transmission, there are three most popular and commonly used methods, namely, permanent magnet coupling (PMC), capacitive power transfer (CPT), and inductive power transfer (IPT). Regarding capacitive WPT technology, it employs the alternating electric field to wirelessly deliver power. Also, the CPT has less electromagnetic interference (EMI) than the conventional electromagnetic-field-based technologies since the electric flux has a tendency to through within the conductive plate. On the other hand, the magnetic flux has a tendency to go from the coil in all directions, forming a closed magnetic flux loop. Another benefit of the CPT is that the wireless power could still be delivered even through metallic barriers. In other words, the lower and upper surfaces of the inserted metallic barrier can be regarded as conductive plates in an electric field, which will split the previous electric field but not interfere with the wireless power transmission. Although CPT technology has its specific advantages, a big realistic challenge it faces is the small coupling capacitance. Furthermore, any misalignments of the coupling plates or the existing air gaps will greatly reduce the capacitance, which turns the CPT wireless charging impractical for EV charging. In terms of PMC technology, it employs mechanical force as the energy carrier, which was firstly proposed to replace traditional contact gears. But as the air gap distance increases, the system efficiency and delivered power of the PMC method are very low.

The IPT technology is a widespread near-field technology that has been commonly applied in traditional induction motors and transformers. The non-resonant IPT technology has been widely employed in low-power wireless charging consumer electronics, such as smartphone and electric toothbrush. Nevertheless, the transferred power decays rapidly ($1/r^3$) as the distance increases. Thus, the effective operation range of the non-resonant IPT technology is always limited to several centimeters. In order to achieve a longer transmission range as well as higher transmission efficiency, the magnetic resonance coupling (MRC) IPT technology is developed, which can be

Table. 1. Comparison and Classification of the WPT Technologies for Wireless EV Charger

Energy-carrier	Classification		Range	Power	Efficiency	Remarks
Electromagnetic field	Near-field	Conventional non-resonant IPT	Low	High	High	Not capable for EV charging due to short range
		MRC IPT	Medium	High	High	Most widely used in wireless EV charging
	Far-field	Microwave or Laser	High	Low	Low	Request complicated tracking mechanisms, large antennas, direct line-of-sight path
		Radio wave	High	Low	Low	Not capable for EV charging due to low power and efficiency
Electric field	Capacitive WPT		Medium	Medium	Low when large air-gap	Proposed for wireless EV charging but not popular
Mechanical force	PMC- Magnetic gear		Medium	Medium	Medium	Proposed for wireless EV charging but not popular

classified as the near-field transmission but is enhanced by using the LC resonators. As a result, the power delivery range can be extended, as well as the efficiency. Until now, the MRC IPT topology has been the most widely used in wireless EV charging system due to its high efficiency and wide air-gap range with appropriate compensation topologies. The comparison is presented to make a summary as listed in Table 1.

1.2 Equivalent Circuit for IPT System

In general, the magnetic pads in IPT systems can be regarded as a loosely coupled transformer with the primary and secondary windings with self-inductances L_1 and L_2 (respectively), and a mutual inductance M, as plotted in Fig. 3. The loosely coupled

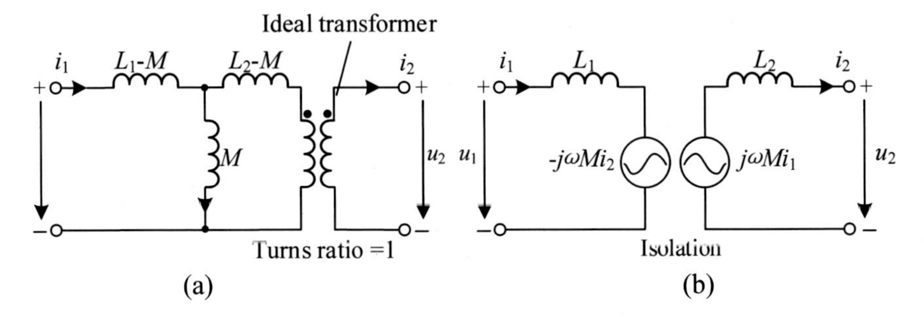

Fig. 4 Equivalent circuits of the two loosely coupled coil pads. **a** T-model. **b** Isolated circuit model (M-Model)

inductors in the IPT system have been simplified in Fig. 4. The relation between voltages u_i and currents i_i in the primary and secondary coil pads is given by

$$u_1 = L_1 \frac{di_1}{dt} - M \frac{di_2}{dt}$$
$$u_2 = L_2 \frac{di_2}{dt} + M \frac{di_1}{dt}$$

where $i = 1, 2$ represents the transmitter and receiver sides, respectively.

The ratio between the self- and mutual inductances $k = M/\sqrt{L_1 L_2}$ is known as the coupling factor and determines the portion of the magnetic flux linking both pads compared to the total magnetic flux produced by the coils. Power transformers tend to have high coupling factors closing to 1, which ensures high efficiency and minimum leakage inductance. For IPT coil pads, however, k is usually from 0.15 to 0.5 subject to the magnetic core layout of the coil pads. The coupling factor is particularly sensitive to the coil topology as well as to the ratio of the distance between coil pads and their size. The coupling factor is considered as a measure of the effectiveness of the coil pad's design. Higher coupling factors result in better efficiency and less leakage flux.

The magnetic couplers can be graphically represented with different models. The T-model seen in Fig. 2a is the most used. An alternative representation is shown in Fig. 2b. Here, both pads are represented as isolated circuits each one with one current-controlled voltage source representing the coupling between them.

1.2.1 Equivalent Load

Generally, a high frequency AC inverter is required to produce an alternating current for the transmitter coil pad, and a high frequency rectifier is required to convert the AC input to DC output. Just as the primary converter can be approximated by its fundamental components, the secondary converter and load can be approximated by an equivalent impedance. Since in an uncontrolled bridge rectifier the converter

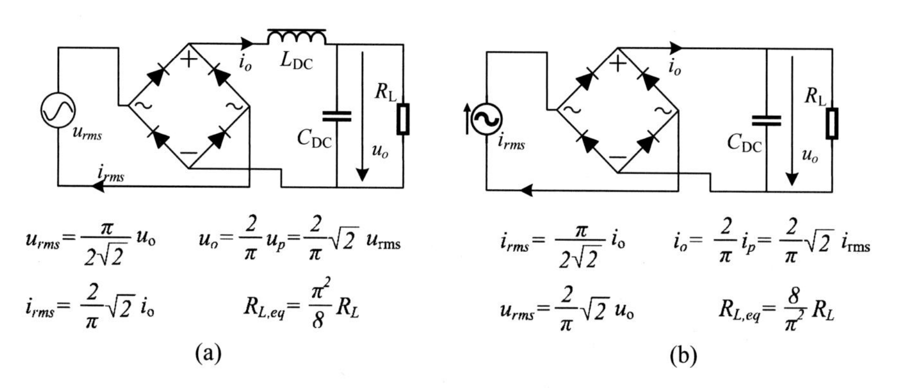

Fig. 5 Equivalent circuits for different driven sources **a** voltage driven source. **b** Current driven source

voltage (u_2) and current (i_2) are in phase, the equivalent impedance is purely resistive. The selection of the appropriate equivalent resistance depends on whether the receiver is driven as a voltage or a current source as shown in Fig. 5 [8].

The equivalent resistance values are given by

$$\text{Square} i_2 \text{ waveform: } R_{\text{L,eq}} = \frac{\pi^2 u_{\text{DC},2}^2}{8P_2}$$

$$\text{Square } u_2 \text{ waveform: } R_{\text{L,eq}} = \frac{8u_{\text{DC},2}^2}{\pi^2 P_2}$$

where P_2 is the power at the receiver estimated as the product between the DC output voltage $u_{\text{DC},2}$ and current $i_{\text{DC},2}$ respectively. This expression is only valid for uncontrolled converters. If the receiver side is active, the equivalent load will be controlled by the converter.

1.2.2 Operating Frequency for the WPT System

A high operating frequency f is required, since the induced voltage for the receiver coil from the transmitter coil and the capability of power transmission are proportional to the frequency f. Furthermore, high operating frequencies result in higher power density with smaller magnetic coil pads. However, some challenges are introduced with the high operating frequency. First, it is not suitable for all kinds of semiconductor transistors at high operating frequencies. Besides, the switching losses of the transistors rise with the operating frequency unless the transistors are switched with the patterns of zero voltage switching (ZVS) and zero current switching (ZCS). Moreover, the AC loss problems caused by the skin effect and proximity effect are introduced positively with the increasing frequency. Therefore, the Litz wire consisting of multiple thin strands is demanded, which reduces the AC resistance but increases the cost of coil pads. Also, high operating frequency for the magnetic core results in higher eddy current loss and hysteresis loss, which can reduce the transmission efficiency due to the higher core losses. Finally, higher operating frequency brings higher impedance and then results in a higher voltage across the coil or compensated capacitor, as well as a higher VA rating. However, over the past few years, the SAE J2954 standard has set a recommended frequency range for EV applications, to be between 81.39 kHz and 90 kHz, typically set at 85 kHz. Therefore, the operating frequency of 85 kHz is widely applied in recent academic publications and the relevant industrial wireless EV charger.

1.2.3 Harmonics of Current

In general, a typical IPT system with a voltage source inverter (VSI) H-bridge inverter produces a square-wave voltage output for the compensation network at the referred operating frequency (e.g., 85 kHz from the SAE standard). This is accomplished by operating the two legs of the inverter with 180 degree shift. In a square voltage waveform, the magnitude of the n-th harmonic is calculated by $u_i(n) = 4u_{\mathrm{DC},i}/(\pi n)$. Even though the harmonics drop at a frequency of around 20 dB/dec, the harmonic components of this voltage waveform are still high. Aiming to ease the harmonic components of the voltage, multi-level converters can be deployed with the cost of higher complexity and more electronic components. Generally, another method, pulse width modulation (PWM), is another solution. Nevertheless, the PWM requests a much higher operating frequency up to MHz level, which introduces higher switching losses and the system's complexity. Fortunately, such complex solutions are not required for most IPT systems. This is because the impedance of the coil pads limits the harmonic content of the voltage from being passed into the current. The equivalent load impedance in a series-compensation IPT system is illustrated in Fig. 6a. Due to the resonant circuit between the coil and the converter, the impedance reaches a minimum at the resonant frequency (frequency of operation) and increases at around 20 dB/dec elsewhere. The frequency response of the current is calculated from the ratio between the voltage and the impedance at each specific frequency. Hence, the magnitude of the current harmonics decreases rapidly at a rate of 40 dB/dec, as shown in Fig. 6b. Consequently, the harmonic content of the square voltage waveform produces almost negligible current harmonics [9]. As a result, the inverter can be represented as an ideal voltage source for most practical purposes. The same principle applies to the converter on the receiver side.

1.3 Maximum Transmission Efficiency

The standard SAE J2954 requires magnetic couplers used in EV applications to sustain efficiencies over 85% during operation. Most commercial units report efficiencies greater than 80% (between the grid and the battery), and often close to 90%.

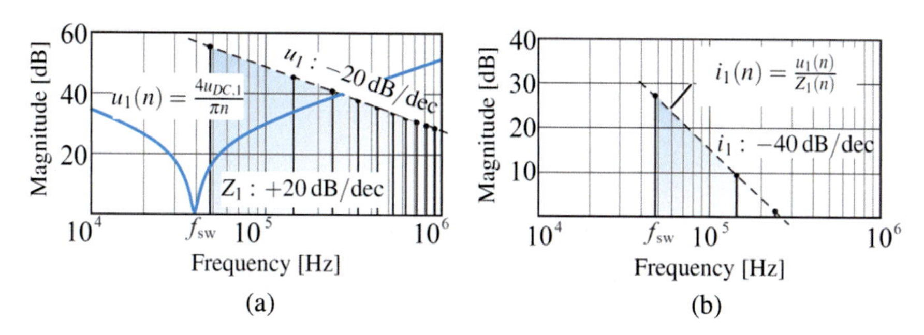

Fig. 6 Frequency responses of the primary converter. **a** Voltage and impedance in primary converter. **b** Current in primary converter

The efficiency achieved with IPT systems is usually around 5% less than that of plug-in chargers. It is worth noting that higher efficiencies (> 99%) are possible for an optimum design; however, efficiency is compromised when interoperability between pad manufacturers and tolerance to pad misalignment are needed. The power losses in the IPT systems can be attributed to the power electronic components (MOSFETs/diodes, connectors, and passive elements) and the magnetic couplers (copper and core losses). MOSFETS switching losses can be ameliorated by using soft-switching techniques like ZVS or ZCS. Losses in the other power electronic elements, however, are inherent to each component (dissipation factors). Thus, they must be sized appropriately to reduce their impact on the system's efficiency. The same is true for the coil and core in the magnetic couplers. An adequate design is required to ensure high efficiency.

To understand the effect on the transmission efficiency via the design of the magnetic couplers, an expression of the maximum efficiency $\eta_{\max}$ can be derived. Considering only the magnetic couplers, an equivalent model of an IPT system can be constructed, as shown in Fig. 7. A series compensation is implemented by adding a series capacitor C_i to each coil. Each capacitor resonates with its respective coil at the desired frequency of operation, i.e.,

$$f = \frac{1}{2\pi}\frac{1}{\sqrt{L_1C_1}} = \frac{1}{2\pi}\frac{1}{\sqrt{L_iC_i}}(i = 1, 2, ..., n)$$

The losses in the coil can be modeled by an equivalent resistor connected in series with the coil inductor. Note that this loss equivalent resistance represents both the copper and magnetic losses of the coil. Thus, the power transfer efficiency can be estimated from the ratio: $P_2/(P_2 + P_{\text{loss},1} + P_{\text{loss},2})$, where $P_{\text{loss},1}$ and $P_{\text{loss},2}$ correspond to the losses in the coil as depicted in Fig. 7. An equation for the maximum efficiency can be calculated from this equivalent circuit:

$$\eta_{\max} = \frac{k^2Q_1Q_2}{\left(1 + \sqrt{1 + k^2Q_1Q_2}\right)^2} \approx 1 - \frac{2}{kQ}$$

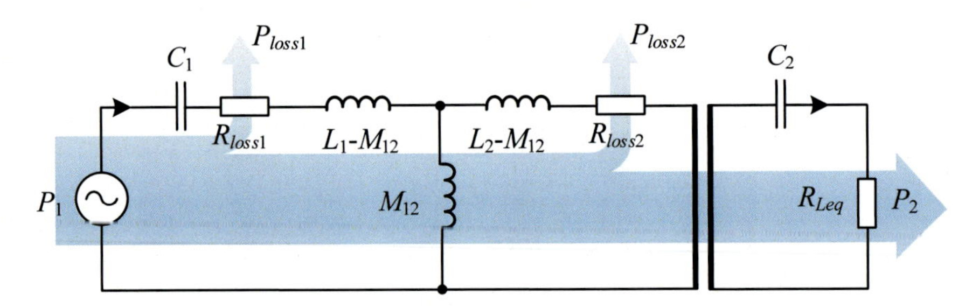

Fig. 7 Power flow path including the power losses in an IPT system with series compensation

where Q_i is the quality factor of the individual coil. The quality factor Q_i represents the ratio between the coil reactance and resistance: $Q_i = \omega L_i/R_{i,\mathrm{ac}}$. The quality factor can also be understood as the ratio between the power stored by the coil (inductance) and the power loss in the coil. It is clear that the efficiency of the magnetic coupler is mainly dependent on k and Q. The product kQ is one of the figures of merit of IPT technology. Quality factors in IPT technology are typically kept over 100 to ensure higher efficiencies.

To improve the coil pad's efficiency, Q and k should be increased. By raising the frequency of operation or by reducing the overall AC resistance, the quality factor Q can be further enhanced. Generally, the AC resistance of the coils can be reduced by increasing the cross-section of the coil at the expense of more cost, volume, and weight. On the other hand, the coupling coefficient k can be increased by optimizing the design of the pad. The factor k is particularly dependent on the ratio between the coil area and the distance between pads. Consequently, coils that occupy larger areas yield higher coupling factors. Larger pads imply lower power density as well as an increment in the copper losses due to the extra coil length. A compromise is usually required during the design. On the other hand, k can also be enhanced by decreasing the distance between coil pads. However, this is determined by the specified application. For EVs, the clearance between pads ranges from 75 mm to over 200 mm. For EVs, typical designs result in coupling factors below 0.3.

The maximum pad efficiency is only achieved for a specific load $R_{\mathrm{L,eq}}$. Deviations from this value produce a drop in efficiency. This decay is shown in Fig. 8 for two different compensation networks: series–series (SS) and series–parallel (SP) [10]. Here, γ represents the equivalent load normalized by the coil reactance, i.e., $\gamma = R_{\mathrm{L,eq}}/(\omega L_2)$. In IPT systems, γ is usually referred to as the load matching factor. An optimal pad design must ensure maximum efficiency at the rated load. At any other loads, the efficiency of the system will be sub-optimal unless the equivalent impedance of the load is controlled to keep the load matching factor constant at every operating point. For this, a synchronous rectifier or a DC-DC converter between the rectifier and the battery is required to further reduce the power losses.

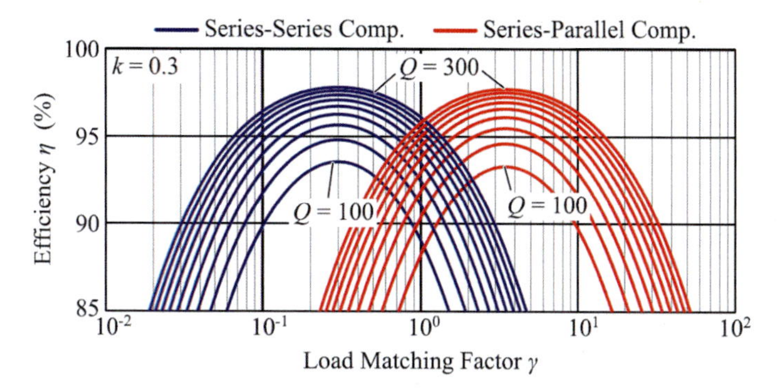

Fig. 8 Efficiency comparison with respect to load matching factor for the series–series and the series–parallel compensated systems

1.4 Power Transfer Capability

The amount of power transmitted between coil pads could be evaluated from the open-circuit voltage V_{oc}, the short-circuit current I_{sc}, and the load quality factor Q_{R} as follows

$$P_2 = V_{\mathrm{OC}} I_{\mathrm{SC}} Q_R = \omega^2 i_1^2 \frac{M^2}{L_2} Q_R = P_{\mathrm{su}} Q_R$$

where Q_{R} refers to the ratio between the apparent power in the secondary coil VA_2 and the power output P_2. In a series compensated system, Q_{R} is equivalent to the ratio between the secondary coil reactance (ωL_2) and the equivalent load resistance ($R_{\mathrm{L,eq}}$); i.e., $Q_{\mathrm{R}} = \omega L_2 / R_{\mathrm{L,eq}}$. In parallel compensated systems, Q_{R} is given by $R_{\mathrm{L,eq}}/(\omega L_2)$. Higher Q_{R} values result in higher power transfer. In practice, however, this value is usually limited to 10.

Meanwhile, P_{su} refers to uncompensated power and only considers the magnetic circuit independently of the load and power electronics. For this reason, P_{su} is used as a metric when comparing different designs of magnetic couplers. P_{su} can also be written as

$$P_{\mathrm{su}} = (\omega L_1 i_1) i_1 \frac{M^2}{L_1 L_2} = u_1 i_1 k^2 = \mathrm{VA}_1 k^2$$

where VA_1 represents the apparent power of the unloaded primary coil pad. Effectively, the VA rating of the magnetic pad couplers is inversely proportional to the coupling factor. Lower VA ratings are preferred as they lead to lower losses and lower component voltage/current stress.

2 Compensation Networks for IPT System

Compensation circuits are essential elements in an IPT system. Because of the large air gaps, the leakage flux of the coil pads is considerably high. Thus, the power factor of an uncompensated IPT system is low. Moreover, large reactive currents cause unacceptable losses. Resonant tanks are therefore used for reactive power compensation to achieve a unity power factor. Several compensation circuits have been analyzed during the last decades. The number of passive components (capacitors and inductors) as well as their performance related to power flow control, current sharing, and robustness to the misalignment is different for different topologies. Taking the mutual inductances and the coil leakage inductances in different resonant circuits into account, different compensation networks can be applied to enhance the desired system performance. In general, there are four basic compensation networks for the MRC IPT system [11], specifically, series-series (SS) compensation, series–parallel (SP) compensation, parallel-series (PS) compensation, and parallel-parallel (PP) compensation, as shown in Fig. 9. Furthermore, the LCC-compensation network [12] and the LCL-compensation network were investigated in detail, which aim to enhance the system performance [13].

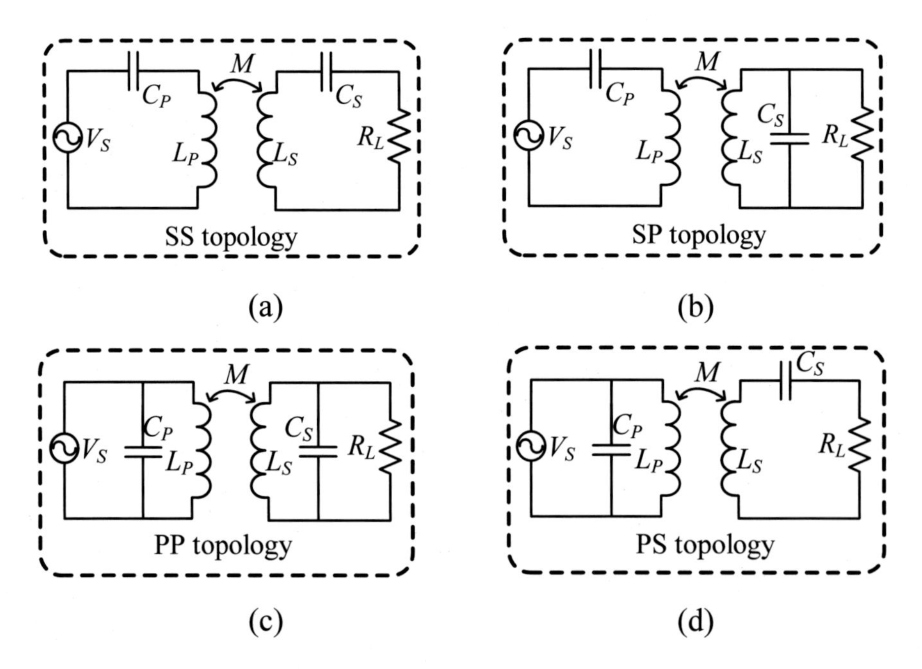

Fig. 9 Basic compensation networks. **a** Series-series compensation. **b** Series–parallel compensation. **c** Parallel-parallel compensation. **d** Parallel-series compensation

2.1 Compensation Networks on the Secondary Side

Figure 9a and b show the equivalent resonant circuits of the secondary side with series (-S) and parallel (-P) compensation networks, respectively, where C_P, L_P are the compensated capacitance and the inductance of the primary coil pad; C_S, L_S are the compensated capacitance and the inductance of the secondary coil pad; and R_L is the load equivalent resistance. Generally, the equivalent series resistance of the compensated capacitor is ignored to simplify the mathematical calculations. When the current flowing in the primary coil is I_P while the current flowing in the secondary coil is I_S, the reflected impedance to the primary side from the secondary side can be calculated as

$$Z_r = \frac{-j\omega \mathrm{MI_S}}{I_P} = \frac{\omega^2 M^2}{Z_S}$$

where M is the mutual inductance between the primary and secondary coil pads, which is related to the coupling coefficient k, $M = k\sqrt{L_P L_S}$.

The equivalent impedance Z_S of the secondary side is determined by the compensation network, which can be calculated as

$$Z_S = \begin{cases} j\omega L_S + \dfrac{1}{j\omega C_S + 1/R_L} \text{ (in parallel)} \\ j\omega L_S + \dfrac{1}{j\omega C_S} + R_L \text{ (in series)} \end{cases}$$

Typically, the operating frequency of the primary side equals the resonant frequency of the secondary side provided by $\omega_0 = 1/\sqrt{L_S C_S}$.

Then,

$$\text{for parallel-secondary} \begin{cases} \mathcal{R}e\,(Z_r) = \frac{R_L M^2}{L_S^2} \\ \mathcal{I}m\,(Z_r) = \frac{-\omega_0 M^2}{L_S} \end{cases}$$

$$\text{for series-secondary} \begin{cases} \mathcal{R}e(Z_r) = \frac{\omega_0^2 M^2}{R_L} \\ \mathcal{I}m(Z_r) = 0 \end{cases}$$

Usually, the power delivered to the secondary side from the primary side is considered as the power consumed by the real component of the reflected impedance, which has

$$P = \mathcal{R}e\,(Z_r) I_P^2$$

Since that, it is noticed that the power harvested by the secondary side will drop quadratically with the decrease of mutual inductance; consistently, drop with the increase of transmission air gap.

2.2 Compensation Networks on Primary Side

Figure 9a and c show the equivalent resonant circuits of the primary side with series (-S) and parallel (-P) compensation networks, respectively. From the view of the AC power source, the equivalent load impedance can be defined by different combinations of compensation networks on the primary side and secondary side. The equivalent load impedance Z_P in the primarily series-compensated system can be achieved from

$$Z_P = j\omega L_P + \frac{1}{j\omega C_P} + Z_r$$

The equivalent load impedance Z_P in the primarily parallel-compensated system can be achieved from

$$Z_P = \frac{1}{j\omega C_P + 1/(j\omega L_P + Z_r)}$$

With the aim of reducing the reactive power of the AC power supply, generally, the resonant circuits should be operated at the nominal resonant frequencies, where the real part (resistance) of the load impedance Z_P is equal to zero. Therefore, the zero-phase-angle (ZPA) between the output current and voltage of the inverter can be realized. However, the ZPA operation is not so superior when considering the switching loss of the inverter under hard-switching mode. In practice, the operating frequency of the primary side can be shifted away slightly from the nominal resonant frequency to accomplish a small part of reactive power that can help the power switches of the AC inverter be operated under ZCS or ZVS mode.

Table 2 shows the comparison of four basic compensation networks. It is worth noting that the imaginary part (reactance) of the reflected impedance from the series-compensated secondary is zero. Therefore, the nominal resonant frequency of the series-compensated primary will not be altered by the variations of the mutual inductance and load. On the other hand, for SP and PP compensation networks, variation in mutual inductance will change the nominal resonant frequency of the primary side. Thus, the SS compensation network is currently the most popular choice due to its high tolerance of system parameters. From the scattering parameters, the transmission efficiency can be expressed as $\eta = |S_{21}|^2$, where the network is matched at both ports. It can be noticed from this expression that the transmission efficiency drops rapidly as the distance increases. Usually, the SS compensation is desired once $\omega^2 M^2/R_L < M^2 R_L/L_S{}^2$, while PP compensation is desired once $\omega^2 M^2/R_L > M^2 R_L/L_S{}^2$.

Table 2 Comparison of four basic compensation networks

Compensation network	Reflected resistance Re (Z_r)	Reflected reactance Im (Z_r)	Secondary quality factor (Q_S)
SS	$\frac{\omega_0^2 M^2}{R_L}$	0	$\frac{\omega_0 L_S}{R_L}$
SP	$\frac{R_L M^2}{L_S^2}$	$-\frac{\omega_0 M^2}{L_S}$	$\frac{R_L}{\omega_0 L_S}$
PS	$\frac{\omega_0^2 M^2}{R_L}$	0	$\frac{\omega_0 L_S}{R_L}$
PP	$\frac{R_L M^2}{L_S^2}$	$-\frac{\omega_0 M^2}{L_S}$	$\frac{R_L}{\omega_0 L_S}$
LCC-S[a]	$\frac{\omega_0^2 M^2}{R_L}$	0	$\frac{\omega_0 L_S}{R_L}$
LCL-P[a]	$\frac{R_L M^2}{L_S^2}$	$-\frac{\omega_0 M^2}{L_S}$	$\frac{R_L}{\omega_0 L_S}$

[a] The compensation capacitor C_{PP} and the inductance L_{SP} are determined by the load condition and the output voltage of the inverter. Generally, the operating resonant frequency will be set at or close to that of the secondary

2.3 LCC Compensation Network

Aiming to realize a more flexible operation of ZPA, ZCS, ZVS, etc., the LCC compensation network was investigated, as indicated in Fig. 10a and b, by adjusting the parameters of the compensated capacitors. In the symmetrical T-type LCC compensation network, shown on the primary side, constant voltage output or constant current output can be accomplished irrespective of the load impedance.

LCC compensation topologies are often applied in multi-load IPT systems such as road-powered EV (RPEV) systems. With the aim of realizing low switching stress and low turn-off loss, the near ZCS operation under ZPA was investigated for the LCC compensation network by introducing the parallel-connected capacitor C_{PP} and the series-connected inductor L_{SP} [12]. The parameter design process under ZCS operation is different from traditional topologies. First, the nominal resonant frequency ω_0 of the secondary side and the primary power P_P should be resolved. After that, the current of primary coil I_P and the parallel compensated capacitor C_{PP} are determined as

$$I_P = \sqrt{\frac{P_P}{R_F}}$$

$$C_{PP} = \frac{I_P}{\omega_0 V_{\text{inverter}}}$$

where V_{inverter} is output voltage of the inverter, R_F is the reflected resistance from the secondary side.

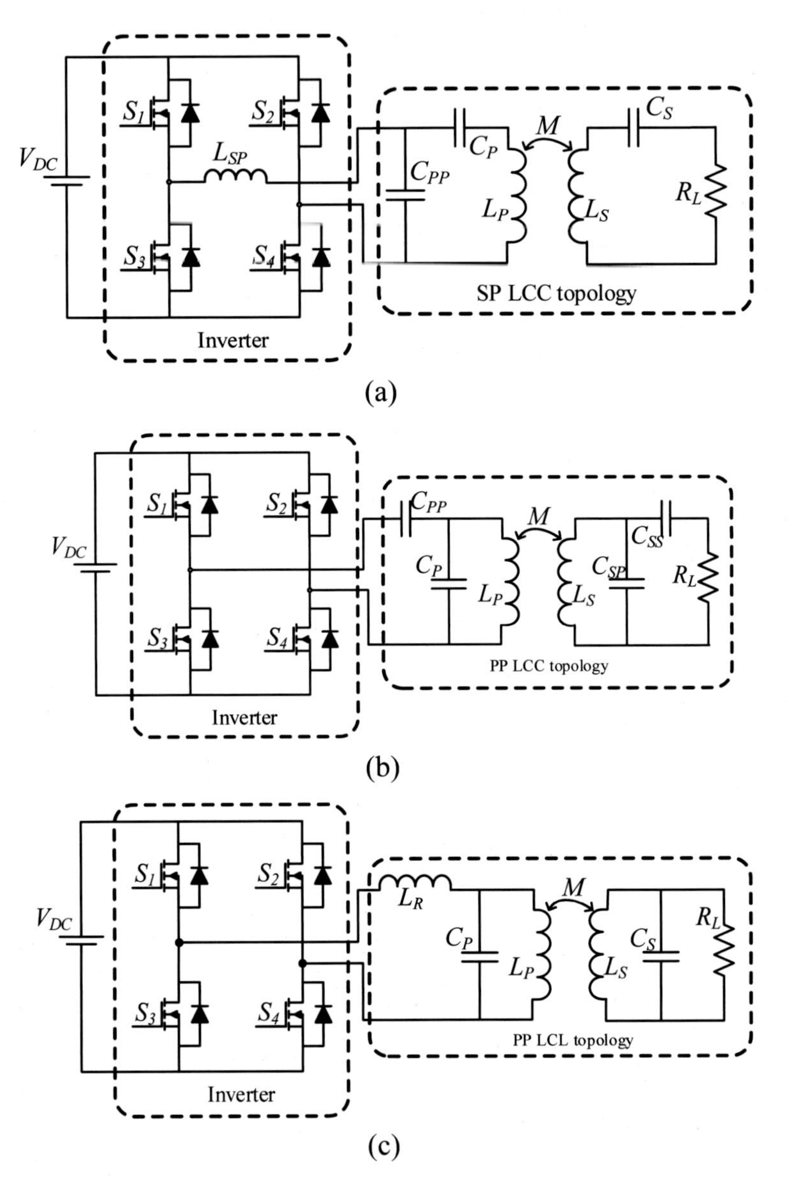

Fig. 10 Other two commonly used compensation networks. **a** Compensation network—SS LCC. **b** Compensation network—PP LCC. **c** Compensation network—PP LCL

Accordingly, the resonant capacitor C_{P} and the series-connected inductor L_{SP} can be calculated as

$$C_{\mathrm{P}} = \frac{C_{\mathrm{PP}}}{L_{\mathrm{P}}/L_{\mathrm{SP}} - \pi^2/8}$$

$$L_{SP} = \frac{1}{\omega_0^2 C_{PP}}$$

In order to improve the transmission distance, a PP LCC topology was proposed. By altering the ratio of series compensation capacitance and parallel compensation capacitance, the maximum transmission efficiency can be obtained within a certain distance. However, the main disadvantage is that the system performance is sensitive to the variations of parameters.

2.4 LCL Compensation Network

Furthermore, by applying the series inductance, the reflected capacitive reactance from the secondary side in the PP compensation can be tuned, as shown in Fig. 10c. The LCL compensation network has several advantages [13]. One is that when the WPT system is set under the nominal resonant frequency, the inverter for the LCL network only provides the active power needed by the load. The LCL resonant network is powered by a DC voltage source, and its main advantage is that the output current is determined by the input voltage source while regardless of load changes. As a result, the controller design can be simplified and easy to control the output power. Furthermore, the LCL compensation can operate under both discontinuous current and continuous current conditions. In addition, the inverter that operates close to unity power factor (UPF) can be accomplished through the control method of variable frequency. The features of several compensation networks are compared in Table 3.

Since the LCL compensation can still maintain high transmission efficiency under a low quality factor Q, it is more favored in high power applications. Aiming to have the operation closer to the UPF, as shown in Fig. 10c, an extra series compensated capacitor can usually be placed in series with the L_R to form an LCLC network, in which the DC current can be prevented from flowing through the inductor. However, for high-power applications, it is easy to cause inductor saturation due to excessive high-frequency and high current [14]. Figure 11 shows the efficiency comparison with respect to Q values of different networks. It can be noticed that LCL and LCLC compensations are more efficient than LC compensation. In this LCL compensation, the main consideration is that the position of receiver coil should be fixed when aligning to the transmitter coil, that is, the system is less tolerant of position changes.

Table 3 Features of different compensation networks for MRC IPT system

Network	Features	Network	Features
SS network	• No reflected reactance • High tolerance of parameter variations • Most commonly applied, preferred when $M^2R_L/L_S{}^2 > \omega^2M^2/R_L$	SP network	• Capability of stable current output • Reflected reactance associated to the operating frequency
PS network	• Capability of stable voltage output • No reflected reactance	PP network	• Reflected reactance associated to the operating frequency • Preferred when $M^2R_L/L_S{}^2 < \omega^2M^2/R_L$
LCC network	• Highly sensitive to the variations of capacitance and inductance • Constant current input for the transmitter coil irrespective of the existence of secondary • High tolerance capability of misalignment with properly matched parameters • Lower power transfer efficiency • Typically, LCC SP compensation network is for multi-load WPT • Can reach ZPA and ZCS operations simultaneously	LCL network	• Continuous or discontinuous current • The series inductance helps to tune out reflected reactance • Applying control method of variable frequency to have UPF operation • Low quality factor Q still have high transmission efficiency • Low tolerance capability of misalignment • Help to remove VAR loading in high power applications

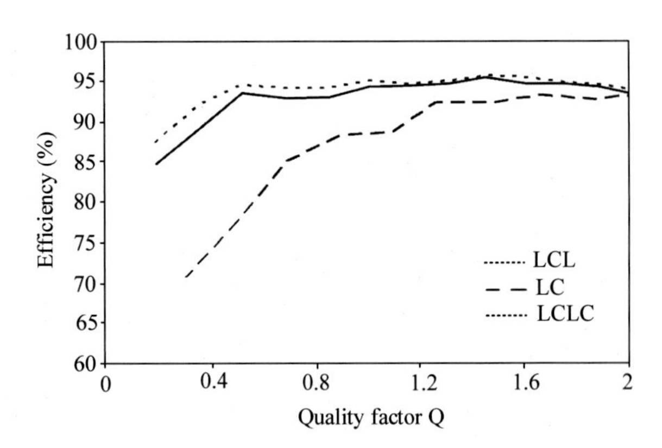

Fig. 11 Efficiency comparison in terms of different networks, LC, LCL and LCLC compensations

3 Magnetic Coupler Pads

3.1 Coil Pad Design for Wireless EV Charging

The transmitter and receiver coil pads, as the energy carriers in the WPT system, perform an essential role in generating and harvesting the spatial magnetic fluxes. The magnetic couplers, namely transmitter and receiver coil pads, comprise the coil windings, magnetic cores, and shielding plates. The design of magnetic coupler pads determines the system's transmission efficiency, the VA rating, and the capability of power transfer. Figure 12 shows the typical coil pad topologies for transmitter and receiver designs, which can be applied for wireless EV charging for both move-and-charge and park-and-charge [15, 16]. In the early stage of WPT development, the circular/square pads (CP/SP) were proposed for many years. This topology generally helps to improve the flux area; and can reduce edge flux leakage. However, low transmission efficiency and large total flux leakage are inevitable. Therefore, it is generally acceptable to design the transmitter coil according to specific requirements with a simple constructure. To reduce flux leakage and achieve low reluctance flux path, soft ferrite bars are usually added to the backside of the coil pad, as shown in Fig. 12a. And an aluminum shielding plate is set at the back of the coil pad to support the magnetic flux distribution and shield the magnetic leakage flux. The advantage of this topology is the easy establishment of symmetrical distribution of flux lines around the central. However, the transmission distance, the height of main flux path, is constrained by the size of the coil pad.

Based on the CP/SP coil pads, as shown in Fig. 12b, an enhanced prototype named Double-D pad (DDP) was proposed, in which the current directions in the dual coil windings are opposite. Because of the parallel magnetic field distribution along the ferrite bars or plates, the magnetic flux path can be taller and narrower. Thus, the transmission distance is extended, and the transmission efficiency is improved. However, when the receiver pad is centrally aligned, DDP topology exhibits poor interoperability characteristics, as does the CP topology. In order to well address the issue of interoperability, the Double-D quadrature pad (DDQP) topology was designed to offer parallel magnetic field and perpendicular magnetic field simultaneously, as shown in Fig. 12c, in which one additional coil is placed on the DDP coil pad. As the DDQP topology can offer non-polarized field and polarized field by adjusting the current's direction flowing in the coil, the WPT system is much more flexible than other coil pad topologies. But the disadvantage is that the number of coils increases, resulting in additional coil and control costs.

Based on the DDQP topology, another high-compliance topology named bipolar pad (BPP), as shown in Fig. 12d, is introduced to offer parallel field and perpendicular field, where the two coil windings partially overlap. At the same time, by eliminating one coil, the complexity and cost of the WPT system are reduced. Depending on the diverse combinations of the current direction, the BPP topology can conduct single coil type, SP type, and DD type. Therefore, the BPP topology can be actually applied to the design of multi-mode secondary charging pads, and then it has great potential in wireless EV charging. The development of the WPT widespread topologies of the magnetic couplers discussed above is listed in Fig. 13. Besides, the factors and features of commonly used coil pads are compared and summarized in Table 4.

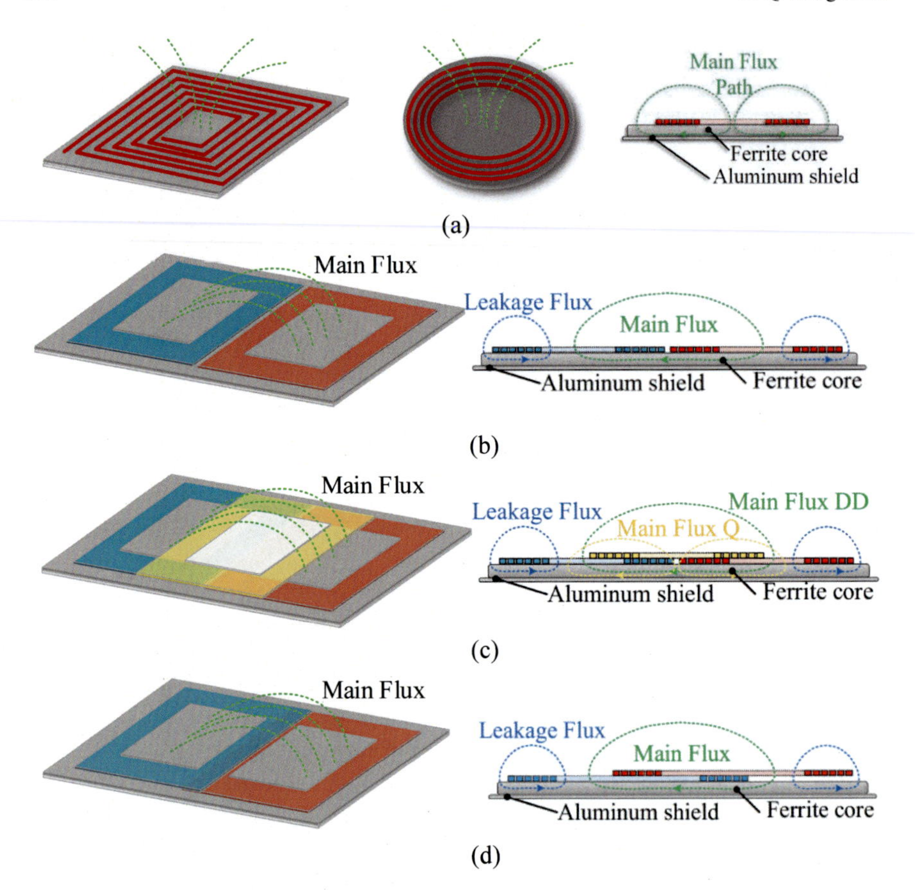

Fig. 12 Typical magnetic couplers for wireless EV charging. **a** CP/SP single-coil topologies with the non-polarized field. **b** DDP topology with the polarized field. **c** DDQP topology with non-polarized and polarized fields. **d** Polarized BPP topology

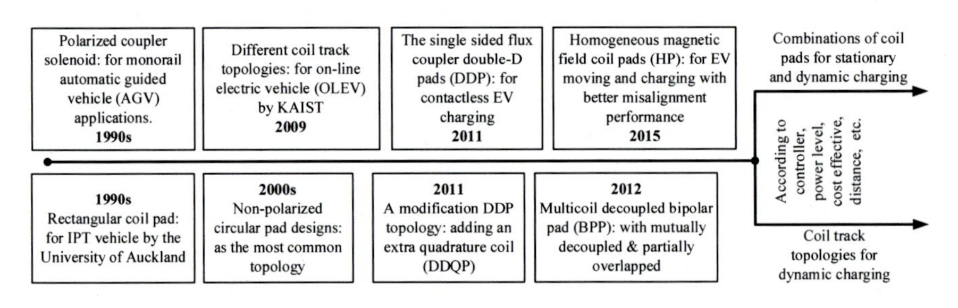

Fig. 13 Development of WPT typical coil topologies

Table 4 Comparison of the widely applied magnetic couplers for wireless EV charging

Topology	CP/SP topology	DDP topology	DDQP topology	BPP topology
Factors	• System weight • Power level • Types of electric vehicle. • Limited cost and size • Chassis structure • Distance between the primary and secondary	• System weight. • Unwanted flux leakage. • Power level. • Thickness and length of ferrite bars. • Polarity of magnetic flux direction to be coupled. • Limited coil pad size and cost	• System weight • Thickness and length of ferrite bars • Power level • Control schemes • Limited coil pad size and cost • Chassis structure. • Types of EVs.	• Overlap of the central area • Thickness and length of ferrite bars • Power level • Chassis structure • Limited coil pad size and cost • Air-gap between the transmitter and receiver
Features	• Most widely applied in the transmitter side or receiver side • Flexible applications. Non-polarized perpendicular magnetic field pattern by CP • Poor interoperability characteristics • Flux symmetric around center of the CP • Generate and couple perpendicular flux	• Commonly applied in the primary side • Higher interoperability to different coil pas of the secondary side • ingle sided flux generation • No reverse flux to conduct the cancellation of the unwanted rear flux • Easy to be saturated in ferrite bars for high power applications	• Higher interoperability to different coil pas of the secondary side • Commonly applied in the secondary side • Providing effective charge zone three times than DDP when used for the secondary • Regulable excitation modes • Versatile design for the middle coil to suit the air-gap • Inferior material usage efficiency	• Mostly the same features of DDQP topology applied in the secondary • Mutually decoupled partially overlapped coil structure • Using less coil than DDQP • Commonly used in the secondary • Higher interoperability to different coil pas of the secondary side • Deliver almost the same power for given size of the DDQ topology

3.2 Magnetic Cores for Coupler Pads

Soft-magnetic materials are added to the IPT pads to improve their magnetic performance. The addition of magnetic cores results in higher P_{su} and k, and lower VA ratings. High resistivity, permeability, and saturation point are some of the requirements for the core material. High resistivity is required to constrain the eddy current loop occurred in the magnetic core, while a slim hysteresis loop is required to ensure minimum magnetic losses. For IPT applications, manganese-zinc soft ferrites, such as the K2004 or the N87 from EPCOS, are the most used ferrite materials. Ferrite is used because of its adequate magnetic properties, large availability, and low cost. Ferrite cores, however, are brittle and make the practical fabrication of small-scale or geometrically complex cores impractical. Ferrite cores come up with a relatively low magnetic saturation, limiting the power rating of the magnetic coupler. Moreover, the performance of ferrite deteriorates rapidly as the operating temperature rises. Ferrite-less pads or alternative materials are the current focus of research.

When both the transmitter and receiver pads are non-polarized, plates or radially oriented bars are preferred, as shown in Fig. 14a [17]. Generally, discrete core bars achieve similar performance to a continuous core (plate). Optimal sizing and placement within the coil pad region have been the prominent topics of research over the past few years. In addition, polymer-based nanoparticles ferrite were also proposed to enhance the mechanical properties of the coil pads and reduce the usage of ferrite [18]. Even the obtained permeability is low, it effectively improved the mechanical robustness of the material.

For polarized pads, both ferrite bars and plates are used, as shown in Fig. 14b. In most cases, the maximum flux is restricted below the saturation point to reduce the losses in the core. The optimum spacing for discrete bars was determined to be twice the core bar thickness. The thickness of the bars was determined according to the maximum allowed flux density within the core. Due to the low magnetic saturation

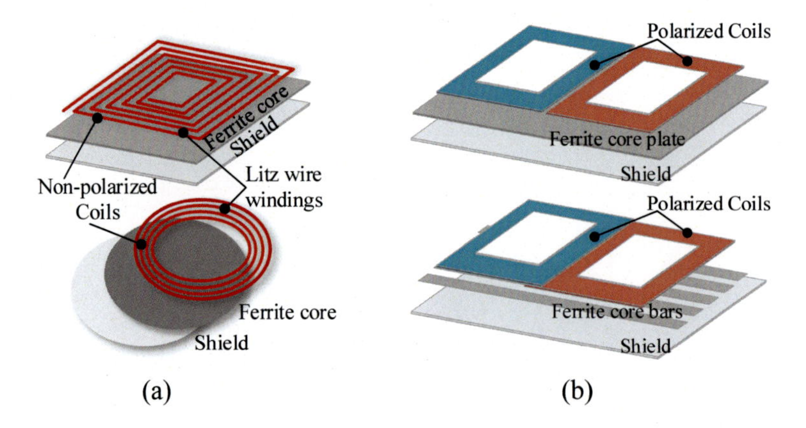

Fig. 14 Typical wireless EV charging core designs. **a** Non-polarized and **b** polarized coil pads

point of the ferrite, the maximum magnetic flux density is constrained, normally under 0.45 T. It is worth noting that large ferrite plates are not commercially available and bespoke designs are required. Moreover, large ferrite pieces are difficult to manufacture due to the brittleness of the material. Thus, ferrite plates are usually formed by several smaller units. The interface between units leads to several problems, such as partial saturation and localized heating in the cores. These problems are more probable as the number of units used in the core increases.

3.3 Magnetic Cores for Coupler Pads

Power losses within the magnetic core are mainly of two types: eddy-current and hysteresis losses. The former results from the circulation of induced currents within the core while the latter results from the stiffness of the magnetic domains. Both are proportional to the magnitude and frequency of the magnetic field density B. In general, iron losses in the magnetic core are evaluated by employing the Steinmetz equation

$$P_v = C_m C(\vartheta) f^{\alpha} B_{pk}^{\beta}$$

where α, β, and C_m are constants obtained heuristically for each material; f stands for the operating frequency, B_{pk} for the peak value of magnetic flux density while $C(\vartheta)$ is a correction factor due to changes in temperature (ϑ). It is worth noting the Steinmetz's units are kW/m^3. Thus, to calculate the total losses, volume integration is required. This is commonly done with finite element method (FEM) packages after solving for the magnetic field distribution within the core.

Ferrite losses can also be estimated directly from FEM simulations by specifying the permeability as a complex quantity. Complex permeability (μ') can be understood as follows. When an AC excitation $H = H_m e^{jwt}$ is applied to a ferromagnetic material, a flux density $B = B_m e^{j(\omega t - \delta)}$ is obtained. Here, δ represents the phase angle between H and B, and it is related to the power loss in the ferrite due to the AC magnetization. The complex permeability can be characterized as the ratio between H and B as

$$\mu' = \frac{B}{H} = \mu_r + j\mu_i = \frac{B_m}{H_m}\cos(\delta) - j\frac{B_m}{H_m}\sin(\delta)$$

The complex permeability μ' reflects the combined effect of eddy-current, hysteresis, and relaxation losses [19]. Complex permeability values can be obtained from the material datasheets. The measurement of complex permeability is commonly performed by using toroidal cores. Thus, the measured values are considered unidirectional uniform flux. When the core is exposed to traverse fluxes, alternative methods of estimating losses are required, particularly for anisotropic and/or laminated cores.

4 Stationary and Dynamic Wireless EV Charging

4.1 Park-And-Charge for Stationary Charging

To facilitate the PAC process for electric vehicles, the IPT technology is extended to a plugless technology, where the transmitter coil pad is installed on the floor of the garage or parking lot, and the receiver coil pad is installed at the bottom of the vehicle. Drivers don't need to bother with those bulky and dangerous charging cables. The wireless charging system is very simple to use and the charging process takes place automatically once the driver has parked correctly [20]. This plugless PAC scheme not only improves the convenience of use, but also provides a way to overcome the standardization of charging plugs. This plugless PAC not only enhances the convenience of users' utilization, but also presents a mean to overcome the standardization of EV charging interfaces. Based on the MRC IPT, the transmitter side and receiver side are with the same resonant frequency to ensure the system can wirelessly deliver power with high efficiency and power density, while the energy dissipating in non-resonant objects, such as drivers or metal bodies, is quite low [21].

The up-to-date development and research of WPT for PAC are vigorous and diverse, such as achieving bidirectional WPT between EVs, compensating the misalignment among magnetic couplers, and combining information transfer and power transfer within the same channel. Some state-of-the-art researches are briefed below.

(1) In practical PAC system, the design of magnetic couplers plays an important role in an effective WPT system. For example, a compound transmitter coil pad with a non-uniform pitch spiral coil winding can provide an uniform distribution of magnetic flux density over the majority of the charging zone, thus resolving the misalignment tolerance problem. Also, bipolar transmitter coil pad was developed to interoperate with simple receiver coil pads for wireless power receiving with large lateral misalignment.

(2) Vehicle-to-grid (V2G) technology has a promising future as electric vehicles can act as mobile energy routers to support and stabilize the grid with renewable energy storage and supply. By integrating WPT into V2G, the system with bidirectional power interface has been investigated to enable charging and discharging for multiple EVs simultaneously. In recent years, various bidirectional resonant converters have been investigated for V2G via the WPT technique, aiming to enhance power flow control, power rating, and capability of fault tolerance.

(3) The wireless EV charging is currently actively developing the simultaneous wireless power and information transfer (WPIT) technology, which requests power to be transferred from the ground to the vehicle, as well as the data transmission between the transmitter and the EV battery management system. For instance, the fundamental component of the current waveform in a WPT

system can be applied for transferring power, while its third-order harmonic component can be deployed to exchange data.

4.2 *Move-And-Charge for Dynamic Charging*

Instead of parking or stopping, EV charging prefers wireless charging without plug-in action when moving. In other words, a set of transmitters are placed in charging zones or lanes under the roadway, while the receivers are mounted on the chassis of multiple EVs. Thus, the moving EV could pick up wireless energy from the charging zones or lanes, which is entitled MAC technology or dynamic charging. Fundamentally, it has a high potential to resolve the longstanding issue of the EV charging process. That is to say, EVs do not need to place so many batteries, which can greatly reduce the cruising cost. At the same time, EVs can be conveniently charged during driving in the charging zones, thereby the cruising range can be automatically extended. Unlike the PAC, the MAC system is much more challenging in the infrastructure, and control complexity, etc.

Regarding the type of transmitters beneath the lanes, it could be either separate coil pads or a single rail design. The separate coil pad design involves a large number of coil pads, each of which is the same size or smaller than that of an EV. As shown in Fig. 15a, the separate coil pads can be individually activated by detection sensors and power inverters. However, this type of transmitter inevitably involves many coil pads, detection sensors, and power inverters, resulting in high installation complexity and enormous investment costs. In contrast, the single rail design only requires one primary rail or actually a long transmitter coil and one power inverter to wirelessly energize several vehicles, as shown in Fig. 15b, which has a much lower installation complexity and much less investment cost than the separate coil pad design. Aiming to have more flexible scalability and maintainability, single rail design is often arranged in segmented lanes, each segment uses one power inverter to wirelessly energize several EVs.

4.3 *Developments of Move-And-Charging EVs*

As so far, many research institutes are focusing on the move-and-charging technologies. Especially, the Korea Advanced Institute of Science and Technology (KAIST) is elaborating the online electric vehicle (OLEV) project, which has well resolved many problems such as continuous power transfers, high-frequency current-controlled inverters, and cost-effective enhancement [22]. The OLEV project was launched in 2009. Pioneering techniques for coil pad designs and roadway construction helped the transmission efficiency reach up to 83% with an output power of 60 kW at a nominal resonant frequency of 20 kHz. The transmission distance is set at 200 mm and lateral tolerance of 240 mm can be attained. Besides, the construction cost of

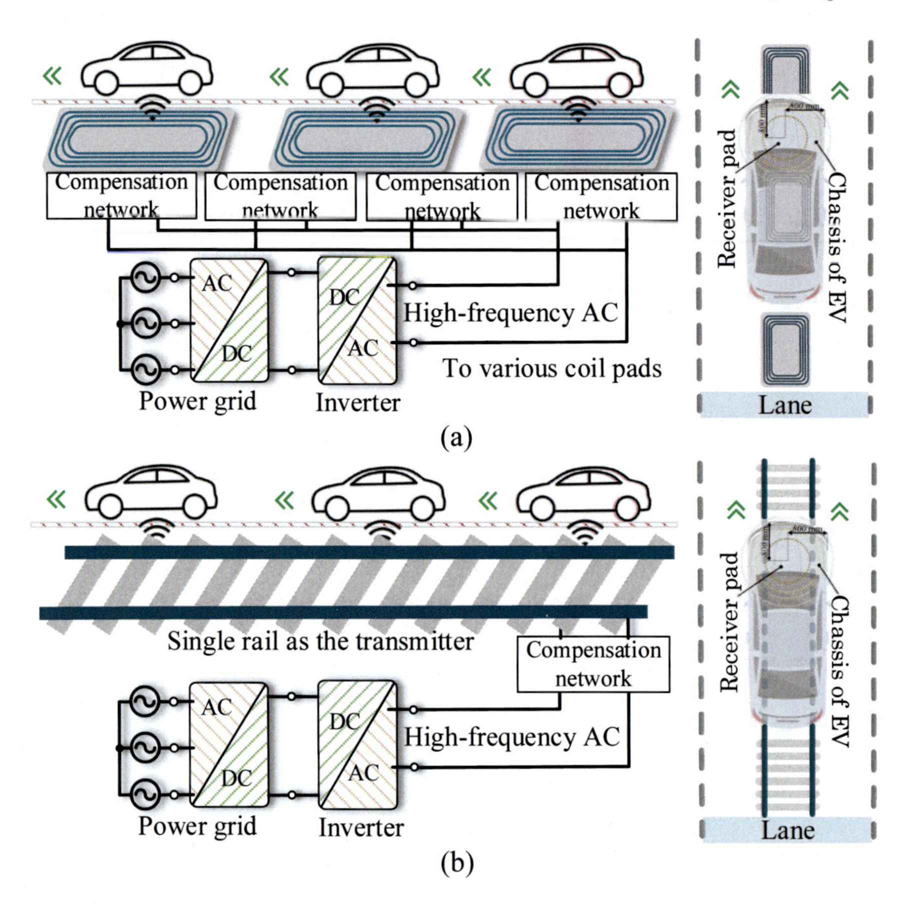

Fig. 15 Configuration of different types of MAC systems. **a** Many separate coil pads as the transmitter. **b** Single rail as the transmitter

power rail has been significantly lowered to at least one-third after the optimization of the rail design. The battery capacity has been reduced to 20 kWh, and the current in the transmitter has also been effectively mitigated as low as 200 A during high power operation. Until now, the OLEV system has been implemented at five locations. As shown in Fig. 16, the project of OLEV has been proceeding to the forth-generation. Besides, the development of the fifth-generation and the sixth-generation OLEV are in progress, where the ultra slim S-type transmitter rail of 4 cm was proposed to further enhance the cost-effectiveness of the construction.

Ever since the 1990s, researchers from Auckland University have been proposing different IPT topologies for wireless EV charging [24]. For example, the Auckland University team proposed an IPT system for RPEV that includes a number of small power coil pads, in which the length of a coil pad is much shorter than that of a vehicle's body to get rid of unnecessary powering and loading. However, there are many factors to be considered with this approach, such as increased deployment and

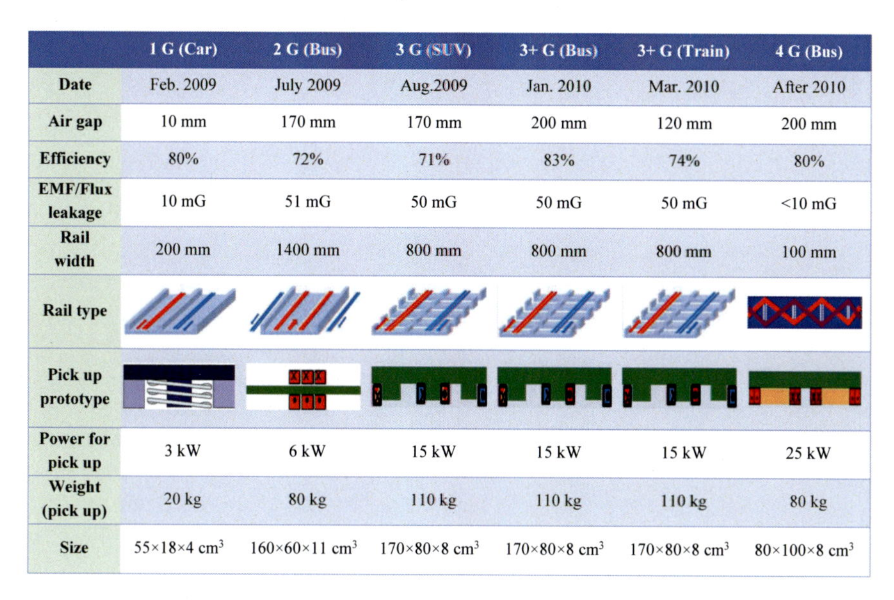

	1 G (Car)	2 G (Bus)	3 G (SUV)	3+ G (Bus)	3+ G (Train)	4 G (Bus)
Date	Feb. 2009	July 2009	Aug.2009	Jan. 2010	Mar. 2010	After 2010
Air gap	10 mm	170 mm	170 mm	200 mm	120 mm	200 mm
Efficiency	80%	72%	71%	83%	74%	80%
EMF/Flux leakage	10 mG	51 mG	50 mG	50 mG	50 mG	<10 mG
Rail width	200 mm	1400 mm	800 mm	800 mm	800 mm	100 mm
Rail type						
Pick up prototype						
Power for pick up	3 kW	6 kW	15 kW	15 kW	15 kW	25 kW
Weight (pick up)	20 kg	80 kg	110 kg	110 kg	110 kg	80 kg
Size	55×18×4 cm³	160×60×11 cm³	170×80×8 cm³	170×80×8 cm³	170×80×8 cm³	80×100×8 cm³

Fig. 16 An evolution map of developed OLEV from KAIST [23]

control complexity, as well as the maintenance cost of the ground placed coil pads, which should be able to survive in harsh road conditions.

Besides, the research team from Bombardier has investigated different IPT systems for the PAC and MAC schemes of trams and buses since 2010. Both the stationary and dynamic charging were developed with a peak output power of 250 kW, transmission air-gap of 60 mm, and operating resonant frequency of 20 kHz. From April 2013, the research team from Endesa has been actively involved in the vehicle initiative consortium for transport operation & road inductive application (VICTORIA) project and has applied three charging schemes: traditional plug-in, stationary charging, and dynamic charging. With the purpose of demonstration, the RPEV with a maximum power of 50 kW was implemented with U-type rail and operated on a 10 km bus route with a self-guided control system in 2014. Moreover, a team from Integrated Infrastructure Solutions (INTIS) has developed a 25 m U-type transmitter rail in its own test center. Based on a single-phase power system, the double U-type transmitter rail can provide a maximum power of 200 kW at the operating resonant frequency of 35 kHz. As so far, the INTIS has developed two IPT systems: stationary wireless EV charging and RPEVs, which deployed the operating frequency of 30 kHz at an air-gap of 10 cm for the maximum output power of 30 kW.

4.4 Energy Encryption for Dynamic Charging

The dynamic charging technique for wireless EV charging has drawn increasing attention in recent years. Indeed, the dynamic charging system will be one of the most important development directions for EV charging in the foreseeable future. Typically, for busy roads with a long-scale single transmitter rail WPT system, there may be several vehicles being charged at the same time. This means that all participating vehicles can harvest the wirelessly transmitted energy through the electromagnetic field generated by the single transmitter rail. In such a way, the energy is easily picked up by some unauthorized vehicles, such as unstable, unpaid, and illegal vehicles. Therefore, the energy security of wireless EV charging system has drawn more and more attention. Since a large impedance will be presented once the operating frequency is far away from the nominal resonant frequency, then the resonant frequency can be employed to conduct the energy encryption via a capacitor array [25]. In an energy encryption system, both the primary and authorized receiver have the same encryption algorithm based on a secret key. Figure 17 shows the flowchart of the energy encryption for the wireless EV charging system. The authorized EV can get the correct operating frequency by switching the capacitor array based on the authorized key. The main drawbacks of this capacitor array topology are that the number of sections number is limited by the number of capacitors and the power variation that happens during switching capacitors. Thus, the continuously variable capacitor is required to accomplish continuously adjustable capacitance to replace the capacitor array [26].

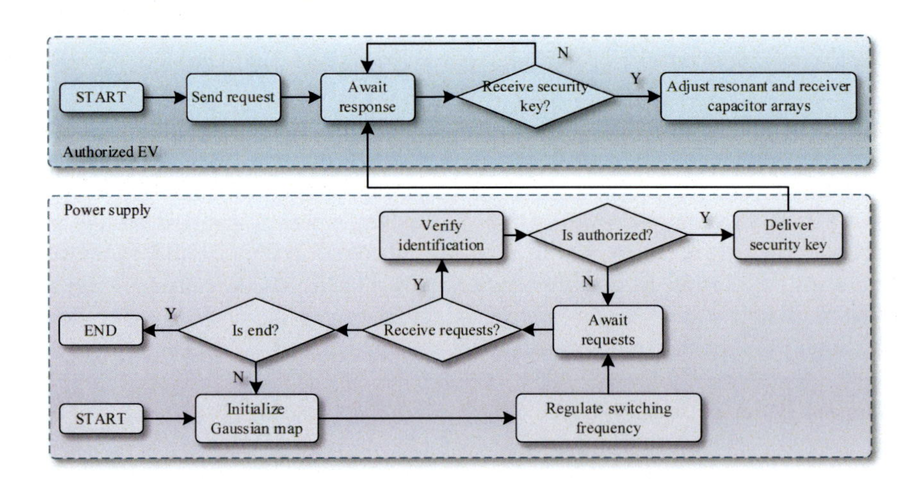

Fig. 17 Flowchart of energy encryption for wireless EV charging system [27]

5 Flux Leakage Regulations and Magnetic Shielding

According to guidelines and regulations from the International Commission of Non-Ionizing Radiation Protection (ICNIRP), the public generally should not be exposed to more than $B_{\text{rms}} = 27\ \mu\text{T}$ of magnetic flux among the frequency range from 3 kHz to 10 MHz. This value is slightly relaxed for occupational exposure to about $B_{\text{rms}} = 100\ \mu\text{T}$. For many IPT systems, the general public limit is considered and, therefore, a limit of $B_{\text{rms}} = 27.3\ \mu\text{T}$ is usually selected. Contrary to the ICNIRP, the IEEE.C95 standard defines a limit of $205\ \mu\text{T}_{\text{RMS}}$ for exposure to head and torso and $1130\ \mu\text{T}_{\text{RMS}}$ for limbs [28]. The lack of agreement between these two standards is attributed to differences in statistical and biological models, stated objectives, and specification of safety limits in specific body parts. In literature, limits vary across research institutes. In recent years, however, more authors seem to favor the pacemaker limit of $15\ \mu\text{T}_{\text{RMS}}$ as it considers a worst-case scenario. To ensure code compliance, the magnetic flux is measured at different points around the pad, and the safe zone is defined accordingly. A distance of ~800 mm from the center of the receiver pad is usually considered as the limit of the safety area (considering the vehicle width as 1.6 m), as shown in Fig. 18.

The magnitude of the flux leakage depends on the frequency of operation, the primary apparent power, and the coupling factor between pads [29]. High power ratings complicate the compliance of safety codes as the flux leakage increases with the power level. Therefore, most IPT systems use magnetic shielding to reduce leakage flux. Two main shielding methods can be found in literature: metallic shielding and reflection coils [30]. A metallic shield is a metal sheet placed in strategic locations within the pad, usually below the pad. The magnetic flux reaching the shield induces currents which, in turn, creates a magnetic field that opposes the main one. The currents in the shield produce losses that can be estimated with FEM simulations. A reflection coil, on the other hand, is an additional winding placed below the pad. When excited, it produces a magnetic field that cancels the flux generated by the main windings. The reflection coil can be active (driven by a power source) or passive (driven by the main winding). Active coils require more power electronics

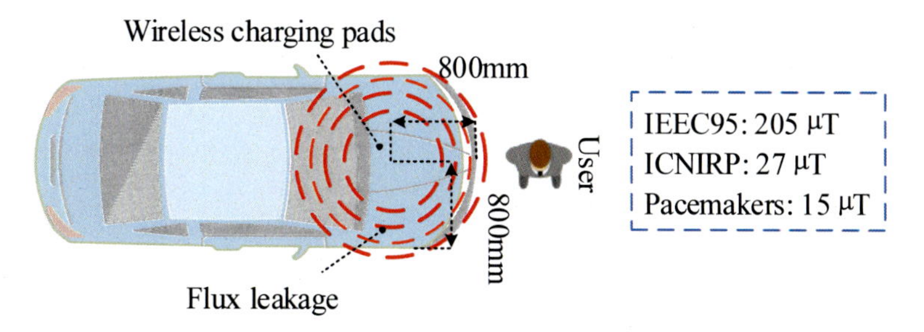

Fig. 18 Different regulations of the stray flux leakage measured out of the coil position with a radius of 0.8 m

and control units but are more versatile and easier to design as compared to passive coils.

6 Conclusion

Wireless EV charging is a crucial technology to accelerate the promotion of EV utilization. First, it increases safety by limiting user interaction with high-voltage galvanic terminals. Second, wireless EV charging technology allows for automation of the charging process. This not only increases the user's comfort but also allows for vehicles-to-grid (V2G) and opportunity-charging schemes. Through V2G, the battery of every EV works as a storage unit for the grid, facilitating peak-shaving and the integration of renewable generation. Opportunity-charging schemes, on the other hand, can reduce the range anxiety of EV users, reduce the depth of discharge of batteries, and increase their lifespan.

In this chapter, the fundamentals of WPT techniques have been introduced, including the classifications, equivalent circuits, and the basic calculations of power and efficiency. Then, the different compensation networks and magnetic coupler pads have been described for applying for wireless EV charging. After that, the pack-and-charge and move-and-charge technologies have been introduced with different coil pad configurations, followed by the flux leakage regulations and magnetic shield methods. For the next generation of wireless EV charging, the magnetic core will be very critical to improve the power density and then reduce the core loss by exploring new magnetic materials, such as Nanocrystalline ribbon core with high saturation, high permeability and low core loss. Moreover, the main power converters, including grid-connected AC-DC converter, DC-AC inverter before the compensation network, and the synchronous AC-DC rectifier, are taking parts of the total system loss. It will be challenging to develop new IPT system adaptable converter topologies, power transistor switching patterns, and new power devices to further enhance the power density, the conduction loss and switching loss, as well as system reliability.

Acknowledgements This work was supported in part by a grant (Project No. 52107011) from the Natural Science Foundation of China (NSFC), China; in part by a grant (Project No. 9211302) from the Environmental and Conservation Fund (ECF), Hong Kong SAR; a grant (Project No. 9610529) from City University of Hong Kong; in part by a grant (Project No. SGDX20210823104003034) from the Science Technology and Innovation Committee of Shenzhen Municipality, China. The authors would like to thank the editor-in-charge, Dr. Christopher H. T. Lee, for his invitation and effort to improve this chapter.

References

1. Covic GA, Boys JT (2013) Inductive power transfer. Proc IEEE 101(6):1276–1289

2. Pries J, Galigekere VPN, Onar OC, Su G-J (2019) A 50-kW three-phase wireless power transfer system using bipolar windings and series resonant networks for rotating magnetic fields. IEEE Trans Power Electron 35(5):4500–4517
3. Hui SYR, Zhong W, Lee CK (2014) A critical review of recent progress in mid-range wireless power transfer. IEEE Trans Power Electron 29(9):4500–4511
4. Jiang C, Gaona DE, Shen Y, Zhao H, Chau KT, Long T (2020) Low-frequency medium power capacitor-free self-resonant wireless power transfer. IEEE Trans Industr Electron 68(11):10521–10533
5. Jiang C, Chau KT, Leung YY, Liu C, Lee CH, Han W (2017) Design and analysis of wireless ballastless fluorescent lighting. IEEE Trans Industr Electron 66(5):4065–4074
6. Liu C, Jiang C, Song J, Chau KT (2018) An effective sandwiched wireless power transfer system for charging implantable cardiac pacemaker. IEEE Trans Industr Electron 66(5):4108–4117
7. Jiang C, Chau KT, Liu C, Lee CH (2017) An overview of resonant circuits for wireless power transfer. Energies 10(7):894
8. Steigerwald RL (1988) A comparison of half-bridge resonant converter topologies. IEEE Trans Power Electron 3(2):174–182
9. Gaona Erazo D (2021) Enhancement of inductive power transfer technology: Iron-based nanocrystalline ribbon cores. University of Cambridge
10. R. Bosshard, "Multi-objective optimization of inductive power transfer systems for EV charging," ETH Zurich, 2015.
11. Wang CS, Stielau OH, Covic GA (2005) Design considerations for a contactless electric vehicle battery charger. IEEE Trans Industr Electron 52(5):1308–1314
12. Zhu Q, Wang L, Guo Y, Liao C, Li F (2016) Applying LCC compensation network to dynamic wireless EV charging system. IEEE Trans Industr Electron 63(10):6557–6567
13. Wang CS, Covic GA, Stielau OH (2004) Investigating an LCL load resonant inverter for inductive power transfer applications. IEEE Trans Power Electron 19(4):995–1002
14. Keeling NA, Covic GA, Boys JT (2010) A unity-power-factor IPT pickup for high-power applications. IEEE Trans Industr Electron 57(2):744–751
15. Zaheer A, Hao H, Covic GA, Kacprzak D (2014) Investigation of multiple decoupled coil primary pad topologies in lumped IPT systems for interoperable electric vehicle charging. IEEE Trans Power Electron 30(4):1937–1955
16. Pearce MGS, Covic GA, Boys JT (2018) Robust ferrite-less double D topology for roadway IPT applications. IEEE Trans Power Electron 34(7):6062–6075
17. Bosshard R, Kolar JW, Mühlethaler J, Stevanović I, Wunsch B, Canales F (2014) Modeling and Pareto optimization of inductive power transfer coils for electric vehicles. IEEE J Emerg Sel Top Power Electron 3(1):50–64
18. Delgado A, Salinas G, Rodríguez J, Oliver JA, Cobos JA, Finite element modelling of litz wire conductors and compound magnetic materials based on magnetic nano-particles by means of equivalent homogeneous materials for wireless power transfer system, pp 1–5
19. Füzer J, Dobák S, Kollár P (2015) Magnetization dynamics of FeCuNbSiB soft magnetic ribbons and derived powder cores. J Alloy Compd 628:335–342
20. Lee CH, Hua W, Long T, Jiang C, Iyer LV (2021) A critical review of emerging technologies for electric and hybrid vehicles. IEEE Open J Veh Technol
21. Chau KT, Jiang C, Han W, Lee CH (2017) State-of-the-art electromagnetics research in electric and hybrid vehicles. Prog Electromagn Res 159:139–157
22. Mi CC, Buja G, Choi SY, Rim CT (2016) Modern advances in wireless power transfer systems for roadway powered electric vehicles. IEEE Trans Industr Electron 63(10):6533–6545
23. Choi SY, Gu BW, Jeong SY, Rim CT (2014) Advances in wireless power transfer systems for roadway-powered electric vehicles. IEEE J Emerg Sel Top Power Electron 3(1):18–36
24. Boys JT, Covic GA (2015) The inductive power transfer story at the University of Auckland. IEEE Circuits Syst Mag 15(2):6–27
25. Zhang Z, Chau KT, Qiu C, Liu C (2015) Energy encryption for wireless power transfer. IEEE Trans Power Electron 30(9):5237–5246

26. Zhang Z, Pang H (2019) Continuously adjustable capacitor for multiple-pickup wireless power transfer under single-power-induced energy field. IEEE Trans Industr Electron 67(8):6418–6427
27. Chau KT (2016) Energy systems for electric and hybrid vehicles. The Institution of Engineering and Technology (IET)
28. Bosshard R, Iruretagoyena U, Kolar JW (2016) Comprehensive evaluation of rectangular and double-D coil geometry for 50 kW/85 kHz IPT system. IEEE J Emerg Sel Top Power Electron 4(4):1406–1415
29. Lin FY, Covic GA, Boys JT (2015) Evaluation of magnetic pad sizes and topologies for electric vehicle charging. IEEE Trans Power Electron 30(11):6391–6407
30. Kim J, Kim J, Kong S, Kim H, Suh I-S, Suh NP, Cho DH, Kim J, Ahn S (2013) Coil design and shielding methods for a magnetic resonant wireless power transfer system. Proc IEEE 101(6):1332–1342

Advanced Vehicle-to-Grid: Architecture, Applications, and Smart City Integration

Albert Y. S. Lam

Abstract Electric vehicles (EVs) will become prevalent in this decade. An EV is equipped with a battery, which enables its interaction with the power grid in many interesting ways. An EV not only gets charged from but also discharges its energy back to the grid. Due to their mobility, an aggregation of EVs turns into a significant energy resource with high flexibility and convenience, constituting the vehicle-to-grid (V2G) system. In this chapter, we investigate some advanced V2G technologies in terms of architecture, applications, and smart city integration. We first introduce various system components and discuss a hierarchal structure, which is generalized to most V2G architecture. Then several innovative applications are reviewed, including frequency regulation, voltage regulation, unit commitment, and energy trading. After that, we study how V2G can be enhanced with new vehicular technologies and integrated into a smart city. EVs with self-driving capability allow us to drive a fleet of vehicles to strive for various V2G objectives. With the support of dynamic wireless power transfer, we can turn a road network into a big energy buffer, which can function like V2G. It can be seen that V2G technologies continue to evolve, become smarter, and cultivate new applications which are impossible before.

1 Introduction

According to Global EV Outlook 2021 [1], 10 million electric vehicles (EVs) ran on the roads in different parts of the world by the end of 2020. In spite of the global economic downturn due to COVID-19, there was a 41% increase in EV registrations in 2020. We expect that the trend continues and the growth will be even more compelling after the pandemic and economic recovery. With such a significant growth of EVs in key markets and the EV-favored mandates of many nations, it is not hard to imagine EVs will dominate in all kinds of vehicle populations in the future.

A. Y. S. Lam (✉)
Fano Labs and The University of Hong Kong, Hong Kong, China
e-mail: albert@fano.ai; ayslam@eee.hku.hk

C. H. T. Lee (ed.), *Emerging Technologies for Electric and Hybrid Vehicles*,
Green Energy and Technology, https://doi.org/10.1007/978-981-99-3060-9_11

The compelling EV adoption allows reducing our dependency on fossil oil. EVs get charged from different energy sources. While some energy is generated by independent renewable energy sources, a large portion is acquired through the power grid. On the other hand, EVs can be considered as moving batteries and they can supply energy to the grid at appropriate locations. Here comes the Vehicle-to-Grid (V2G) system [2]. V2G refers to a system that allows EVs to communicate with the power system and provide demand response services. This can be done by discharging excessive energy of EVs to the grid or by charging the EVs with excessive power from the grid. The core idea is to maintain the grid with appropriate power levels with the assistance of EVs. When compared to the size of energy shortfall and excess in the grid, the amount of contribution from a single EV is insignificant. To make the system practical, V2G usually involves a fleet of EVs of considerable size. How to organize and manage the EVs makes a huge impact on effectiveness of V2G.

V2G presents in the form of EV aggregation. As constituted by independent entities, a mature system is always on with swift response. In other words, it is easy to gather a number of EVs to support the grid by contributing their batteries in a short period of time. V2G is particularly good for providing ancillary services to the power grid, maintaining a reliable electricity system. Some examples of V2G applications include frequency regulation, voltage regulation, unit commitment, and energy trading, especially in the presence of renewable energy generation. As renewables are intermittent, their integration creates uncertainties in the grid, which can be leveraged by the autonomous and swift nature of EVs.

In a smart city, technologies of different sectors are elevated with support of information and communication technologies and their interaction enables synergetic effects for the good of citizens. EVs set an important milestone in vehicular technology, which is continuous to evolve. Some new forms of vehicles, which are also EVs, are enabled with other advanced technologies. One example is the autonomous vehicle (AV), which can drive itself without active human involvement. The autonomy of AVs can upgrade V2G to a new extent with more functionality and capacity. While we usually assume that EVs need to park stationarily in order to participate in a V2G system, there exists a new vehicular system design, which allows moving EVs to interact with the power system, in a way aligned with V2G. These new inventions help make V2G smarter and integrate V2G into the roadmap of smart city development.

In this chapter, we investigate the recent advancement of V2G in terms of architecture, applications, and smart city integration. In Sect. 2, we will introduce some fundamentals of V2G, followed by different V2G architectures in Sect. 3. Section 4 describes several V2G applications and we will investigate how new technologies bring V2G to a new dimension in a smart city in Sect. 5. Section 6 concludes this chapter.

2 Fundamentals of V2G

a. Why V2G

There are several reasons why V2G was proposed and is getting popular:

(i) Decarbonization

The conventional internal combustion engine vehicles rely on fossil fuel to provide energy for operation. As mobility is essential to modern civilization, transportation becomes one of the largest greenhouse gas emitting sectors, accelerating global warming. Most of the energy used to charge EVs is provided from the grid, in which the smart grid initiative promotes high penetration of renewable energy sources in electricity generation [3]. In fact, EV batteries are the most cost-efficient storage to store the renewable energies. We can see that the increasing adoption of EVs can in turn reduce our need of fossil fuel and thus reducing carbon intensity. Moreover, traditionally online generation is used to provide ancillary services. As EVs move around the city, it is relatively easy to gather a sufficient number of EVs at the bus of power network where imbalance between supply and demand takes place. With V2G, EVs become a handy alternative to maintain proper power flow and to recover from power system failures.

(ii) Energy efficiency

Usually, a power system covers a large geographical area. Buses are connected with power lines and each bus provides services to its neighborhood. To avoid unnecessary duplicate of system resources, the power system is not a fully connected but sparse network technically. It is common to transmit power from a source to a destination via multiple hops. The power lines have impedance, which induces power loss. In general, the longer the length of the power lines, the more power is lost. This phenomenon is also true for providing ancillary services, in which the service providers are located in some fixed locations. However, it is possible that any bus in a power system needs services and the providers can be far away. The unnecessary power loss can be reduced if the ancillary service providers can be found nearby. As EVs are mobile, a V2G system can basically be constructed anywhere. We can make use of the closest V2G to support a problematic bus so as to curtail the power loss.

(iii) Electrification

Besides transportation, V2G gives EVs another purpose, i.e., supporting the power grid with their internal batteries. Instead of a pure consumer product, an EV can be used to make profit. The V2G purpose only happens when the EVs are idle and their owners do not need to satisfy their original use of EVs. This encourages the adoption of EVs and electrification of automotives. As electricity is the primary energy source, EVs are made more energy efficient. This in turn suppresses the use of fossil fuels and contributes to environmental protection.

b. Types

V2G is a general term describing an interaction of EVs with the grid. It can be further classified into three types based on the direction of power flow and the scale of grid connectivity:

(i) Unidirectional V2G

Unidirectional V2G is also known as smart charging or V1G. As the name implies, power is only allowed to flow in one direction, from grid to vehicle in this case. We intelligently alter the duration and rate when charging an EV. When applied to a fleet of EVs, power can be drawn from the grid and distributed to the EVs. In this way, the excessive power from the grid can be transferred to the EVs. No discharging of EVs is taken to provide power back to the grid.

(ii) Bidirectional local V2G

This type involves the power interaction between EVs and a closed environment, like a residence, a building, or any other infrastructure, which are also called vehicle-to-home, vehicle-to-building, and vehicle-to-everything, respectively. EVs are used to take up the energy produced on-site, e.g., from renewables, and serve as residential power backup in power outage. No power involved is engaged with the grid.

(iii) Bidirectional V2G

In this type, EVs are employed to support the grid, instead of merely a local environment. The amount of power is usually larger than that required in bidirectional local V2G and thus more EVs are involved. The grid operator makes use of the batteries in EVs to provide ancillary services or acquire energy from EVs in peak demand periods. Without mentioned otherwise, V2G will be referred to this type throughout this chapter.

c. **Implementation by Country**

Here we give some examples of actual V2G implementations in several countries. The following is not exhaustive but to give some ideas of the current development and progress happening in different parts of the world:

- In the United States, at least two school bus producers are electrifying school buses with V2G technology [4]. 50 V2G bidirectional charging stations were built in the campus of the University of California, San Diego [5]. The bidirectional EV charging system developed by Fermata Energy was the first to be certified to the North American safety standard UL9741 in 2020 [6].
- Energy companies EDF Energy and Nuvve drove a campaign to install 1500 V2G charging stations in the United Kingdom in 2018 [7]. This setup facilitates an additional energy storage capacity of 15 MW, which is good enough to support the power required for 4000 households. The stored energy can be traded in the energy market and to enhance the grid flexibility during the peak usage hours.
- In Japan, Toyota Tsusho Corporation and Chubu Electric Power jointly demonstrated V2G charging and discharging on EVs and plug-in hybrid vehicles in 2018. Two V2G charging stations have been set up at a parking infrastructure in Toyota City to assess the EV capacity of power balancing between EVs and the grid [8].

3 Architecture

In this section, we will first introduce different system components and then the conventional V2G architecture. After that, we will discuss some new V2G designs.

a. **General architecture**

The general V2G system model is illustrated in Fig. 1, which consists of a power grid with power generation, and aggregators to accommodate the contributing EVs.

(i) Power system

In general, the power system, or the power grid, is composed of generators, the transmission network, distribution network, and the loads [9]. The generators supply power to the system, ranging from the traditional power plants, which generate electricity from primary energy, to renewable energy sources, e.g., solar and wind farms. The transmission network brings electrical energy in high voltage from a generator to a substation, in which the voltage is transformed from high to low. The distribution network distributes power from a substation to a load in low voltage. A load can be a factory, commercial/residential building, or even an EV. Multiple distribution networks are connected to the transmission network to acquire power from the generation. In the traditional power system, the power is flown from bulk generators in the transmission network to the loads in the distribution network. With the advent of smart grid [10], the power flow can be bidirectional. In other words, it is possible to have distributed energy sources in the distribution network to supply power to other

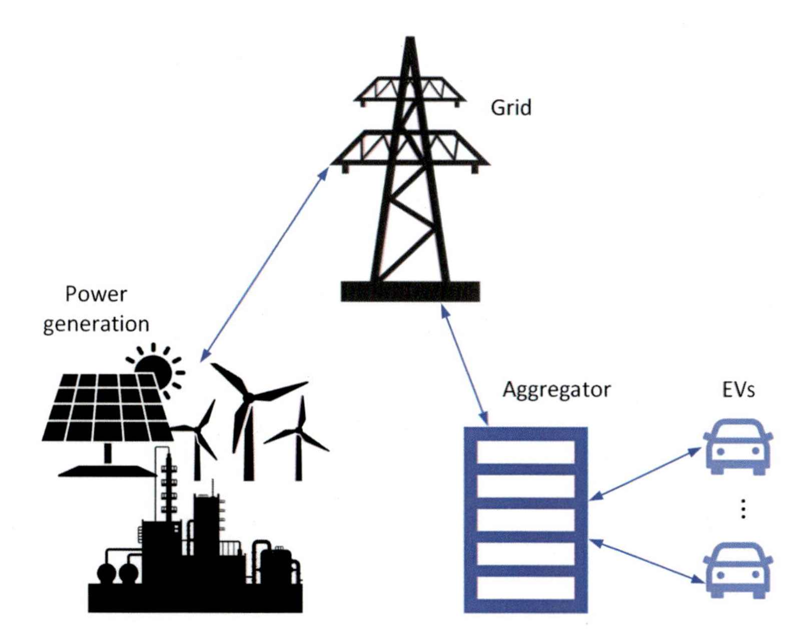

Fig. 1 General V2G model

loads in the same distribution network, or even to serve the loads in another distribution network via the transmission network. This encourages distributed renewable energy generations to contribute to the grid or other new inventions, like V2G, which were impossible in the traditional power system.

(ii) Electric vehicles

EVs can be generally classified into battery EVs (BEVs), hybrid EVs (HEVs), and plug-in hybrid EVs (PHEVs). A BEV is solely motorized by an electric battery. An HEV acquires its energy externally from gasoline but it has a battery to store energy gained through regenerative braking. An HEV cannot be plugged to recharge from a charging facility. A PHEV is similar to HEV but its battery can be recharged with a charger. The connection between a vehicle and a charger also exists in multiple forms. The majority of EVs get charged through plug-in cables. Nowadays, (static) wireless charging is getting popular in which an EV is charged automatically when being driven over a charger without the need of cable. Dynamic wireless charging is a new form of wireless charging which allows an EV to get charged on the move. When an EV supports V2G, it is capable of charging and discharging its energy from and to the grid via a cable or wirelessly. In this chapter, without loss of generality, we consider an EV a vehicle equipped with a battery, in which the energy can transferred back and forth with the grid in any connection form.

Each EV has its own battery of different sizes, from 17.7 kWh of Smart EQ for four to 107.8 kWh of Mercedes EQS [11]. Its state-of-charge (SoC) indicates the amount of energy stored in its battery normalized by its maximum capacity. The charging rate depends on the equipment used at the charging facility. The charging time may be required as little as half an hour or half a day, based on the battery size and charging rate. With a 50kW rapid charger, energy equivalent to 100 miles of range can be charged for many EV models [12].

(iii) Aggregator

In V2G, EVs interact with the grid to provide those services (see Section 4 for more details) which become more convenient with the presence of EVs. As the battery size of a single EV is very small, when compared to the power need at the grid, the contribution of a single EV is negligible. Even when there are large number of EVs connected to the grid simultaneously, without proper coordination, the total amount of contributive power from the EVs can still be far from the required power level. For example, some EVs get charged while some are discharging. The contributions of certain EVs may be canceled out each other and net contribution by the EV fleets may become smaller than expected. As another example, in a certain period of time, some EVs connect to the grid earlier while some connect later. For some time-sensitive V2G applications, the amount of EV contribution precisely at a certain period of time may not satisfy the power requirement at the grid. Therefore, in order to make V2G effective, we need to coordinate a certain number of EVs with appropriate charging and discharging behaviors so that the aggregate contributive power of the EVs meet the power level required by the grid. The power system per se is a very complex system and it implements many measures to address different (con-)contingency situations. From the traditional power system perspective, an EV is simply a load and many EVs are logically considered as many loads. EVs are traditionally not considered to be a core part of the power system and there is no control on the grid side to manage a fleet of EVs as loads. An aggregator acts as the middleperson between the EVs and the grid and we implement different control algorithms at an aggregator to control the charging and discharging behaviors of the governed EVs. A parking facility is a natural location to act as an aggregator for V2G accommodation.

b. **Single-aggregator architecture**

The general system model given in Fig. 1 follows the single-aggregator architecture. In this architecture, only one aggregator is employed to coordinate the fleet of EVs for V2G support. When the fleet size is small, centralized control at the aggregator is sufficient for the EV management [13]. However, the computational complexity of the control algorithms grows exponentially with the number of EVs. When the EV population becomes large, distributed control mechanisms are developed to overcome the scalability issue [14, 15]. Each EV takes up a part of the computation and the aggregator turns to coordinate and combine the distributed EV computations to provide an orchestra effect. However, these distributed algorithms [14, 15] require iterative message passing between the aggregator and EVs and they may suffer from enormous communication delays.

c. **Hierarchical architecture**

When it comes to a large geographical region, we need to have multiple aggregators, each of which corresponds to a specific location managing its own local EVs. Then we have some higher-level controllers to coordinate the local aggregators with the top level connecting to the grid. A hierarchical architecture is designed with multiple levels of aggregators to coordinate large-scale EVs [16, 17]. [18] generalizes the idea and proposes a general tree-like structure depicted in Fig. 2. Similar to the general V2G architecture, this model consists of an EV fleet, a set of V2G aggregators, and a grid operator. There are multiple levels: EV, charging station, district, and grid levels. EVs are at the lowest EV level and they are connected to some aggregators, called EV aggregators (EVAs). As those aggregators are naturally realized by charging stations, they form the charging station level. With one level up, called the district level, the EVAs are grouped together based on proximity and other economic and technical factors. Another type of aggregator, called aggregator of aggregators (AoA) is used to coordinate certain EVAs. A possible AoA is a distribution substation in the district level. Be noted that there exist multiple sub-levels of AoAs connected together constituting an aggregator structure of AoAs in the district level. In other words, an AoA can be used to coordinate both EVAs and AoAs in the sub-levels underneath. On the top grid level, the grid operator serves as the point of operation between the V2G and the grid. It works with AoAs or even directly with EVAs when they are close to the grid operator and the need of AoA is unnecessary. It is possible to have multiple grid operators in the power system, each of which works with its connected hierarchical V2G structure. In this case, each hierarchal V2G structure operates independently.

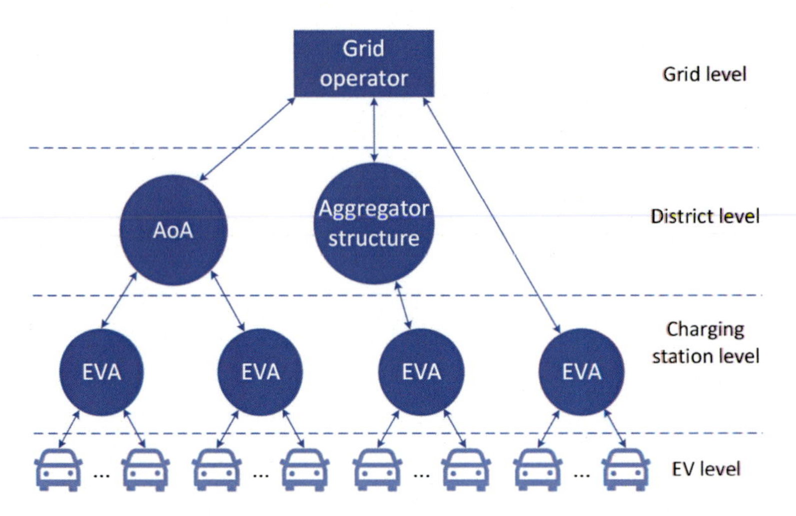

Fig. 2 Hierarchical V2G

d. **Operation protocols**

V2G aims to provide services to the grid with the governed EVs. The grid operator given in Fig. 2 is the only entity representing the grid in the V2G architecture. The grid operator conveys the service requests to the aggregators, each of which in turn needs to coordinate with its contributing EVs for the service commitment. An entity needs to communicate with one other and exchange information. All these are achieved with some kinds of operation protocols.

An operation protocol sets the procedure for the entities to achieve the centralized V2G scheduling objectives. As EVs are dispersed relative to the fixed aggregators, long-distance communication is needed for data exchange. It is realized through the modern communication technologies. For example, [19] makes use of software-defined networking technology to overcome the problems due to high mobility of vehicles and high volume of data transmission required. An operation protocol also facilitates the participation of new aggregators in the system without unnecessary restructuring. We consider the general hierarchical structure given in Fig. 2. An entity only communicates with its neighbors to simplify the design. Based on the parties involved, there are three types of operation protocols: the grid operator-aggregator protocol, the aggregator-aggregator protocol, and the aggregator-EV protocol [18]. Basically, a V2G scheduling process is composed of two phases, i.e., information collection phase and V2G commitment phase. The information collection phase is a bottom-up process, in which a child node proposes its contributing V2G capacity, i.e., the amount of power it can provide for V2G, to its parent node. For an EV, its V2G capacity is its power limit decided by its maximum charging and discharging rates, SOC, and other system requirement. For an EVA or AoA, its V2G capacity refers to the upper and lower power limits aggregated from its subordinate EVs or aggregators. The information collection phase is followed by the V2G commitment phase, which assigns the power commitment for each entity based on its capacity in a top-down manner. The details of the protocol designs can be found in [18].

4 Applications

V2G offers a series of applications, which becomes more convenient with V2G than the traditional approaches. V2G also facilitates some new applications which may not be possible without the participations of EVs.

a. **Frequency regulation**

A proper power system should maintain its operating frequency close to its nominal value. e.g., 50 Hz in Europe and most parts of Asia, and 60 Hz in North America. Operating at a frequency away from the nominal frequency can damage the connected generators, equipment, and other infrastructure. Further frequency deviation may cause partial or full blackout in the power system. Frequency shift is mainly due to the imbalance of power supply and demand. When the generated power is larger than the demand, the system frequency will be driven higher than the nominal value. This may happen when there is a sudden rise of power production from renewables or a sudden drop of power consumption collectively, e.g., due to weather change. Similarly, if the power generation is smaller than the load, the system frequency will get lower than the nominal setting. This may take place due to a sudden turn-off of a power plant or a sudden surge of power demands. In a power system, the various generations and loads may be on and off throughout the day. Although the generations are more controllable, it is almost impossible to just rely on the traditional generators to match the power demand at the loads, which are highly unpredictable. When it comes to the smart grid, the renewables are heavily involved, making the generations more unpredictable, and thus further worsening frequency deviation.

We strive for a persistent system frequency by balancing the generations and loads at all times. Frequency regulation is a measure to always ensure power balance, so as to maintain a constant system frequency, in a short period of time, from seconds to minutes. Traditionally, an over-frequency event can be easily handled by reducing the amount of power out at the generators. An under-frequency event can be addressed by spinning reserve or battery energy storage system (BESS) [20]. The former refers to standby generation spinning at grid frequency and it can become online within minutes to bring extra power generation to serve the unexpected loads. The latter is a rapid reserve that becomes active from a standby mode within milliseconds. However, maintaining spinning reserve or BESS, which are very often being standby, is relatively not economical.

Each EV is equipped with a battery, which primarily serves as an energy storage to support various vehicle operations. We take an EV to commute. Besides those used for public transport or for commercial use, it is not hard to imagine that most vehicles are idle most of the time throughout the day. Ideally, a collective of batteries from idle EVs nearby form a BESS. It is handy to connect all these EVs to provide frequency

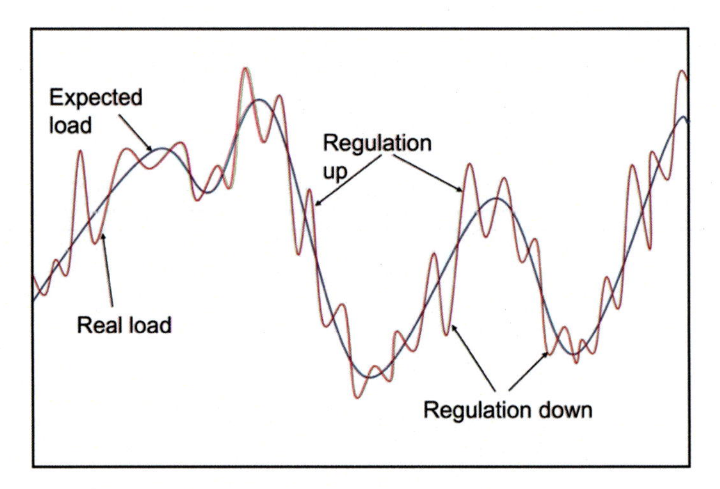

Fig. 3 Regulation-up and -down

regulation service through V2G. Many works demonstrated how to configure a V2G in a centralized or distributive manner, e.g., [21–23]. Most assume that EVs are willing to participate and their schedules of participation are all known. However, V2G is a dynamic system, in which EVs join and leave independently. To make V2G more reliable, it is important to tell how much power a V2G can support for both regulation-up and -down incidents during the day. Regulation-up provides extra power for the power shortfall in the grid while regulation-down refers to the absorption of excessive power from the grid. They are illustrated in Fig. 3 in which the generation follows the expected load to provide power to the grid. Determination of regulation-up and -down capacities for a V2G can facilitate setting up a frequency regulation contract between an aggregator and the grid operator.

In [24], both capacities for regulation-up and regulation-down of a V2G system for an aggregator is estimated by the queueing theory. As shown in Fig. 4, an aggregation of EVs is modeled with a queueing network, in which the regulation-down queue (RDQ), regulation-up-and-down queue (RUDQ), and regulation-up queue (RUQ) are connected to one another. Consider an EV i gets parked in a parking facility, gets charged during the stay, and participates in V2G. The arrival of an EV follows a Poisson process at rate λ. One participates in which queue or moves from one queue to another depends on its SoC $x_i(t)$ at time t. Let $\underline{x}_i$ and $\overline{x}_i$ be the lower and upper limits of the target SoC of EV i. If $x_i(t) \leq \underline{x}_i$, EV i will enter RDQ and only participate in regulation-down since the EV can take up extra energy from the grid. If $x_i(t) \geq \overline{x}_i$, EV i will enter RUQ and only participate in regulation-up since it can supply instant energy to the grid. If $\underline{x}_i < x_i(t)\overline{x}_i$, the EV will enter RUDQ and support both regulation-up and -down simultaneously. During its stay, the EV gets charged and $x_i(t)$ increases with time according to the charging rate and it will move to another queue according to the progress its SoC. It is allowed for an EV to quit the system at any time and thus it will leave its participating queue. Let q_1

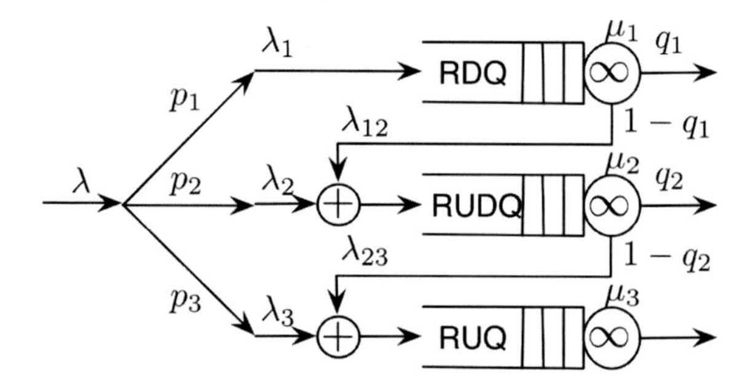

Fig. 4 A queueing model for regulation-up and -down capacity estimation (adopted from [24])

and q_2 be the probabilities of leaving RDQ and RUDQ, respectively. Assume that an EV joins RPQ, RUDQ, and RUQ with probabilities p_1, p_2, and p_3, respectively. The durations of the stay in the three queues are exponentially distributed at rates μ_1, μ_2, and μ_3, respectively. All the parameters $\lambda, p_1, p_2, p_3, q_1, q_2, \mu_1, \mu_2$, and μ_3 can be deduced from the historical data of the parking facility. From [24], we can analytically determine the expected numbers of EVs in RDQ, RUDQ, and RUQ to be

$$L_1 = \frac{p_1\lambda}{\mu_1}$$

$$L_2 = \frac{\lambda(p_1 + p_2 - p_1q_1)}{\mu_2}$$

$$L_3 = \frac{\lambda(1 - p_1q_1 - p_1q_2 - p_2q_2 + p_1q_1q_2)}{\mu_3}$$

Respectively. By fixing P_{EV} to be the contribution of an EV in a regulation event, we can determine the regulation-down and regulation-up capacities in the steady state to be $P_{EV}(L_1 + L_2)$ and $P_{EV}(L_2 + L_3)$, accordingly. This queueing model can also be easily generalized to other V2G applications given in the following.

b. **Voltage regulation**

Electronic equipment is designed to be operated within a prescribed voltage range. Deviating from the operating voltage worsens the equipment performance and serious derivation may further damage the electronics therein [25, 26]. In a power system, power is transmitted from the source to load over power lines. For an ideal power line with zero resistance and reactance, there is no change of voltage magnitude between the two ends in the power transmission. However, it is not the case in reality, especially in the distribution network, in which the line conductors have low reactance-to-resistance ratio. Long distribution lines may significantly bring down the system voltages. Different load conditions, e.g., utilization of heavy electric products, may also drag down the voltage level. Some devices can actively intervene the system voltage. For example, on-load tap changers (OLTCs) can be installed at substations to adjust the voltage of the winding terminals. Distributed generator inverters can also provide reactive power support for voltage control [27]. In the smart grid, the distributed renewable energy sources (RESs) impose significant impact on the system voltage. The intermittency of renewables brings voltage variations and deteriorates power quality [28]. They may cause reverse power flow back to the substations, harming the electric utilities [28].

Voltage regulation is the measure of how to maintain the near constant voltage under different load conditions. Similar to frequency regulation, EVs can serve as distributed power sources and storage to supply instant power and absorb excessive power, respectively. Thus V2G is promising to realize voltage regulation when renewable energy sources are widely integrated into the power systems. We can implement voltage management by coordinating EVs with OLTCs [29]. We may control the charging strategy of multiple EV charging stations by satisfying certain power flow and bus voltage constraints based on distributed model predictive control [30]. In these implementations, it is hard to control the parking schedules of EVs, which impose unavoidable impact to the effectiveness of the solutions. It is possible to encourage EV participation by prioritizing the owners' concerns and preferences. [31] developed economic and control mechanisms for EVs to mitigate voltage violation in low-voltage power distribution network.

The more predictive of EVs, the more reliable V2G is. With the advance of artificial intelligence (AI), more EVs are built with self-driving capability and they are known as autonomous vehicles (AVs) or driverless cars. AVs can be actively involved to improve V2G performance. [32] proposed a method to coordinate the charging and parking schedules of a fleet of electric AVs to mitigate voltage fluctuations in a distribution network. AVs are allocated to appropriate charging facilitates to facilitate power exchange between the grid and charging facilities for over- and under-voltage relaxation. The problem is formulated as a mixed integer quadratic program. We confine the voltage violation by minimizing $\sum_{i,t}\left(V_{i,t}-1\right)^{T}\left(V_{i,t}-1\right)$ to regulate the voltage to 1.0 p.u., where $V_{i,t}$ is the voltage at bus i at time t. We bound the reactive power of RESs by the apparent power capability and real power generation and set limits of dispatch power at substations. The aggregated power of AVs at bus i is the total AV power injection at parking facility f. [32] involves new technologies, namely AVs, which are not common found in a typical power system or smart grid nowadays. We will further explore how to exploit new technologies in a smart city in Sect. 5.

c. **Unit commitment**

In electrical power generation, a set of power generators are coordinated to produce a right amount of power for the loads. Due to their geographical distribution, each generator incurs different cost of power generation, e.g., due to different cost of raw materials, extra power to produce for power loss during transmission, etc., to serve a particular location. For excessive generations, storing unused power in energy storage is not economical. Thus the best is to generate the amount of power at the generation just sufficient for the demand together with the power loss during transmission. The unit commitment problem is about determining the on–off schedule for a set of power generators to strive for minimum cost or maximum revenue in power generation.

When a huge population of EVs are attached to the power system, their simultaneous charging and discharging activities will significantly impact the schedule of power generation at generators.

We can address the unit commitment problem as a constrained optimization. We may maximize the energy production profits or minimize the energy production costs, and impose constraints on reliability and emissions, generation limits, must-run and must-off requirement, power balance, spinning reserve, minimal uptime and downtime, and ramp rate limits. To emphasize EVs' impact, V2G operation is incorporated into the model [33] and we include additional constraints related to EVs, e.g., battery capacity limit and charging frequency limit. Due to the non-linearity, the resultant EV-based unit commitment problem is non-convex, and heuristic approaches are commonly adopted for problem solving. For example, [34] made use a nature-inspired metaheuristic called Chemical Reaction Optimization [35], giving outstanding performance.

d. **Energy trading**

In the previous discussion, EVs are generally assumed to participate in V2G ideally. In reality, in order to take part in V2G, an EV needs to contribute its battery capacity and energy at some costs explicitly and implicitly. One major concern is that charging and discharging a battery may accelerate its depreciation and shorten the battery life. Although some research indicates that shallow change of SoC, which usually happens in V2G, is beneficial to battery life [36], most EV owners are not willing to contribute their cars without clear incentive. To motivate EV participation, monetary incentives on energy exchange for EVs foster a new energy market. In a typical V2G system as shown in Fig. 1, EVs and the grid operator are the service providers and the service consumer, respectively. They in turn become the sellers and buyer in the induced energy market. In [37], a multi-layer architectural design was proposed and a double action mechanism was developed to facilitate energy trading. The mechanism is shown to be strategy-proof and converges asymptotically. Topology of active distribution system in terms of microgrids were also considered in the auction energy trading mechanism [38]. [39] formulated energy trading as a large-scale EV charging problem for aggregator profit maximization. These examples illustrate the establishment of a new energy market for V2G, which enables many opportunities for business and energy management.

5 Smart City Integration

We have examined how EVs and V2G contribute to the power system in the previous sections. In this section, we investigate how the concept of V2G is extended to a smart city in the vehicle's perspective. A smart city makes use of modern information and electronic technologies to enhance various city-wise operations, in the areas of energy, mobility, government, environment, etc. EVs naturally bridge smart energy and smart mobility together. When it comes to vehicles with autonomous control,

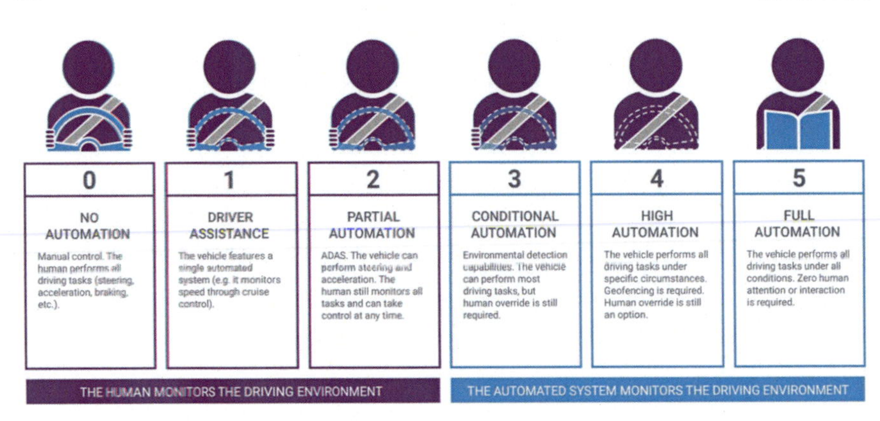

Fig. 5 Levels of driving automation (adopted from [41])

the power and functionality of V2G can be further enhanced. There exists a new form of energy network, called Vehicular Energy Network (VEN), whose functions are highly related to V2G, with more capacity and flexibility. We will review how autonomous vehicles and VEN integrate into the smart city in the V2G perspective.

a. **Autonomous vehicle**

An AV is also known as a driverless car, self-driving car, etc. and it can drive itself without active human intervention with the support of AI. According to the Society of Automotive Engineers, there are six levels of automotive autonomy, ranging from zero automation to full automation with optional steering wheel [40], which are illustrated in Fig. 5. To bring the full capacity of AV, we assume Level-5 autonomy thereafter.

As discussed in Sect. 4a, V2G suffers from uncertainty due to driver behaviors and thus the capacity of a V2G is hard to be guaranteed. It is important that a V2G provider gets paid for the service guarantee even though no actual power transmission between the grid and V2G has taken place. This is known as capacity payment [42]. In general, an EV participates in V2G only when it parks. One normally parks a car in the nearest available parking spot relative to a specific destination. As the main purpose of V2G is to provide the energy needs of the grid from EVs, the energy drive is from the grid while the EVs play a relatively passive role. An aggregator is highly tied to a parking facility, it is normally difficult to control a sufficient number of EVs parked at a particular parking facility to fulfill the timely energy need of the aggregator. Different parking facilities have diverse V2G objectives, e.g., the applications discussed in Sect. 4, resulting in different EV demands at different times. The question becomes how to allocate appropriate numbers of EVs to a given set of parking facilities at a specific time.

After finishing dropping off the passengers at the destination, an AV is basically free to go and park per se. There may be certain parking requirements, e.g., related to range, car owner 's preference, etc., for specific cars but an AV has more flexibility for its actual parking location. In a system-wise point of view, we can allocate a right number of AVs with parking intention to a parking facility according to the energy need of the aggregator subject to certain constraints.

In [43], the coordinated parking problem (CPP) for AVs to support V2G was formulated. A control center is supposed to coordinate the AV parking for a number of parking facilities by sending control signals to the vehicles by means of vehicular communication technologies. The time horizon is given in time slots $\{t = 1, ..., D\}$. We coordinate the set of AVs $\mathscr{R}$ by $t = 0$. CPP is formulated as an integer linear program with binary variables x_{kt}^{f} which indicates if AV k is allocated to Parking Facility f in time slot t. The constraints are defined as follows:

- An AV will be allocated to one parking facility in the time horizon, assuming D is good for a few hours during the day;
- An AV will take time to travel from its original location to its assigned parking facility f before parking and to traverse to another designated location after parking. We should reserve sufficient time for the AV on these two legs of journey.
- An AV will stay at the assigned parking facility for a certain number of time slots sufficient for charging the vehicle itself and for supporting V2G;
- An AV should be allocated to a parking facility within its tolerable travel range;
- An AV is only specified to be parked in $(\underline{t}_k, \bar{t}_k)$ only. It should not be assigned to any parking facility out of this period.
- A sufficient number of AVs are secured to be parking at f based on its demand profile p_f, which indicates the number of vehicles required to support V2G in the time horizon.

CPP is formulated by maximizing $\sum_{k\in\mathcal{R},t\in\mathcal{J},f\in\mathcal{F}} x_{kt}^{f}$ subject to all the constraints listed above. As an integer linear program (ILP), the problem can be solved by a standard ILP solver or the distributed algorithm presented in [43].

AVs can also be coordinated to form a public transportation system, in which AVs are assigned to systematically satisfy transportation requests with riding sharing capability [44]. It provides ride-hailing services (See Fig. 6). By employing AVs, the vehicle travel itineraries can be well-planned such that an AV can be employed to serve multiple groups of passengers with different time and location requirements. The AVs can be arranged so as to maximize the chance of ride-sharing and to minimize the transport resources. When an AV is not engaged in service, it should be parked in an area with high potential service demand to reduce the waiting time for passengers and this is called AV rebalancing. Similar to [43], the idle AVs can be parked at locations to support V2G. In [45], a joint rebalancing and V2G coordination strategy for AV-based public transportation system was proposed. The total arrival time for the AVs is minimized to provide services subject to some rebalancing and V2G requirements. The problem is an ILP, which is addressed by a heuristic based on Genetic Algorithm and Model Predictive Control in low time complexity [45].

Other than serving passengers, AVs can also be employed to convey goods from transportation hubs to destinations for last-mile delivery in a supply chain. [46] presented an autonomous vehicle logistic system to accommodate logistic demands for a smart city. It focuses on determining optimal routes to guide AVs to get charged at

Fig. 6 **Ride**-hailing and ride-sharing services

appropriate locations with distributed renewable generations while accommodating logistic requests. The problem is formulated as a quadratic-constrained mixed integer linear program, which is solved via dual decomposition [46]. The system can be easily modified to support V2G. Instead of merely utilizing the excessive energy at the distributed renewable generations, the charging constraints can be replaced by the V2G-related constraints adopted in [45]. As both types of constraints are linear, the adoption of the V2G constraints will not elevate the difficulty of the optimization problem, which can be tackled by similar techniques given by [45] and [46].

b. **Vehicular energy network**

Recall that V2G aims to stabilize the power system by providing or acquiring instant energy from a fleet of EVs. The EVs are generally assumed to be parked at some suitable locations to provide aggregate energy effects with size relative to the need in the power system. Technically speaking, the definition of V2G does not require the EVs to be stationary. As long as there is supportive equipment, the EVs can still function like V2G on the go. [47] proposed a vehicle system, VEN, and it realizes a similar idea for moving EVs to complement the grid.

Wireless power transfer (WPT) allows a vehicle to get charged wirelessly and dynamic wireless charging elevates the charging mechanism by supporting the charging of moving EVs [48]. By near-field electromagnetic induction, charging and discharging can be realized when a supported EV is passing by a road segment with WPT track laid underground, as shown in Fig. 7. VEN is built upon the transportation network in which multiple WPT tracks are constructed at certain road segments. Consider an EV that picks up an "energy packet," i.e., wirelessly charged with a certain amount of energy, when it passes by a location (see Fig. 8). It then carries the energy packet along its route and discharges the energy packet at another location. Since the WPT happens at the background, the driver does not need to do anything intentionally to support the charging and discharging mechanisms. Consider there exists a control center to coordinate VEN. With advanced vehicular communication technologies, the control center can acquire the travel itinerary of each participating EV and thus the present and near-future traffic flows contributed from the EVs can be determined. Without altering their travel plans, we just need to assign charging and discharging activities on the appropriate EVs at the right locations. It is possible that an energy packet is being carried by multiple EVs along multiple routes undergoing several charging-charging cycles before it can reach the desired location. In this way, we can systematically utilize EVs, as energy carriers, to transport energy over a large geographical region. By having certain locations on VEN as energy sources and destinations, energy can be transmitted over VEN in terms of energy packets conveyed across the region by moving EVs.

We can consider VEN as a huge imaginary energy buffer spreading over a large region. Since both the power system and road infrastructure aim to serve people for daily activities, their service coverages are more or less similar. When there is excessive energy at a bus of the power grid, VEN can take up the energy at a nearby location over the transportation network. Similarly, when an energy deficit takes place, VEN can supply energy to the bus with energy needs. A mature VEN can realize energy transmission for many energy sources and destinations simultaneously and charging and discharging events happen all the times in VEN. Energy is a commodity without identity. In other words, the energy from one source is used in the same way as that from another source. VEN can be designed to be flexible in the sense that energy can be withdrawn in advance of depositing energy at a source or energy can be bumped into VEN before fixing an energy destination. An example is to connect renewable energy sources to VEN. The EVs constituting VEN can always consume

Fig. 7 Dynamic wireless charging (adopted from [49])

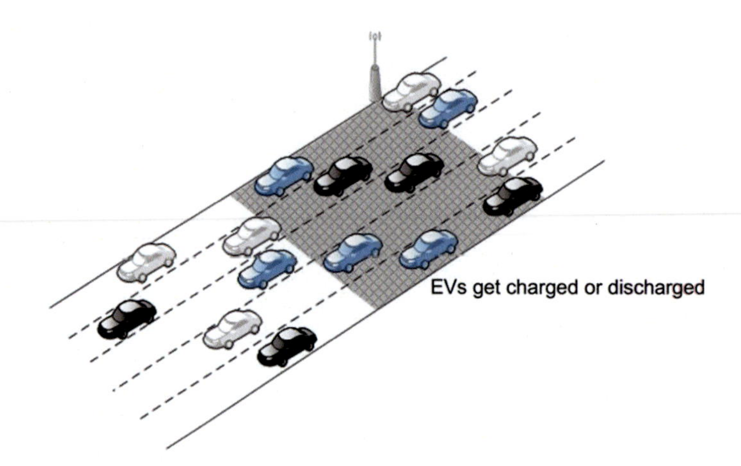

Fig. 8 VEN assigns charging and discharging events along appropriate vehicular routes. EVs of different colors stand for different routes (adopted from [50])

"free" energy for their own use. This exactly reveals the core objectives of V2G. The main difference is that we no longer require the EVs to be parked stationarily to make up V2G.

VEN has a layered architecture, illustrated in Fig. 9. The lowest layer is the road network modeled as a directed graph G(N, A), where N and A stand for the set of road junctions and road segments connecting the road junctions. We construct a vehicular network over the road network. In the vehicular network, a vehicular route $r_i = \langle a_1^i, \ldots a_{r_i}^i \rangle$ is a sequence of road segments, where its j-th segment is given by $a_j^i \in A$. $r_i(n, m) = \langle a_1^i, \ldots a_m^i \rangle$ is the sub-route of r_i connecting the n-th segment to m-th segment. We build an energy network on the vehicular network. In this layer, we define a set of energy sources and destinations as $N_s \subset N$ and $N_d \subset N$, respectively. We connect an energy source $s \in N_s$ to an energy destination $t \in N_d$ via a set of energy path $P(s, t)$. The j-th energy path $p_j(s, t) \in P(s, t)$ is a sequence of segments of vehicular routes as $p_j(s, t) = \left\langle r_1^j(n_1, m_1), \ldots, r_1^j(n_i, m_i), \ldots, r_{|pj|}^j\left(n_{|pj|}, m_{|pj|}\right)\right\rangle$, where $r_i^J(n_i, m_i)$ is the i-th segment of $p_j(s, t)$. In this way, we can transmit energy from s to t via $P(s, t)$.

How to construct the energy paths is called opportunistic routing as a path is implicitly built upon EVs opportunistically. [50] presented a method to construct the energy paths connecting a source to a destination. [51] extended the work of [50] to multi-source multi-destination routing in several variants of VEN of static and dynamic traffic. It takes time for an EV to go along a road segment, which incurs a delay $d(a_k)$ on segment a_k. The propagation delay of $p_j(s, t)$ is given by $d(p_j) = \sum_{i=1}^{|p_j|} \sum_{a_k \in r_i^j(n_i, m_i)} d(a_k)$. The energy transmission rate g_j of $p_j(s, t)$ is governed by $g_j \leq w\, f_i^J$, $i = 1, \ldots, |p_j|$, where w is the energy packet size and f_i^j is the EV flow rate of the i-th segment of $p_j(s, t)$. Over a time window T, the amount

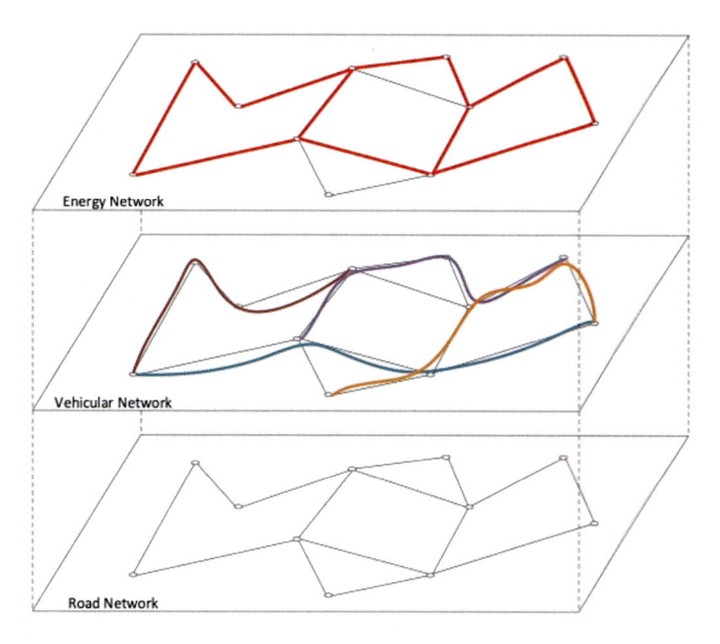

Fig. 9 Layered structure of VEN

of energy x_j transferable along $p_j(s, t)$ is subject to $x_j \leq (T\text{-}d(p_j))z^{|p_j|}g_j$, where z is the energy efficiency in a charging-discharging cycle. Therefore, the total amount of energy $x(s, t)$ transferred from s to t is the sum of energy along all paths, given by $x(s,t) = \sum_{j|p_j \in \mathcal{P}(s,t)} (T - d(p_j))z^{|p_j|_{g_j}}$ and the corresponding energy loss is $L(s,t) = \sum_{j|p_j \in \mathcal{P}(s,t)} \left(\frac{1}{z^{|p_j|}} - 1 \right) x_j$. In this way, we can maximize the energy transfer with the energy transfer maximization problem given in [47]. We can also minimize the total energy loss by solving the energy loss minimization problem given in [52].

The energy path construction discussed in [50] is done based on the assumption that we have perfect information about the traffic flows. This is achievable if we have full control of the vehicles (e.g., as AVs) or the advanced vehicular communication technologies. As the EVs are driven by humans, the traffic flows may deviate from the expected values due to human-induced errors or other conditions. [53] proposed a robust routing scheme based on robust optimization. It shows that the design can cope with the sensitivity of traffic uncertainties without impairing the system efficiency significantly.

6 Conclusion

EVs are getting prevalent in the city. They get charged from the power grid to support their mobility and they can be considered as moving energy storage. Most vehicles are set idle most of the time during the day. An aggregation of these idle batteries become a huge energy resource to support the grid and constitutes a V2G system. A number of EVs are coordinated by an aggregator, which acts as an agent between the EVs and the grid operator. To drive a lot of EVs for a common V2G objective, aggregators can be connected together, forming a hierarchical structure for the ease of coordination. Due to its flexibility, abundance, and convenience, V2G is particularly useful to provide services to frequency regulation, voltage regulation, unit commitment, and facilitate energy trading in a new energy market. EVs are evolving and equipped with new functionalities with advanced vehicular technologies. With the support of AI, an EV can drive itself and turn into an AV. We can manage a fleet of idle and moving AVs more easily to strive for V2G purposes. After-service AVs can be summoned to appropriate parking facilities to enhance the V2G performance. The AVs can further form a public transportation system and a logistic system with the consideration of V2G. With dynamic charging and discharging, moving EVs form VEN to serve as a big energy buffer for a large geographical area, which revolutionizes V2G to a great extent. We can see that V2G technologies continue to evolve, get smarter, and cultivate new applications which are impossible before.

References

1. IEA (2021) Global EV outlook 2021. IEA, Paris
2. Kempton W, Tomic J (2005) Vehicle-to-grid power implementation: from stabilizing the grid to supporting large-scale renewable energy. J Power Sources 144:280–294
3. Chen X, Leung K-C, Lam AYS (2020) Power output smoothing for renewable energy system: planning, algorithms, and analysis. IEEE Syst J 14(1):1034–1045
4. Engle J, Electric school bus charging hub could provide 'blueprint' for grid support". Renewable Energy World. Retrieved 2022-02-06
5. UC San Diego expands triton rides program with vehicle-to-grid service from Nuvve. Accessed 13 Dec 2018. https://nuvve.com/ucsd-triton-rides/
6. Fermata energy receives the first UL certification for 'vehicle-to-grid' electric vehicle charging system. Accessed 22 Jun 2022. https://www.ul.com/news/fermata-energy-receives-first-ul-certification-vehicle-grid-electric-vehicle-charging-system
7. Oil & Gas 360, EDF Energy and Nuvve corporation announce plans to install 1,500 smart electric chargers in the United Kingdom. Accessed 9 Jan 2019. https://www.oilandgas360.com/edf-energy-and-nuvve-corporation-announce-plans-to-install-1500-smart-electric-chargers-in-the-united-kingdom/
8. Market Screener. Toyota Tsusho: and Chubu electric power announce to initiate Japan's first ever demonstration project of charging and discharging from storage batteries of electric vehicles to the electric grid. Accessed 9 Jan 2019. https://www.marketscreener.com/quote/stock/TOYOTA-TSUSHO-CORPORATION-6494673/news/Toyota-Tsusho-and-Chubu-Electric-Power-Announce-to-Initiate-Japan-s-First-Ever-Demonstration-Proje-27561256/
9. Grigsby LL (2018) Electric power generation, transmission, and distribution, 3rd edn. CRC Press, Boca Raton, FL

10. Ekanayake J, Liyanage K, Wu J, Yokohama A, Jenkins N (2012) Smart grid: technology and applications. Wiley, Hoboken
11. Useable battery capacity of full electric vehicles. Accessed 25 Mar 2022. https://ev-database.org/cheatsheet/useable-battery-capacity-electric-car
12. Pod Point, How long does it take to charge an electric car? Accessed 22 Jun 2022. https://pod-point.com/guides/driver/how-long-to-charge-an-electric-car
13. Le Floch C, Kara EC, Moura S (2018) PDE modeling and control of electric vehicle fleets for ancillary services: a discrete charging case. IEEE Transactions on Smart Grid 9(2):573–581
14. Mohammadi J, Kar S, Hug G (2016) Distributed cooperative charging for plug-in electric vehicles: a consensus + innovations approach. In: Proceedings of IEEE global conference on signal and information processing, Washington, DC, USA. pp 896–900
15. Lin J, Leung KC, Li VOK (2014) Optimal scheduling with vehicle- to-grid regulation service. IEEE Internet Things J 1(6):556–569
16. Shao C, Wang X, Shahidehpour M, Wang X, Wang B (2017) Partial decomposition for distributed electric vehicle charging control considering electric power grid congestion. IEEE Transactions on Smart Grid 8(1):75–83
17. Yao W, Zhao J, Wen F, Xue Y, Ledwich G (2013) A hierarchical decomposition approach for coordinated dispatch of plug-in electric vehicles. IEEE Trans Power Syst 28(3):2768–2778
18. Chen X, Leung K-C, Lam AYS, Hill DJ (2019) Online scheduling for hierarchical vehicle-to-grid system: design, formulation, and algorithm. IEEE Trans Veh Technol 68(2):1302–1317
19. Wang J, Zeng P, Jin X, Kong F, Wang Z, Li D, Wan M (2018) Software defined Wi-V2G: A V2G network architecture. IEEE Intell Transp Syst Mag 10(2)
20. Chen T, Lam AYS, Song Y, Hill DJ (2020) Reducing BESS capacity for accommodating renewables in subtransmission systems with power flow routers. In Proceedings of international conference on smart grids and energy systems, Perth, Australia
21. White CD, Zhang KM (2011) Using Vehicle-to-Grid Technology for Frequency Regulation and Peak-Load Reduction. J Power Sources 196(8):3972–3980
22. Liu H, Hu Z, Song Y, Lin J (2013) Decentralized vehicle-to-grid control for primary frequency regulation considering charing demands. IEEE Trans Power Syst 28(3):3480–3489
23. Zhong J, He L, Li C, Cao Y, Wang J, Fang B, Zeng L, Xiao G (2014) Coordinated control for large-scale EV charging facilities and energy storage devices participating in frequency regulation. Appl Energy 123:253–262
24. Lam AYS, Leung K-C, Li VOK (2016) Capacity estimation for vehicle-to-grid frequency regulation services with smart charging mechanism. IEEE Trans Smart Grid 7(1):156–166
25. Petinrin JO, Shaabanb M (2016) Impact of renewable generation on voltage control in distribution systems. Renew Sustain Energy Rev 65:770–783
26. Douglass PJ, Garcia-Valle R, Østergaard J, Tudora OC (2014) Voltage-sensitive load controllers for voltage regulation and increased load factor in distribution systems. IEEE Trans Smart Grid 5(5):2394–2401
27. Juamperez M, Guangya Y, Kjær SB (2014) Voltage regulation in LV grids by coordinated volt-var control strategies. J Mod Power Syst Clean Energy 2:319–328
28. Sun H et al (2019) Review of challenges and research opportunities for voltage control in smart grids. IEEE Trans Power Syst 34(4):2790–2801
29. Cheng L, Chang Y, Huang R (2015) Mitigating voltage problem in distribution system with distributed solar generation using electric vehicles. IEEE Trans Sustain Energy 6(4):1475–1484
30. Zheng Y, Song Y, Hill DJ, Meng K (2019) Online distributed MPC-based optimal scheduling for EV charging stations in distribution systems. IEEE Trans Industr Inf 15(2):638–649
31. Hoque MM, Khorasany M, Razzaghi R, Wang H, Jalili M (2022) Transactive coordination of electric vehicles with voltage control in distribution networks. IEEE Trans Sustain Energy 13(1):391–402
32. Chen T, Chu KF, Lam AYS, Hill DJ, Li VOK (2020) Electric autonomous vehicles charging and parking coordination for vehicle-to-grid voltage regulation with renewable energy. In: Proceedings of IEEE PES general meeting, Montreal, Canada

33. Egbue O, Uko C, Aldubaisi A, Santi E (2022) A unit commitment model for optimal vehicle-to-grid operation in a power system. Int J Electr Power & Energy Syst 141
34. Yu JJQ, Li VOK, Lam AYS (2013) Optimal V2G scheduling of electric vehicles and unit commitment using chemical reaction optimization. In: Proceedings of IEEE congress on evolutionary computation, Cancun, Mexico
35. Lam AYS, Li VOK (2010) Chemical-reaction-inspired metaheuristic for optimization. IEEE Trans Evol Comput 14(3):381–399
36. Jeong S, Jang YJ, Kum D (2015) Dynamic analysis of the dynamic charging electric vehicle. IEEE Trans Power Electron 30(11):6368–6377
37. Lam AYS, Huang L, Silva A, Saad W (2012) A Multi-layer market for vehicle-to-grid energy trading in the smart grid. In: Proceedings of the 1st IEEE INFOCOM workshop on green networking and smart grids, Orlando, FL, USA
38. Zhong W, Xie K, Liu Y, Yang C, Xie S (2019) Topology-aware vehicle-to-grid energy trading for active distribution systems. IEEE Trans Smart Grid 10(2):2137–2147
39. Yu JJQ, Lin J, Lam AYS, Li VOK (2016) Maximizing aggregator profit through energy trading by coordinated electric vehicle charging. In: Proceedings of the 7th IEEE international conference on smart grid communications, Sydney, Australia, Nov. 2016
40. Taxonomy and definitions for terms related to driving automation systems for on-road motor vehicles, April 2021
41. Kirschner M (2022) Dude, where's my autonomous car? Semicond Eng. Accessed 22 Jun 2022. https://semiengineering.com/dude-wheres-my-autonomous-car/
42. Kempton W, Tomic J (2005) Vehicle-to-grid power fundamentals: calculating capacity and net revenue. J Power Sources 144(1):268–279
43. Lam AYS, Yu JJQ, Hou Y, Li VOK (2018) Coordinated autonomous vehicle parking for vehicle-to-grid services: formulation and distributed algorithm. IEEE Transactions on Smart Grid 9(5):4356–4366
44. Lam AYS, Leung Y-W, Chu X (2016) Autonomous vehicle public transportation system: scheduling and admission control. IEEE Trans Intell Transp Syst 17(5):1210–1226
45. Chu KF, Lam AYS, Li VOK (2021) Joint Rebalancing and vehicle-to-grid coordination for autonomous vehicle public transportation system, to appear in IEEE Transactions on Intelligent Transportation Systems
46. Yu JJQ, Lam AYS (2018) Autonomous vehicle logistic system: joint routing and charging strategy. IEEE Trans Intell Transp Syst 19(7):2175–2187
47. Lam AYS, Leung K-C, Li VOK (2017) Vehicular energy network. IEEE Transactions on Transportation Electrification 3(2):392–404
48. Li S, Liu Z, Zhao H, Zhu L, Shuai C, Chen Z (2016) Wireless power transfer by electric field resonance and its application in dynamic charging. IEEE Trans Industr Electron 63(10):6602–6612
49. KAUST Discovery, Charging ahead for electric vehicles. Accessed 22 Jun 2022. https://discovery.kaust.edu.sa/en/article/1073/charging-ahead-for-electric-vehicles
50. Lam AYS, Li VOK (2018) Opportunistic routing for vehicular energy network. IEEE Internet Things J 5(2):533–545
51. Chow CCT, Lam AYS, Liu W, Chau KT, Multi-source multi-destination optimal energy routing in static and time-varying vehicular energy network, submitted for publication
52. Lam AYS, VOK Li (2016) Energy Loss Minimization for Vehicular Energy Network Routing. In Proceedings of ACM workshop on electric vehicle systems, data and applications, Waterloo, Canada
53. Lam AYS, Yu JJQ (2017) Robust Routing for Vehicular Energy Network. In: Proceedings of the ACM eighth international conference on future energy systems, Hong Kong

Recent Development of Electric and Hybrid Vehicles

Fawen Shen, Shuangchun Xie, C. S. Teo, and Christopher H. T. Lee

Abstract With the increasing concern of environmental protection and energy sustainability, as well as the rapid advancement of high-performance electric motors, power electronic converters, and energy storage technology, electric and hybrid electric vehicles (EV/HEVs) are perceived as the viable substitution for conventional gas-fueled automobiles. EVs were attracting wide attention in the early 1900s but failed in the competition with gas-powered cars during the following decades. The 1973 gasoline crisis sparked new interest in EVs. Later in 1976, the US Congress recommended EVs as a solution to reduce oil dependency. However, electric vehicles didn't attract worldwide attention until the appearance of the Toyota Prius in 1997. Afterward, famous car manufacturers including Honda, Ford, Nissan, GM, BMW, Mercedes, Land Rover, and Audi made great efforts in improving the performance of EVs. Among them, Tesla is the pioneer pushing the frontiers of innovation in this area. Fueled by the continuous advancement in electric motor, power electronics, and battery technology, the sales of EVs in 2021 reached a new record (U.S. Department of Energy, Energy Vehicle Technologies Office in Transportation Energy Data Book. Oak Ridge National Laboratory, https://tedb.ornl.gov/data/, 2022), as shown in Fig. 1.

The focus of this chapter is to provide a brief review of the state-of-the-art EV/HEVs, with emphasis on technology advancement. In this chapter, a few popular commercial products of EV/HEVs are introduced to show the development status of modern EV/HEVs.

F. Shen · S. Xie · C. H. T. Lee (✉)
School of Electrical and Electronic Engineering, Nanyang Technological University, Singapore 639798, Singapore
e-mail: chtlee@ntu.edu.sg

C. S. Teo
Singapore Institute of Manufacturing Technology, Agency for Science, Technology and Research, Singapore 138634, Singapore

C. H. T. Lee (ed.), *Emerging Technologies for Electric and Hybrid Vehicles*,
Green Energy and Technology, https://doi.org/10.1007/978-981-99-3060-9_12

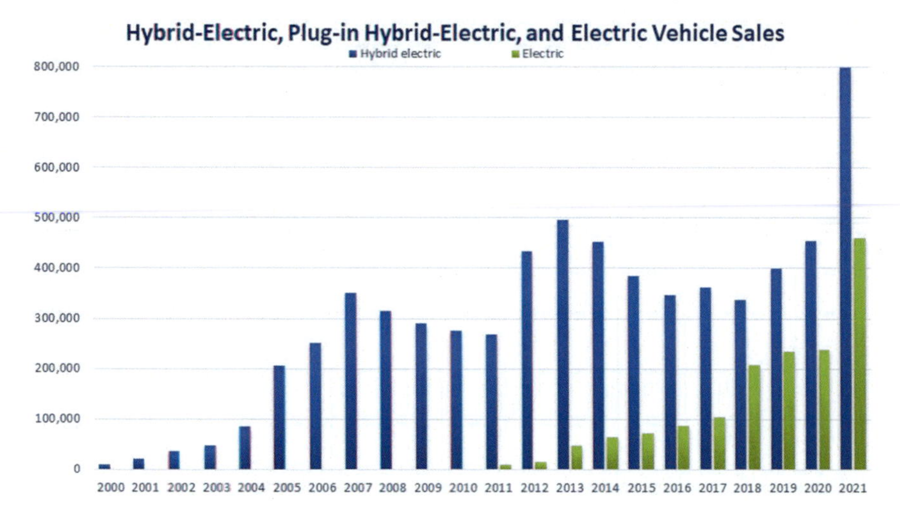

Fig. 1 The sales of electric and hybrid vehicles in the United States [1]

1 Hybrid Electric Vehicles

1.1 Toyota Prius Generation 4

Due to the prominent fuel efficiency, Toyota Prius is regarded as one of the most typical models in HEV family. The Toyota Prius Generation 4 is shown in Fig. 2. It not only exhibits the advantages of the previous Toyota Prius models but also adopts all-wheel drive (AWD) technology for better driving performance. By redesigning the engine, evolving transaxle, improving battery structure, and miniaturizing control unit, the Toyota Prius Generation 4 can achieve an 18% improvement in fuel efficiency.

Figure 2b shows the drivetrain of Toyota Prius Generation 4, where the evolved hybrid transaxle is adopted. Particularly, the generator and drive motor are arranged on different axles, and the new parallel gears replace the previous planetary gear. As a consequence, a 20% mechanical loss reduction and a 12% unit length reduction are simultaneously achieved. Another high spot is the redesigned engine, which can achieve 40% thermal efficiency. Moreover, by improving the structure and components of control unit, the size is reduced by around 33%. The key specifications of the Toyota Prius Generation 4 are listed in Table 1.

(a)

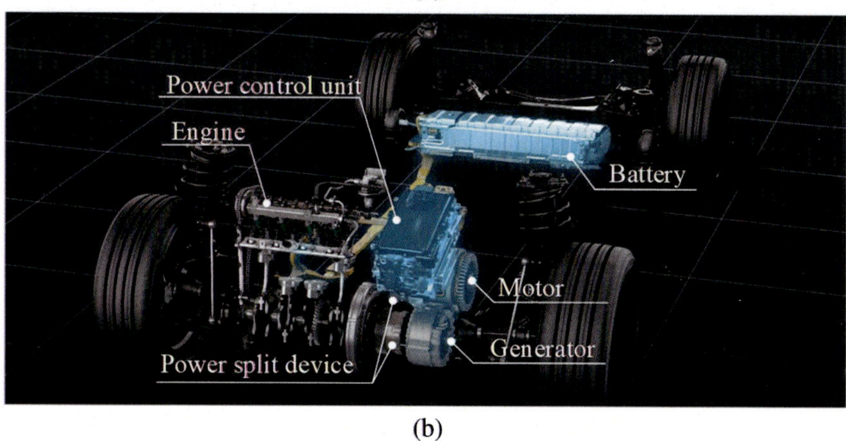

(b)

Fig. 2 Toyota Prius Generation 4 [2]. **a** Exterior view. **b** Drivetrain

1.2 Honda Insight

Honda is a rising star among hybrid electric vehicle manufacturers, challenging the dominant position of Toyota. Different from Toyota hybrid system (THS) with complex planetary gear, Honda developed an extremely simple hybrid transmission system, i.e., Honda i-MMD system. The Honda i-MMD hybrid system consists of two electric motors, a power control unit, an Atkinson-cycle gasoline engine, lithium-ion battery, and an innovative fixed gear transmission. The Honda Insight which is designed as a rival to Toyota Prius, and the i-MMD drivetrain are shown in Fig. 3.

The main working principle of the Honda i-MMD hybrid system is to improve engine fuel economy through series mode at low-speed operation and through parallel (or engine-only) mode at high-speed operation. To be more specific, the Honda i-MMD system is a three-mode system, namely EV mode at low speed, hybrid

Table 1 Specifications of Toyota Prius Generation 4 [2]

Parameters	Toyota Prius Generation 4
Weight (kg)	1461
Drivetrain	Hybrid transaxle
Base price ($)	29,575
Engine	
Type	1.8-L, 4-Cyliner, DOHC
Power (hp @ rpm)	96 @ 5200
Torque (Nm @ rpm)	142 @ 3600
Electric motor	
Category	PM AC motor
Power (hp)	71
Torque (Nm)	163
Traction Battery	
Category	Nickel-metal hydride (NiMH)

drive mode at middle speed, and engine-only mode at high speed. As compared with THS, the unique engine-only mode of Honda i-MMD system contributes to further improving fuel economy at high-speed cruising operation. Table 2 shows the specifications of Honda Insight model. It can be seen that the output power of electric motors in Honda Insights is significantly higher than that in Toyota Prius, i.e., 129 hp versus 71 hp. This is attributed to the extremely simple structure of Honda i-MMD system where there is only one clutch to separate ICE and electric motors. In addition, the high power-density lithium-ion battery is employed in Honda Insight to reduce the weight and increase driving range. Last but not least, the Honda Insight can achieve comparable fuel economy with Toyota Prius, namely 52 mpg versus 56 mpg under combined driving conditions.

2 Electric Vehicles

2.1 Tesla EVs

The pure EVs didn't attract much attention worldwide until Tesla Motors released the Tesla Roadster in 2008. The Tesla Roadster achieved a driving range of 394 km on a single charge, a range unprecedented for pure EVs at that time. Furthermore, the Tesla Roadster could accelerate from 0 to 100 km/h in less than four seconds and reach a top speed of 200 km/h, which is comparable to gasoline-powered vehicles. Although the performance was attractive, the high cost impeded its market penetration.

Afterward, Tesla has developed four consequences of EVs including Model 3, Model X, Model Y, and Model S Plaid, as shown in Fig. 4. The Model X employs

(a)

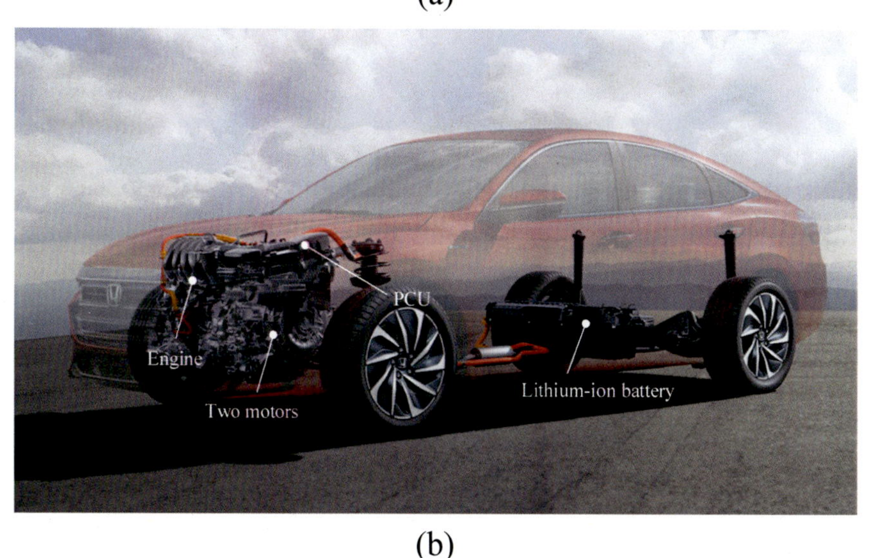

(b)

Fig. 3 Honda Insight [3]. **a** Exterior view. **b** i-MMD drivetrain

both permanent magnet synchronous motor and induction motor with a maximum output power of 670 hp. Based on this engine, the acceleration time from 0 to 100 km/h is shortened to 3.9 s. Moreover, the top speed can be up to 241 km/h. The Model Y also employs both permanent magnet synchronous motor and induction motor with maximum output power of 535 hp. This engine can provide a 0 to 100 km/h time of 3.7 s and top speed of 250 km/h. The Model 3 integrates three permanent magnet synchronous motors with a maximum output power of 221 hp. Based on this engine, the acceleration time from 0 to 100 km/h is shortened to 3.3 s. Moreover, the top speed

Table 2 Specifications of Honda Insight [3]

Parameters	Honda Insight
Weight (kg)	1396
Drivetrain	Hybrid transaxle
Base price ($)	29,790
Engine	
Type	2.0-L, 4-Cyliner, DOHC
Power (hp @ rpm)	107 @ 6000
Torque (Nm @ rpm)	134 @ 5000
Electric Motor	
Category	PM AC motor
Power (hp)	129
Torque (Nm)	267
Traction Battery	
Category	Lithium-ion battery

can be up to 261 km/h. The Model S Plaid exhibits the best performances within the pool. As shown in Fig. 5, it artfully employs three permanent magnet synchronous motors with a maximum output power of 1020 hp. This engine can provide a 0 to 100 km/h time of 2.1 s and top speed of 322 km/h. All of these EVs utilize the Lithium-ion battery. The battery capacity of Model X, Model Y, Model 3, and Model S Plaid are 100 kWh, 82 kWh, 82 kWh, and 95 kWh, respectively. The range of Model 3, Model X, Model Y, and Model S Plaid are 560 km, 514 km, 547 km, and 637 km, respectively. The key specifications of four EVs are summarized and listed in Table 3.

2.2 *BMW EVs*

As one of the largest manufacturers of luxury cars, BMW has been pouring effort and leading the way in making electric cars for over 40 years. In particular, the BMW i3 is the first mass-production battery electric car with carbon fiber reinforced plastic and the second bestselling EV across Europe. Afterward in 2017, BMW announced to make a wholesale shift to electric cars and released i and iX series as sedan and sports activity vehicles (SAVs), as shown in Fig. 6. By adopting the fifth-generation BMW eDrive train as demonstrated in Fig. 7, the electric motor, transmission system, and power electronics are incorporated into a single component, achieving a 30% improvement in power density, and making it much easier to be integrated into different vehicle architectures. Furthermore, the battery pack with an aluminum alloy casing is located in the vehicle chassis, reducing the risk of direct impact on the battery pack in an accident. Last but not least, two electrically-excited synchronous motors (ESMs) are employed in the fifth-generation BMW eDrive train system.

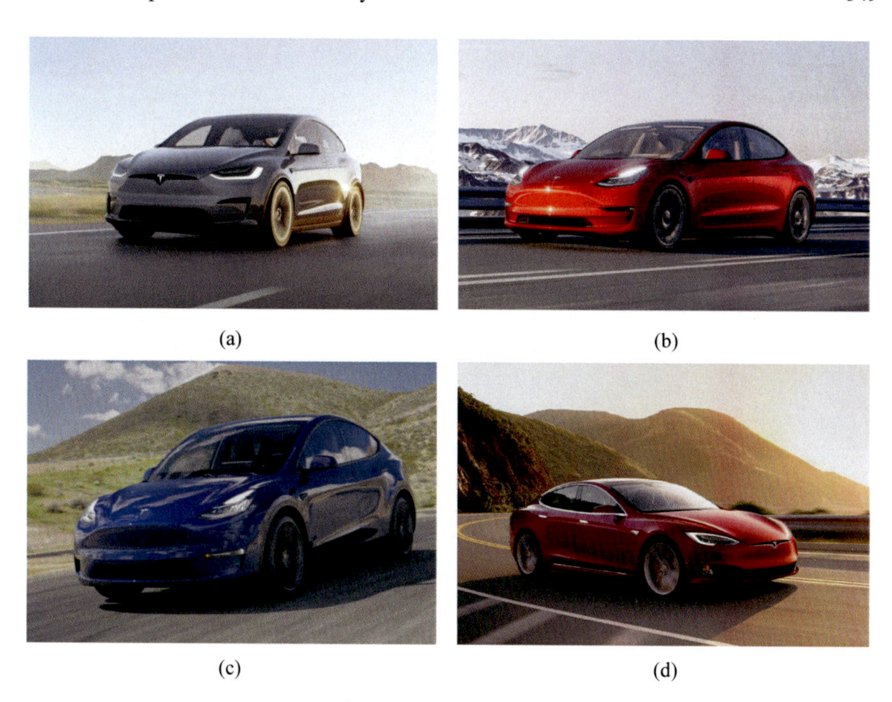

Fig. 4 Exterior view of four Tesla electric vehicles [4]. **a** Model X. **b** Model Y. **c** Model 3. **d** Model S Plaid

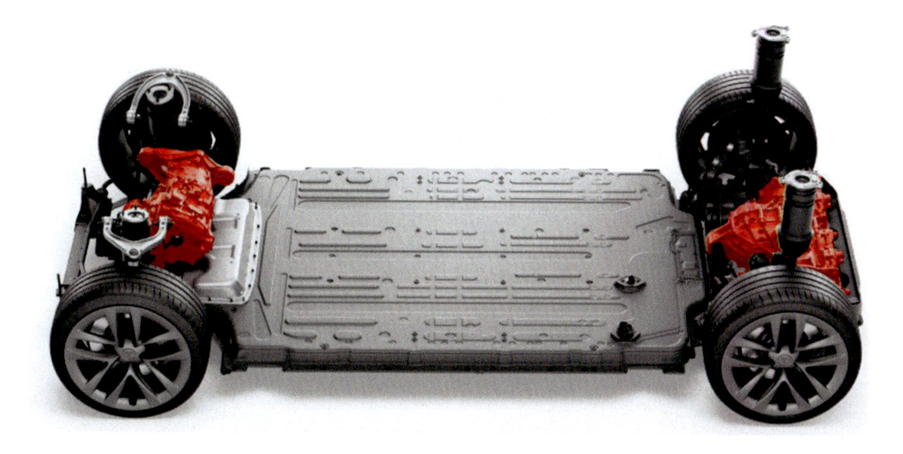

Fig. 5 Tesla Tri-motor AWD system [4]

Table 3 Specifications of Tesla EVs [4]

Parameters	Model X	Model Y	Model 3	Model S Plaid
Peak power (hp)	670	535	515	1020
Drivetrain	Dual-motor AWD		Tri-motor AWD	
Motor type	PM AC motor and IM		PM AC motor	
Top speed (km/h)	241	250	261	322
Range (km)	560	514	547	637
Acceleration 0–100 km/h (s)	3.9	3.7	3.3	2.1
Battery pack	Lithium-ion			
Battery capacity (kWh)	100	82	82	95
Base price ($)	122,440	142,271	64,440	131,190

The employment of non-rare-earth permanent magnet motor contributes to reducing the system cost, improving high-speed cruising performance, and conserving the environment.

Among the BMW i models, BMW i4 and i7 exhibit the most breathtaking performance. With dedicated thermal management system, the maximum output power and operating efficiency of BMW i4 xDrive50 are up to 544 hp and 95%, respectively. As a result, the top speed is 226 km/h and the acceleration time from 0 to 100 km/h is within 3.9 s. With more powerful lithium-ion battery technology, BMW

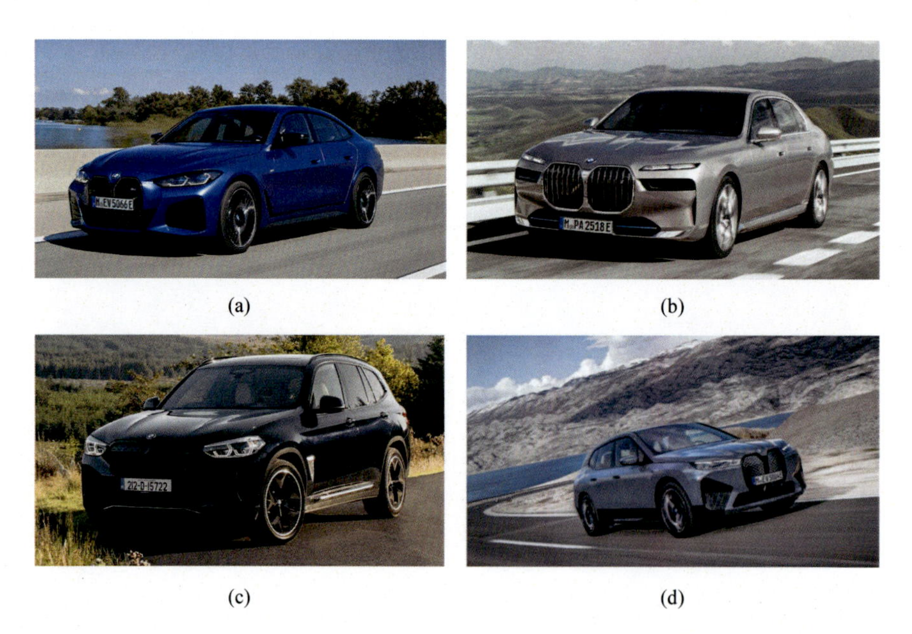

Fig. 6 Exterior view of four BMW electric vehicles [5]. **a** BMW i4. **b** BMW i7. **c** BMW iX3. **d** BMW iX

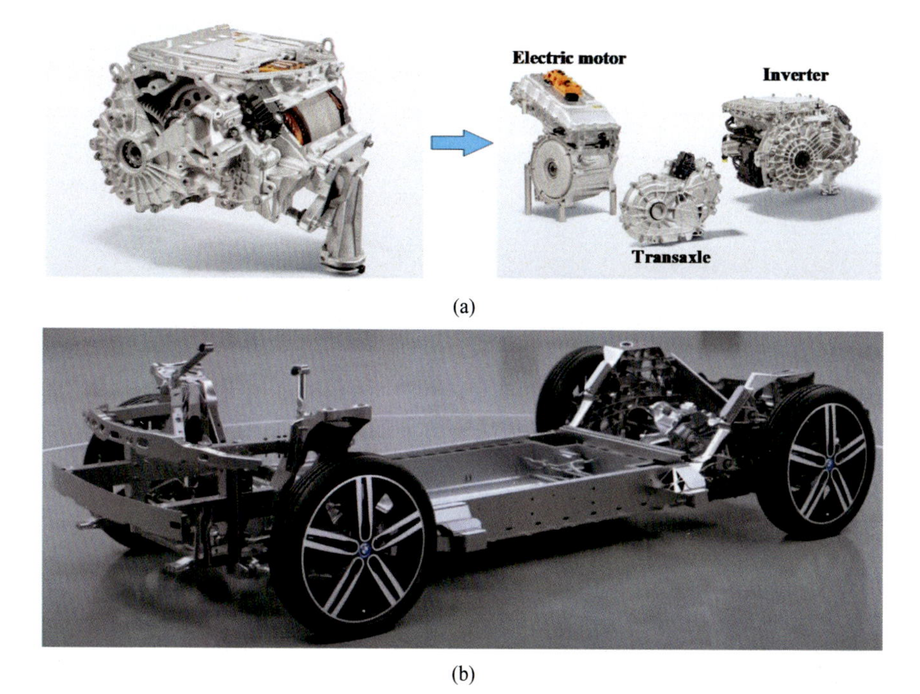

Fig. 7 The fifth-generation BMW eDrive train [5]. **a** Integrated electric powertrain. **b** Overall structure

i7 xDrive60 can achieve a top speed of 242 km/h and driving range of 625 km. The iX3 model is the first attempt of BMW at electric SAVs. The BMW iX3 can achieve a top speed of 180 km/h, an acceleration time from 0 to 100 km/h of 6.8 s, and a driving range of 460 km. The BMW iX is a flagship model of electric SAV, employing six-phase control to achieve increased output power, higher efficiency, and more accurate motion control. The output torque of BMW iX xDrive50 is 1000 Nm, reducing the acceleration time from 0 to 100 km/h to 3.8 s. Furthermore, the driving range is increased to 566 km due to the adopted high-power-density Lithium-ion battery pack. The key specifications of the four EVs are compared and summarized in Table 4.

2.3 Audi EVs

As one of the most famous car brands, Audi has made increasing efforts in electric vehicles since the first Audi e-tron electric vehicle (EV) was launched in 2018. Nowadays, the Audi e-tron has four series of products including e-tron, Q4 e-tron, e-tron S, and e-tron GT. In this part, the latest Audi EVs are introduced. These four

Table 4 Specifications of BMW i and iX Series Models [5]

Parameters	BMW i4	BMW i7	BMW iX3	BMW iX
Peak power (hp)	544	544	286	610
Powertrain	The fifth-generation BMW eDrive train			
Motor type	Dual electrically-excited synchronous motors			
Top speed (km/h)	226	242	180	250
Range (km)	590	625	460	566
Acceleration 0–100 km/h (s)	3.9	4.7	6.8	3.8
Battery pack	Lithium-ion			
Battery capacity (kWh)	83.9	105.0	80.0	111.5
Base price ($)	67,300	120,295	67,680	108,900

types of Audi e-tron are shown in Fig. 8. It can be found that the Audi EV covers both sedan and SUV types, which can fulfill the requirements of different customers.

The Audi e-tron employs two permanent magnet synchronous motors with maximum output power of 402 hp. Based on this engine, the acceleration time from 0 to 100 km/h is shortened to 5.5 s. Moreover, the top speed can be up to 207 km/h. Different from Audi e-tron, Audi Q4 e-tron adopts rear permanent magnet synchronous motor and front asynchronous motor with a maximum output power of 295 hp. This engine can provide a 0 to 100 km/h time of 5.8 s and top speed of

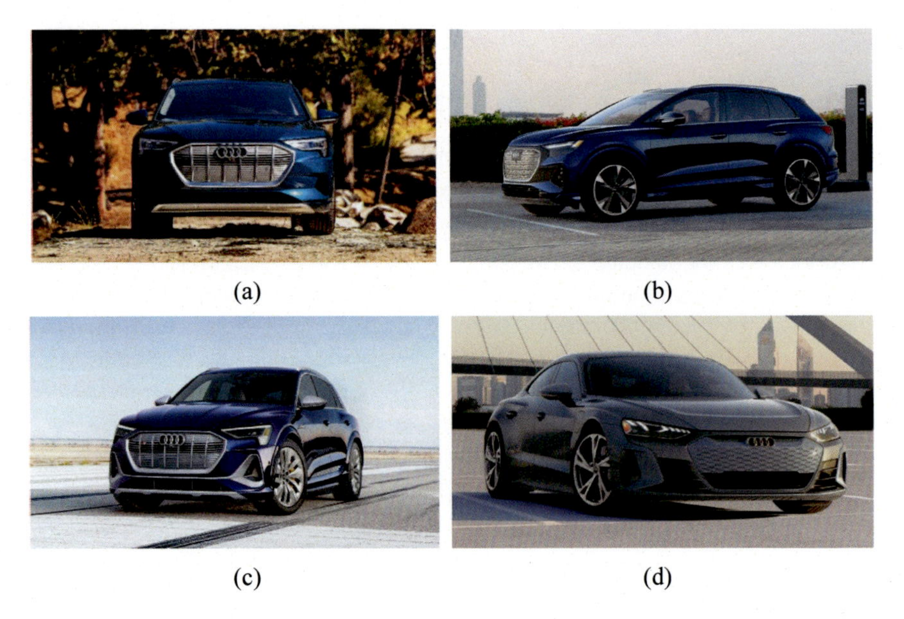

Fig. 8 Exterior view of four Audi e-tron electric vehicles [6]. **a** Audi e-tron. **b** Q4 Audi e-tron. **c** Audi e-tron S. **d** Audi e-tron GT

Table 5 Specifications of Audi EVs [6]

Parameters	e-tron	Q4 e-tron	e-tron S	e-tron GT
Peak power (hp)	402	295	496	522
Motor type	PM AC motor	PM AC motor and induction motor	PM AC motor	PM AC motor
Top speed (km/h)	207	187	217	253
Range (km)	376	402	347	397
Acceleration 0–100 km/h (s)	5.5	5.8	4.3	3.9
Battery capacity (kWh)	95	82	95	93
Base price ($)	70,800	53,300	88,200	104,900

187 km/h. The Audi e-tron S artfully integrates three permanent magnet synchronous motors with a maximum output power of 496 hp. Based on this engine, the acceleration time from 0 to 100 km/h is shortened to 4.3 s. Moreover, the top speed can be up to 217 km/h. The Audi e-tron GT exhibits the best performances within the pool. It employs two permanent magnet synchronous motors with a maximum output power of 522 hp. This engine can provide a 0 to 100 km/h time of 3.9 s and top speed of 253 km/h. All of these EVs utilize the Lithium-ion battery. The battery capacity of e-tron, Q4 e-tron, e-tron S, and e-tron GT are 95 kWh, 82 kWh, 95 kWh, and 93 kWh, respectively. The range of e-tron, Q4 e-tron, e-tron S, and e-tron GT are 376 km, 402 km, 347 km, and 397 km, respectively. The key specifications of four EVs are summarized and listed in Table 5.

The latest Audi EVs are equipped with the quattro drive technology, with which the EVs can achieve the so-called e-torque vectoring. Take the Audi e-tron as an example. The drivetrain of Audi e-tron is shown in Fig. 9. It can be seen that the two motors are accommodated in the front and rear parts, respectively. The e-torque vectoring is illustrated in Fig. 10. When driven normally, only the rear motor provides propulsion. If more power than the rear motor can supply is required, the front motor will be activated. The highly flexible and reliable all-wheel drive system can distribute power to each axle individually, thus providing desirable stability in most driving situations.

2.4 *Porsche EVs*

Owing to the rapid development of EVs, Porsche has also been engaged in this area. Porsche Taycan 2019 was the first attempt from Porsche in the EV market. Currently, there are several Porsche EVs including Taycan, Taycan 4 Cross Turismo, Taycan 4S, Taycan 4S Cross Turismo, Taycan GTS, Taycan GTS Sport Turismo, Taycan Turbo, Taycan Turbo Cross Turismo, Taycan Turbo S, Taycan Turbo S Cross

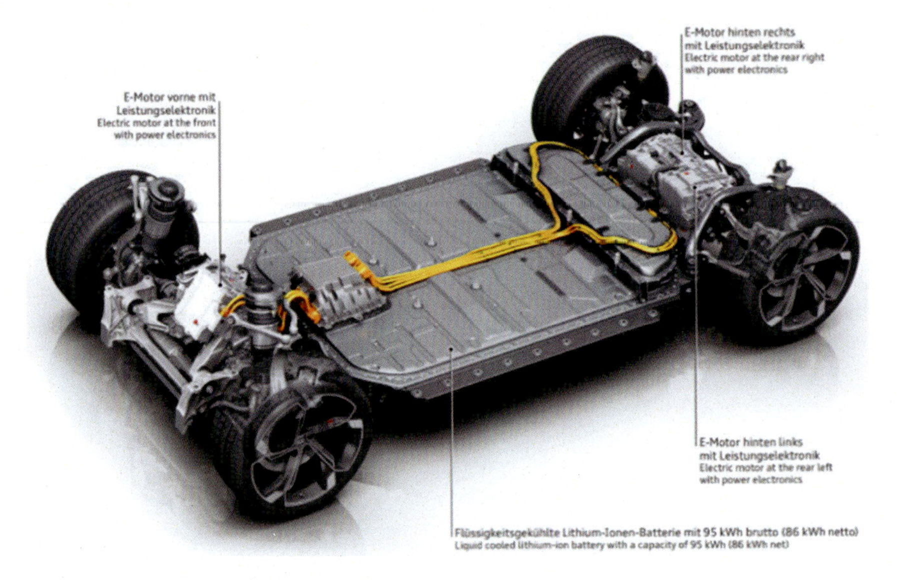

Fig. 9 Drivetrain of Audi e-tron [6]

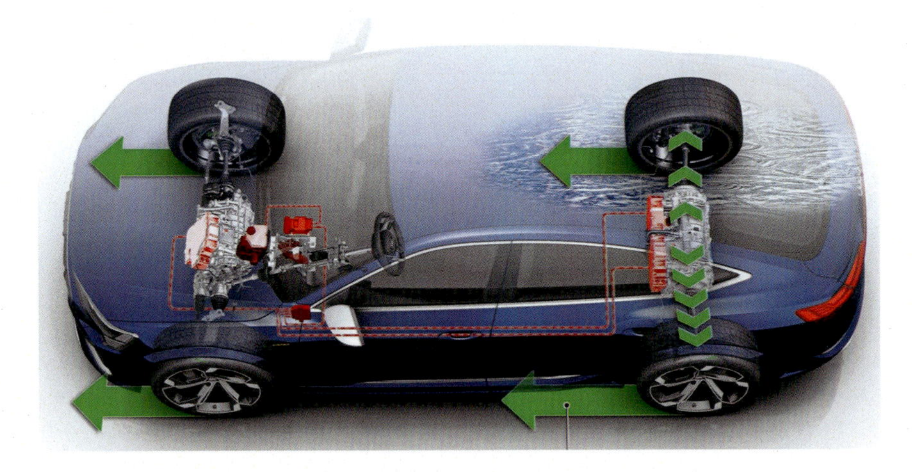

Fig. 10 Illustration of all-wheel drive system in Audi e-tron [6]

Turismo. The typical models of Taycan, Taycan 4S, Taycan GTS, and Taycan Turbo S are introduced in this section. As shown in Fig. 11, these four EVs share the same exterior design.

The Taycan employs permanent magnet synchronous motors with a maximum output power of 402 hp. Based on this engine, the acceleration time from 0 to 100 km/h is shortened to 5.1 s. Moreover, the top speed can be up to 239 km/h. The Taycan 4S also adopts permanent magnet synchronous motors with a maximum output power

Fig. 11 Exterior view of Porsche EVs [7]

of 522 hp. This engine can provide a 0 to 100 km/h time of 3.8 s and top speed of 258 km/h. The Taycan GTS utilizes permanent magnet synchronous motors with a maximum output power of 590 hp. Based on this engine, the acceleration time from 0 to 100 km/h is shortened to 3.5 s. Moreover, the top speed can be up to 258 km/h. The Taycan Turbo S exhibits the best performances within the pool. It employs permanent magnet synchronous motors with a maximum output power of 750 hp. This engine can provide a 0 to 100 km/h time of 2.6 s and a top speed of 260 km/h. All of these EVs utilize the Lithium-ion battery. The battery capacity of Taycan, Taycan 4S, Taycan GTS, and Taycan Turbo S are 79.2 kWh, 79.2 kWh, 93.4 kWh, and 93.4 kWh, respectively. The range of Taycan, Taycan 4S, Taycan GTS, and Taycan Turbo S are 323 km, 323 km, 324 km, and 324 km, respectively. The key specifications of four EVs are summarized and listed in Table 6.

As the best EV model of Porsche, Taycan Turbo S employs cutting-edge technology. Figure 12 shows the drivetrain of Taycan Turbo S. It can be noted that two

Table 6 Specifications of Porsche EVs [7]

Parameters	Taycan	Taycan 4S	Taycan GTS	Taycan Turbo S
Peak power (hp)	402	522	590	750
Motor type	PM AC motor	PM AC motor	PM AC motor	PM AC motor
Top speed (km/h)	239	258	258	260
Range (km)	323	323	324	324
Acceleration 0–100 km/h (s)	5.1	3.8	3.5	2.6
Battery capacity (kWh)	79.2	79.2	93.4	93.4
Base price ($)	86,700	106,500	134,100	187,400

Fig. 12 Drivetrain of Taycan Turbo S [7]

permanent magnet synchronous motors are placed in the front and rear parts for all-wheel drive. Different from the conventional winding using round wires, the hairpin winding with rectangular wires is adopted in these two motors. The wire shape is similar to hairpins, thus being named as "hairpin." The hairpin winding can significantly improve the slot-filling factor of motors. For conventional winding forms, the slot-filling factor is about 45 percent, while the slot-filling factor with hairpin winding can be up to 70 percent. Hence, the torque density and power density of electric motors can be effectively enhanced. Moreover, the cooling of electric motors with hairpin winding can be more efficient.

The Taycan Turbo S adopts two-speed transmission structure shown in Fig. 13. In this system, two gears are employed for different functions. One of the gears is used for initial speed acceleration and the other one is employed to keep the high power and efficiency at high speeds.

Fig. 13 Two-speed transmission structure of Taycan Turbo S [7]

3 Conclusion

Electric motors in EV/HEVs are required to exhibit high efficiency, high torque density, high constant power speed range (CPSR), high power density, fast dynamic response, good flux weakening capability at high speeds, high reliability, and good fault tolerance characteristics. The increasing requirements of EV/HEVs bring about a preference for permanent magnet motors.

In existing motor systems for EV/HEVs, mechanical gearbox is usually adopted to achieve high-torque low-speed output. However, the gearbox suffers from inevitable mechanical loss and frequent maintenance, thus reducing the system efficiency and improving the cost. Hence, direct-drive electric motors without gearbox become promising candidates in future EV/HEVs. On the other hand, the electric motors and corresponding power electronics controllers are separately accommodated, two independent cooling systems are required, thus enlarging the system volume and weight. Therefore, integrated-power-electronics motors will be adopted in future EV/HEVs.

References

1. U.S. Department of Energy (2022) Energy Vehicle Technologies Office, Oak Ridge National Laboratory, Transportation Energy Data Book, Edition 40, table 6.2. https://tedb.ornl.gov/data/
2. Toyota Prius Generation 4, available at https://global.toyota/en/.
3. Honda Insight, available at https://global.honda/.
4. Tesla electric vehicle Models, available at https://www.tesla.com.
5. BMW electric vehicle Models, available at https://www.bmw.com/en/index.html.
6. Audi electric vehicle Models, available at https://www.audi.com/en/models.html.
7. Porsche electric vehicle Models, available at https://www.porsche.com/usa/models/taycan/.

Printed in the United States
by Baker & Taylor Publisher Services